THE ALCHEMY

OF THE

HEAVENS

KEN CROSWELL

ILLUSTRATIONS BY PHILIPPE VAN

ANCHOR BOOKS

DOUBLEDAY

NEW YORK   LONDON   TORONTO   SYDNEY   AUCKLAND

THE ALCHEMY OF THE HEAVENS

SEARCHING FOR MEANING
IN THE MILKY WAY

AN ANCHOR BOOK
PUBLISHED BY DOUBLEDAY
a division of Bantam Doubleday Dell Publishing Group, Inc.
1540 Broadway, New York, New York 10036

ANCHOR BOOKS, DOUBLEDAY, and the portrayal of an anchor are
trademarks of Doubleday, a division of Bantam Doubleday Dell
Publishing Group, Inc.

Book Design by Gretchen Achilles

Library of Congress Cataloging-in-Publication Data

Croswell, Ken.
   The alchemy of the heavens: searching for meaning in the Milky Way /
by Ken Croswell; illustrations by Philippe Van. —1st Anchor Books ed.
      p.   cm.
   Includes bibliographical references and index.
   1. Milky Way—Popular works. 2. Dwarf galaxies—Popular works.
I. Title.
QB857.7.C76   1995
523.1'13—dc20   94-30452
                                                                CIP

CONTENTS

INTRODUCTION:
A NEW GALAXY

IT WAS HARDLY the astronomical discovery of the century. Indeed, the new galaxy that turned up on February 8, 1990, was so dim that the first to signal its presence was a computer. While scanning photographic plates of the sky, a team of British astronomers—and their faithful computer—uncovered the Sextans dwarf, one of the faintest galaxies ever found, right on the Milky Way's doorstep, orbiting our Galaxy just as the Moon circles the Earth.

When he discovered the new galaxy, Mike Irwin, an astronomer at the University of Cambridge, was not thinking about nearby dwarf galaxies, and he certainly was not trying to find any. Instead, Irwin was searching photographic plates for exactly the opposite—luminous quasars at the edge of the observable universe that can radiate a hundred times more light than the entire Milky Way Galaxy. Three years earlier, his scientific team had discovered what was then the most distant known object in the universe, a quasar some 10 billion light-years from the Earth.

Each of the photographic plates Irwin was examining showed a quarter million of the Milky Way's stars, a similar number of galaxies, and possibly a new quasar at a record-breaking distance. Any such quasar lay far beyond the Galaxy's stars, but on the plate the stars and the quasars all looked the same, a myriad of tiny specks. As a first step toward distinguishing the stars from the quasars, Irwin used a computerized machine to scan the plates, count the images, and measure their positions. The plate that he gave to the computer this February morning was of a small constellation called Sextans—the Sextant—which lies just south of Leo. To Irwin, the plate had seemed ordinary, one of the thousands that his team was scanning.

In the afternoon, though, the output from the computer revealed an abnormality. "On this particular plate," said Irwin, "the computer detected an excess of objects in one corner. Having worked on dwarf galaxies before, I was immediately alerted to the fact that it might well be a dwarf galaxy, previously unknown, because I knew there weren't any in that part of the sky." Irwin therefore suspected that the excess images were stars beyond the Milky Way that belonged to this dwarf galaxy. In the log that day, he wrote down the object's coordinates, along with the comment, "New Local Group galaxy???" To verify the apparent discovery, the next day Irwin gave another plate of Sextans to the computer, and six hours later the computer reported that the second plate bore the same enhancement of stars as the first. The new galaxy was real.

Because the galaxy in Sextans appeared large, and because its stars were spread out from one another, Irwin believed that it was nearby, in orbit around the Milky Way. At the time, nine other Milky Way satellite galaxies were known. Two were large and bright, but the other seven were much fainter. The last of these seven dwarfs had been found in 1977. By examining the new galaxy's individual stars, Irwin determined that Sextans was indeed nearby, in orbit around the Milky Way. It was the Milky Way's tenth satellite galaxy and its eighth dwarf, 295,000 light-years away.

The galaxy in Sextans is home to just a few million stars, whose total output of light does not match that of even the single brightest star in the Milky Way. "Without the computer," said Irwin, "there's no way Sextans would have been found. It's virtually impossible to find by eye. In fact, if you look under the microscope at the region of the plate where this object is, it's still very difficult to see. I've asked lots of people to try the test: here's the plate; the galaxy is on here somewhere; can you tell me where it is? And they can't find it."

Exciting though the discovery was, Irwin said that today he has almost forgotten it. "I've gotten very blasé about this galaxy now," he said, noting that Canadian astronomer Sidney van den Bergh had jokingly criticized him for it. "He said that I'd ruined the nice symmetry of the Galaxy and its seven dwarfs—as in *Snow White and the Seven Dwarfs*. Now we've got eight, and it's one too many."

THE ENTIRE UNIVERSE, of course, harbors far more than just eight galaxies. Imagine that you could sail through space at a trillion times the speed of light: galaxy after galaxy would pass you by, each a conglomeration of innumerable stars. Galaxies litter the cosmos on every side, strewn across space like grains of sand over an endless beach. In the vast compass of the universe, each galaxy may seem insignificant.

But one galaxy stands out from every other: our own. The Milky Way Galaxy is the most important galaxy in the universe. We live in it, and we owe our lives to it; we circle one of its stars, and are made of material forged in countless others.

In recent decades, astronomers have delved into the intricacies of the Milky Way and painted a vivid portrait of our Galaxy's structure, evolution, and origin. Indeed, with the plethora of recent discoveries, the Milky Way itself almost seems to be a new galaxy, as new as the recent find in Sextans. But until now, no book for the general reader has presented these accomplishments or told the full story of our Galaxy—following the path from ancient times, when some viewed it as a river in the sky, to the present, when astronomers use it as a laboratory for the study of the formation of galaxies and even of the entire universe.

The Alchemy of the Heavens aims to be that book. Although other books describe how astronomers early in the twentieth century deduced the basic features of our Galaxy, *The Alchemy of the Heavens* also provides a close look at the equally exciting developments that have occurred since 1950. Moreover, the astronomers who made these recent discoveries here provide revealing behind-the-scenes accounts of how the discoveries were made, how well or poorly other scientists received them, and how one finding provoked the next. In this way the story of our Galaxy is much more than just one of science; it is also a deeply human story, full of colorful and controversial characters who sometimes struggled as much with one another as they did with nature.

Deciphering our Galaxy has been one of humanity's greatest—and most difficult—achievements. At first glance, the Milky Way might seem impossible to fathom. Because we live inside it, it is the only galaxy astronomers cannot see from the outside; its billions of stars superimpose themselves on one another like the leaves of trees tangled together in a huge tropical forest. Furthermore, the Galaxy is full of gas and dust that prevent astronomers from seeing most of its stars. For this reason, the stars that do light the night sky represent a tiny fraction of the Milky Way, making ours the one galaxy in the universe that astronomers will never be able to see in its full glory.

Yet these obstacles make the story of the Milky Way richer and more colorful, forcing scientists to draw on a wide array of fields, from stellar evolution and nuclear physics to galactic dynamics and cosmology, in order to achieve their goal. Even today many of the greatest questions in Milky Way research have yet to be answered.

For these reasons, *The Alchemy of the Heavens* tells the history of our exploration of the Galaxy as it actually happened: as one discovery led to another, as one idea triggered another, as one observation overturned another. Like a symphony with unexpected chord changes, the story sometimes veers off in unexpected directions. For example, it might seem that this book should open with the beginning of the universe, with the big bang and what the Milky Way reveals about it. But astronomers' knowledge of this ancient era has come only recently, so this subject appears toward the end of the book.

A brief word, then, about the journey this book embarks on. *The Alchemy of the Heavens* begins with an overview of the modern conception of the Milky Way. The narrative then shifts back in time, to when the Milky Way was the stuff of myths and legends, and moves forward as astronomers began to take notice of it. The first revolution came during the 1910s and 1920s, when astronomers deduced the size of the Milky Way and our place within it.

The journey grows more complex as astronomers learned that nearby stars of different ages had different characteristics and held clues to past epochs in the Milky Way's life. These developments, which originated in the 1940s, stimulated a number of findings during the 1950s and early 1960s. Among them was the realization that most elements on Earth had been forged by the Milky Way's stars; and with that came the first detailed model of the Galaxy's origin and evolution. More recently, during the 1980s and 1990s, astronomers have used improved instruments to scrutinize the nearby stars and deduce far-reaching conclusions concerning the evolution of the entire Milky Way Galaxy.

Meanwhile, in the last few decades, scientists have attempted to augment this view by peering into other parts of the Galaxy. Using infrared and radio telescopes, some astronomers have studied the Galactic center, the point about which all else in the Milky Way revolves and the likely site of a massive black hole. Other astronomers have voyaged in the opposite direction, to the outer fringes of the Galaxy, which harbor nearly a dozen galaxies that orbit ours and abound with mysterious material that no one can see.

Finally, the Milky Way is in many ways a mirror of the entire cosmos, for our Galaxy has much to say about the universe beyond. In particular, since the oldest stars in the Galaxy contain elements that were manufactured just minutes after the big bang, these stars help cosmologists probe the origin of the universe. But the same stars also pose a paradox, for according to some estimates they are older than the universe itself—a logical impossibility that has yet to be resolved. As astronomers on Earth contemplate these issues, there may—or may not —be astronomers elsewhere doing the same, and the final chapter of this book draws on what astronomers now know about the Milky Way to discuss the possibility that intelligent life exists elsewhere in the Galaxy.

THE ASTRONOMERS I interviewed provided unique and often personal insights into how our understanding of the Milky Way has developed. I would like to thank each scientist who spoke with me: Lawrence Aller, John Bahcall, Timothy Beers, Michael Bolte, Geoffrey Burbidge, Margaret Burbidge, Alastair Cameron, Bruce Carney, Joseph Chamberlain, Olin Eggen, William Fowler, Gerard Gilmore, Fred Hoyle, Mike Irwin, Jeffrey Kuhn, Donald Lynden-Bell, William Morgan, John Norris, Donald Osterbrock, Carlton Pryor, Michael Rich, Nancy Roman, Allan Sandage, David Schramm, Martin Schwarzschild, Leonard Searle, Christopher Sneden, Lyman Spitzer, Don Vanden-Berg, Stan Woosley, Farhad Yusef-Zadeh, and Robert Zinn.

I am also extremely grateful to the people who read the manuscript and offered their comments: Marcus Chown, Jeff Kanipe, Laurence Marschall, Richard Pogge, and Farhad Yusef-Zadeh. Of course, these people are not responsible for any errors I have committed in this book.

I thank Philippe Van for his illustrations, and Kitt Peak National Observatory and Paul Hodge for their photographs.

For their support of this project, I thank my agent, Lew Grimes, and my editors, Roger Scholl and Rob Radick.

During the writing of this book, nighttime inspiration was provided by Arcturus, Aldebaran, Altair, Antares, Betelgeuse, Capella, Deneb, Fomalhaut, Polaris, Procyon, Regulus, Rigel, Sirius, Spica, and Vega; and musical accompaniment by Anderson, Bruford, Wakeman, Howe; The Blue Nile; The Dream Academy; Emerson, Lake and Palmer; Genesis; Gypsy; Illusion; Jethro Tull; Kansas; King Crimson; Marillion; The Moody Blues; Rare Bird; Renaissance; Rush; Starcastle; The Strawbs; Rick Wakeman; and Yes.

1

WELCOME TO THE MILKY WAY

THE MILKY WAY Galaxy is a celebration of diversity, abounding with hundreds of billions of stars, each different from every other. The Milky Way's brightest stars emit more light in a single day than the Sun will generate for the next two thousand years, while the faintest stars glow so feebly that if one of them replaced the Sun, noon would be darker than a moonlit night. The Galaxy's hottest stars pour out large quantities of ultraviolet radiation and appear blue, while the coolest stars shine a ruddy red. The largest stars, if at the solar system's center, would touch Saturn, whereas the smallest are smaller than the main islands of Hawaii. The Milky Way's oldest stars date back to the Galaxy's formation, 10 to 15 billion years ago; its youngest are younger than you or I.

Yet all these stars—bright and faint, hot and cool, large and small, old and young—belong to the same mighty galaxy, just as billions of different plants, trees, animals, and people inhabit the Earth. If God created the universe, and if one can judge the creator from the creation, then God must love diversity, for both our planet and our Galaxy revel in it.

The Galaxy's diversity stems in part from its enormous size, since the Milky Way is far larger and brighter than most other galaxies in the

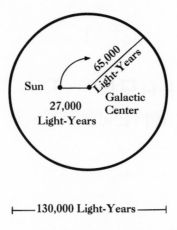

├────130,000 Light-Years────┤

Figure 1. A top view of the Milky Way's disk, which emits most of the Galaxy's light and contains the spiral arms. The Sun lies in the disk, 27,000 light-years from the Galactic center, or about 40 percent of the way from the center to the disk's edge. As viewed from above the disk, the Sun and most other stars revolve clockwise around the Galaxy.

universe. Every star the naked eye can see, including the Sun, is a member of the Milky Way, and the Galaxy is so huge that the Sun requires 230 million years to complete one orbit around the Milky Way's center. An astronomer who tried to map the Milky Way by representing the distance between the Sun and the Earth—93 million miles —as one inch would need a sheet of paper larger than the entire world in order to capture the Milky Way's full extent. On such a map, the Galactic center would be 27,000 miles from the Sun, for the Sun lies 27,000 light-years from the center of the Galaxy.

But even the Sun's distance from the Galactic center is not as extreme as it might appear, because the Sun is actually closer to the center of the Galaxy than to the edge. The Galactic disk, which contains the Sun, is the bright, pancake-shaped part of the Galaxy. It includes the spiral arms that would give our Galaxy the appearance of a pinwheel to an observer outside it. The edge of this disk is some 65,000 light-years from its center, so the disk measures about 130,000 light-years across. But the disk is only part of the Galaxy, and some stars, which do not belong to the disk, lie even farther out.

Beyond these distant stars the Milky Way rules an empire that spans over a million light-years. At least ten other galaxies, all smaller than ours, revolve around the Galaxy like moons orbiting a giant

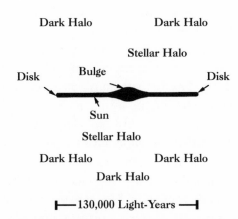

Figure 2. A side view of the Milky Way, showing the disk and the other major components: the bulge, which lies at the Galaxy's center; the stellar halo, which surrounds the bulge and the disk; and the dark halo, which surrounds everything else and extends beyond this page.

planet. These satellite galaxies are colonies in the Milky Way's empire, ruled by its gravitational force. The two largest and most famous of them, the Large and Small Magellanic Clouds, contain billions of stars, while the most remote outpost in the Galactic empire, a dwarf galaxy some 890,000 light-years from the Milky Way's center, harbors a few million stars.

The Milky Way is as dynamic as it is powerful. Every second, a bright orange star named Arcturus moves 3 miles closer to the Sun, a bright white star named Fomalhaut moves 4 miles farther away from the Sun, and the Sun itself moves 140 miles along its orbit around the Galaxy. Like all the Milky Way's citizens, these stars revolve around the center of the Galaxy on their individual orbits the way planets circle the Sun. As a result, the Galaxy is constantly changing. Today, a bright triple star called Alpha Centauri, 4.3 light-years away, is the Sun's nearest neighbor, but millions of years from now Alpha Centauri will be over a hundred light-years distant.

The Galaxy is changing in other ways as well. Every year, in the stellar nurseries scattered throughout its spiral arms, the Milky Way gives birth to about ten new stars. Hundreds of newborn stars speckle the most famous stellar nursery, the Orion Nebula, which is visible to the naked eye as a fuzzy light in the sword of the constellation Orion. Full of gas and dust, the Orion Nebula is set aglow by the very stars it has created. The Sun was born in such a nursery, as were the other stars now shining throughout the Galaxy.

But stars also die. A few times every century, somewhere in the Galaxy, a large bright star exhausts its fuel and explodes, casting off debris in a supernova. The last supernova observed in the Milky Way occurred in the year 1604, before astronomers were using telescopes, but other supernovae have probably exploded since then in sectors of the Galaxy so remote that the supernova's light was obscured by gas and dust. The Galaxy's next supernova could appear at any time.

Most stars, though, die in a less spectacular way, by shedding their outer atmospheres and forming bubbles of expanding gas around themselves. The expanding gas is called a planetary nebula, not because it has anything to do with planets but because, seen through a small telescope, a planetary nebula looks round and extended, like a planet's disk, rather than sharp and starlike. The most famous planetary nebula is the beautiful Ring Nebula in the constellation Lyra. The Sun will end its life this way, as will most of its neighbors. Whether a star dies in this gentle manner or in a violent supernova depends on the mass the star was born with. A star that starts life with less than eight times the mass of the Sun forms a planetary nebula, whereas a star born with more than eight solar masses explodes in a supernova.

Though such objects and processes seem far away, both in space and in time, we owe our lives to the stars that died long ago, before the Sun and Earth were even born. Take a deep breath: 21 percent of the molecules you breathe are oxygen, without which human life could not survive. Oxygen is in the air, the water, our blood; yet when the universe began, there was no oxygen at all. Nearly every oxygen atom on Earth came from high-mass stars that exploded. The heat inside these stars transformed lighter elements into oxygen, which the stars ejected into the Galaxy when they exploded. High-mass stars are rare, however. Rigel, a bright blue star in Orion, is the best example, but it lies some 900 light-years away.

Other types of stars have enriched the Galaxy with other heavy elements, such as carbon and nitrogen. Carbon is the basis of terrestrial life, and nitrogen makes up 78 percent of the Earth's atmosphere. Most carbon and nitrogen atoms were formed in stars that did not explode, stars that ended their lives by creating planetary nebulae.

When the planetary nebulae expanded into space, they enriched the Galaxy with their carbon and nitrogen atoms, some of which joined the cloud of gas and dust that eventually gave birth to the Sun and Earth. You are made of those atoms today.

Another vital element, iron, has two sources. Some of the iron in your blood came from high-mass stars that exploded—the same stars that produced oxygen. But most iron has a humbler source: the white dwarf. White dwarfs make up 10 percent of the Galaxy's stars, and the nearest white dwarf lies just 8.6 light-years away, in orbit around the bright star Sirius. Common though they are, white dwarfs are so small and faint that none is visible to the naked eye, and most lead quiet lives. But a white dwarf can receive material from another star and annihilate itself in a supernova explosion. These white dwarf supernovae differ greatly from the explosions of high-mass stars. While the latter cast large quantities of oxygen into the Galaxy, white dwarf supernovae produce large amounts of iron. Most of the iron now flowing through your veins came from white dwarfs that exploded long ago, perhaps on the other side of the Galaxy. These iron atoms may have drifted through the Milky Way for billions of years before some of the atoms enriched a promising cloud of gas and dust. That cloud gave birth to the Sun and Earth, which 4.6 billion years later gave birth to us.

A Galaxy for All Seasons

To see our Galaxy, it doesn't matter what season you choose, for each season shows a different face of the Milky Way, just as each season reveals a different aspect of a great tree: winter accentuates the bareness of its branches, spring the beauty of its blossoms, summer its leaves of green, autumn its crimson and gold.

Suppose it is winter, when the sky bristles with bright stars. Though winter may be a season of cold and snow, for a stargazer it is the most spectacular. The disk of the Milky Way is high in the sky, and we look away from the Galactic center, toward the Galaxy's edge. This is ideal, because the Sun lies on the inner part of a spiral arm. Spiral arms have bright stars, so when we look away from the Galactic center we stare into the splendors of our own spiral arm. This arm is called the Orion arm, for it contains the bright stars of that constellation, such as the blue supergiant Rigel and the red supergiant Betelgeuse.

These stars lie near a faint band of white light—what some people call "the Milky Way"—that is produced by the combined glow of distant stars in the Galaxy's disk. But every star the naked eye can see, whether or not the star lies in this band, is part of the Milky Way *Galaxy*. How can we both be part of the huge Milky Way Galaxy and still see it as a band of light encircling us? The most luminous stars are in the Galaxy's disk, so when we look at this disk, we see lots of stars— "the Milky Way." When we look away from the disk, we see fewer stars and the sky is darker. It is like being in the midst of a crowd of people. When you look into the crowd, you see people all around you, but when you look up or down, you see none.

As the Earth revolves around the Sun, winter turns into spring, and a darker face of the Galaxy emerges. During spring, we look not into the Galactic disk but above it, so the sky has fewer bright stars, and those which appear bright are fairly nearby, such as Arcturus, which is only 34 light-years from Earth. But the darkness of spring is the Galaxy's greatest illusion, for surrounding the Milky Way's bright disk is a halo of ancient stars. Most halo stars lie closer to the Galactic center than does the Sun, but a second halo engulfs the disk and the stellar halo and extends far beyond both. Though the Milky Way's disk emits most of the Galaxy's light, this second halo, the dark halo, contains most of the Galaxy's mass.

During summer, the Milky Way's disk again rises high, as it did

Figure 3. Even though the Sun is part of the Milky Way, we see a band of white light surrounding us that is also called "the Milky Way." This band results from the combined glow of the stars near us in the Galaxy's disk and surrounds us because the disk itself does.

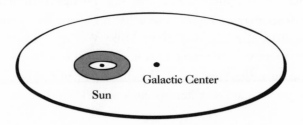

during winter, but we now peer toward the Galactic center and away from the Orion arm. Because we look toward the Galactic center, which lies in the constellation Sagittarius, "the Milky Way" is more spectacular; but because we face away from the Orion arm, we see fewer bright stars.

During autumn, we see another aspect of the Galaxy, for we look below the Galactic plane and the sky is dark. The autumn sky has only one bright star, Fomalhaut, which is 21 light-years away in the constellation Piscis Austrinus. Fomalhaut is the loneliest bright star, more isolated than any other. When it dies, it will shed its atmosphere, forming a planetary nebula whose atoms will drift through the Galaxy, enter star-forming regions, and enrich future generations of stars and planets.

That, in brief, is a tour of the Galaxy, a tour that makes use of information that astronomers have painstakingly gathered during the past hundred years. The struggle to unravel the Galaxy's nature is similar to trying to figure out the shape of a house if one were permanently trapped inside a small closet. So great was the challenge that only in the twentieth century did astronomers deduce the basic properties of the Milky Way. In fact, a hundred years ago, most astronomers believed that our Galaxy was all there was to the universe.

BEYOND THE MILKY WAY

Today astronomers know better. Large and mighty though it is, the Milky Way is just one of billions of galaxies in the observable universe. Our Galaxy is larger than most, but galaxies even bigger than the Milky Way exist, such as the Andromeda Galaxy, the nearest giant galaxy to ours, 2.4 million light-years away. Like the Milky Way, Andromeda rules a vast empire of smaller galaxies.

Andromeda, the Milky Way, and several other nearby galaxies make up a gravitationally bound collection of about thirty galaxies called the Local Group. Of these galaxies, Andromeda and the Milky Way are the largest and rule the rest. Most galaxies in the Local Group circle either Andromeda or the Milky Way, but even those which orbit neither galaxy have their motions affected by the two giants. Indeed, Andromeda and the Milky Way perturb each other. Even though the universe is expanding and tries to carry the two apart, Andromeda and the Milky Way have so much mass—each about a trillion times the Sun's—that their gravitational attraction overcomes the universal expansion. As a consequence, every day Andromeda draws 6.4 million miles closer to the Milky Way. Billions of years from now, the two galaxies may collide and merge, creating a new, gargantuan galaxy.

The Local Group is one among many groups of galaxies throughout the universe. The nearest group beyond the Local Group is the Sculptor Group, which takes its name from the constellation where most of its members lie. These galaxies are 4 to 10 million light-years from the Earth, and the king is a large spiral galaxy, somewhat smaller than the Milky Way, named NGC 253. On the opposite side of the Local Group from the Sculptor galaxies is the M81 group, named for a giant spiral galaxy the size of the Milky Way that lies near the Big Dipper. M81 is 11 million light-years away.

The Local, Sculptor, and M81 groups are themselves members of a still larger structure called the Local Supercluster, a vast, cigar-shaped conglomeration of thousands of galaxies. The Local Group is near one edge of the Local Supercluster, whose heart is the Virgo cluster of galaxies, located between 40 and 70 million light-years away. Beyond the Local Supercluster are other superclusters, many of which take their names from the constellations where their members appear, such as the Pisces-Perseus supercluster, some 250 million light-years away, and the Hercules supercluster, 500 million light-years distant. Superclusters are strung across space like the strands of endless cobwebs.

Viewed in this context, our Galaxy might seem insignificant. But the Milky Way is an ideal laboratory in which to investigate galaxies throughout the universe, because astronomers can see objects in the Milky Way that they cannot in other galaxies. For example, 70 percent of the stars in our Galaxy are faint cool stars called red dwarfs. Most red dwarfs emit less than a hundredth the light of the Sun, and some are among the faintest stars known. After red dwarfs, the most common stars are somewhat hotter and brighter objects called orange dwarfs, which account for 15 percent of the Galaxy's stars. Astronomers cannot see them in most other galaxies, either. Another 10 percent of stars are the white dwarfs, which, again, are invisible in other galaxies. Together, red, orange, and white dwarfs make up 95 percent of all stars in the Galaxy and presumably the universe, but only here in the Milky Way can astronomers actually study them.

One of the major questions facing astronomers today is how galaxies form. The big bang that started the universe expanding is thought to have been smooth, with material spreading out uniformly in all directions, but the present universe, filled with superclusters, is lumpy and uneven. The Milky Way offers unique insight into the origin of one giant galaxy and, by extension, of other large galaxies as well. Especially valuable are the oldest stars in the Galaxy, for they were born when the Galaxy itself was and now serve as time capsules that preserve the Milky Way's ancient past. What they are made of is what the Galaxy must have been made of, billions of years ago; where they lie reveals the shape of the Galaxy when it was born; and the orbits they follow today are relics of the violent era in which the stars arose. Only in the Milky Way can an astronomer study the details of individual stars and, like a celestial archaeologist, piece together the origin and evolution of a galaxy.

The oldest stars in the Galaxy hold secrets to another unanswered question, the age of the universe. The universe must be at least as old as its oldest stars, but astronomers can determine stellar ages only for stars in or near the Milky Way. The oldest of these stars appear to be about 15 billion years old. Cosmologists can also deduce the universe's age in another way, from the rate at which the universe is expanding due to the big bang. The faster the expansion, the younger the universe. In recent years, however, many cosmologists have reached the alarming conclusion that the universe seems to be younger than the Milky Way's oldest stars, a discrepancy that could threaten big bang cosmology.

Another question confronting cosmology is the universe's total mass, which dictates our ultimate fate. Although the universe is still expanding from the big bang, the expansion is slowing, because mass exerts a gravitational pull that brakes the expansion. If the universe is sufficiently massive, it will eventually halt its expansion and begin to collapse. But if the universe does not have enough mass, it will keep expanding forever. The mass associated with visible stars is not enough to stop the expansion, but by studying our Galaxy and others, astronomers have realized that the universe also contains dark, or invisible, matter that slows the universe's expansion. The Galaxy is an ideal place to hunt for the mysterious dark matter that pervades the universe.

As we explore our Galaxy, then, we will see that the Milky Way is more than just the galaxy to which we owe our existence; it is a touchstone for the entire universe.

2

RIVER OF STARS

THOUSANDS OF YEARS ago, before the advent of garish outdoor lighting, the Milky Way was a spectacle from almost anywhere. At that time, the Milky Way was as near as the next clear evening: a ghostly band of stars and glowing light that looked so close one could almost touch it.

Today, such a sight appears only in skies far from cities and suburbs, so most people have never seen the Milky Way and might not recognize it if they did. A few years ago, a young city-raised astronomer was using a telescope atop a remote mountain in Arizona to measure the brightnesses of different stars. Such work is demanding, for even a thin cloud can ruin the observations if it passes in front of a star and dims its light. Consequently, this young astronomer was vigilant, constantly checking the sky for clouds, when she noticed with horror that a band of big white clouds was approaching. She started to close the telescope's dome, but fortunately another astronomer, older and wiser, interceded: the big white clouds were the Milky Way!

THE ANCIENT MILKY WAY

In sharp contrast to today, most people in ancient times knew the Milky Way, and numerous myths sought to explain its existence. To the Greeks, whose myth gave it its name, the Milky Way arose when the infant Heracles—son of Zeus and one of his lovers, Alcmene—bit the breast of Zeus's wife, Hera, and spilled her milk into the sky, milk that the Romans called Via Lactea, or Milky Way. To some of Australia's aborigines, the Milky Way was the smoke from a heavenly campfire that the creator, exhausted following the creation, had lit before going to sleep for the night. To some American Indians, the Milky Way was the road that brave warriors followed to heaven after they died, and the bright stars near this road were campfires where the deceased could rest during their journey.

The most romantic story, though, comes from the Chinese, who saw the Milky Way as a heavenly river. Close to but on opposite sides of this river lie Vega and Altair, the two brightest stars in the summer sky. According to one version of the Chinese legend, Vega, the brighter star, was really a princess, a weaver of exquisite clothes and the daughter of the Sun god, while Altair was a herdsman who tended the imperial cattle. From her side of the river, Vega watched Altair with growing fondness and asked her father to arrange a meeting. Soon after, Vega and Altair fell in love and began spending all their time together; Vega neglected her weaving, Altair his cattle, and the Sun god became angry. After repeated warnings, which Vega and Altair ignored, the Sun god separated the two lovers, placing them once again on opposite sides of the Milky Way, where they remain today. Nevertheless, a tearful Vega extracted a promise from her father. On one night of every year—the seventh night of the seventh moon—a bridge of birds would span the Milky Way's waters, allowing the two lovers a night together. But for the birds to appear, the sky must be clear, and even today, in parts of Asia, some pray for clear skies that night so that the lovers can be reunited.

Just as fanciful, if less poetic, were most of the theories that ancient philosophers proposed for the Milky Way. Aristotle held that it was merely an atmospheric phenomenon, a vapor that the Earth had emitted. He could not admit that the band of light was celestial, because he believed the heavens were perfect, which the Milky Way, with its ragged edges, was not. Other philosophers thought the Milky Way was the path that the Sun once followed, or the zone where the two hemispheres of the sky joined. In the fifth century B.C., the Greek philosopher Democritus, who lived before Aristotle, offered the correct explanation, suggesting that the Milky Way was composed of countless stars, each too faint to be seen individually. Democritus's theory echoed another that he advocated and for which he is better known, that matter consists of tiny particles called atoms.

EARLY MILKY WAY SCIENCE

In 1609, the great Italian astronomer Galileo Galilei turned his newly made telescope on the Milky Way and confirmed Democritus's speculation:

> I have observed the nature and the material of the Milky Way. With the aid of the telescope this has been scrutinized so directly and with such ocular certainty that all the disputes which have vexed philosophers through so many ages have been resolved, and we are at last freed from wordy debates about it. The Galaxy is, in fact, nothing but a congeries of innumerable stars grouped together in clusters. Upon whatever part of it the telescope is directed, a vast crowd of stars is immediately presented to view. Many of them are rather large and quite bright, while the number of smaller ones is quite beyond calculation.

No one cared. Galileo did not visit the subject again, and other astronomers were so obsessed with the solar system that interest in the Milky Way did not resurface until a century and a half later. When it did, it was through the writings of a philosopher rather than the observations of an astronomer. Prussian philosopher Immanuel Kant is best known today for one of the most convincing arguments for the existence of God, the so-called moral argument. God must exist, said Kant, for if he did not, morality would be meaningless, and nothing, not even a massacre of millions of people, would be wrong.

Kant's astronomical reasoning was just as profound. In 1755, after reading a long but misleading review of a book by an Englishman named Thomas Wright, Kant suggested that the Milky Way was a disk of stars. We are in this disk, said Kant, so when we look along its plane, we see numerous stars—"the Milky Way"—and when we look away from it, we see only the few nearby. In his writing Kant credited Wright, but Wright did not believe the Milky Way to be a disk. Like Wright, however, Kant suggested that the Milky Way's stars were revolving around some center, just as the planets of the solar system revolve around the Sun. Kant also speculated that many of the fuzzy celestial objects that astronomers called "nebulae" resembled the Milky Way and were located beyond it. These ideas were all correct, but the world heard little of them, for the publisher of Kant's book went bankrupt and most of the books were impounded to pay off the publisher's debts.

Meanwhile, professional astronomers continued to neglect the Milky Way, and the next advance came a generation later, again from an outsider. German-born musician William Herschel had emigrated to England as a youth and discovered astronomy when he was thirty-four years old. Unlike professional astronomers, Herschel had little interest in the solar system; his goal was to study what lay beyond. Herschel thus embarked on an ambitious attempt to chart the Milky Way. Using large telescopes that he had built himself, he counted the number of stars in hundreds of different directions. Some of the regions he searched presented only a star or two, but others bristled with thousands. In 1785 Herschel used these star counts to publish the first map ever made of the Milky Way's structure. His map revealed that the Milky Way was a disk, as Kant had suggested. According to Herschel, the Sun was near the disk's center, and the disk itself was four times longer than it was thick.

In preparing his map, Herschel had assumed that every star possessed the same true, or intrinsic, brightness. If so, a star's *apparent* brightness indicated its distance: the fainter the star looked, the farther away it was. Modern astronomers are quick to criticize this assumption, knowing that the most luminous stars shine a trillion times more brightly than the least luminous—which suggests that a star's apparent

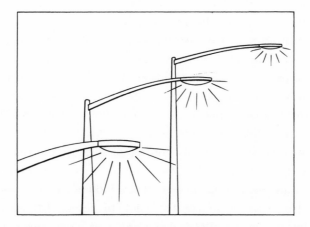

Figure 4. If all stars had the same intrinsic *brightness, as these streetlights do, then astronomers could determine stellar distances from the stars'* apparent *brightness: the fainter the star looks, the farther it must be. Unfortunately, unlike these streetlights, stars differ greatly in intrinsic brightness, so a star's apparent brightness is only a rough measure of distance.*

brightness is almost useless in determining distance. But in fact, Herschel's assumption was not so bad, because intrinsically faint stars—the vast majority—are too dim to contribute much light to the Milky Way, while the intrinsically brightest stars are too rare to do so. As a consequence, most of the Milky Way's light comes from stars lying in a narrow range of luminosity, with intrinsic brightnesses between 10 and 150 times that of the Sun.

Yet Herschel had implicitly made a far worse assumption: that space was free of interstellar material. Astronomers now know that the Milky Way's disk contains gas and dust, which absorb half a star's light within only 2500 light-years of the star. Like a person engulfed in fog, Herschel could see only a fraction of what he should have, and he appeared to be at the center of the little he could see. That is why he erroneously placed the Sun near the Milky Way's center.

Herschel's map had no precise scale attached to it, for he did not know the distance to any of the stars he saw and could not deduce the true extent, in light-years, of the Milky Way. But distance was crucial. Two stars might lie side by side in the same constellation, but there was no guarantee they were the same distance from Earth. So important was this problem that even before Herschel prepared his map he

had tried to measure what astronomers call parallax. As the Earth revolves around the Sun, the apparent position of a star should shift
slightly, because in January astronomers view the star from the opposite side of the Sun as they do in July. The nearer the star, the larger
this shift, or parallax, should be. Unfortunately, although many astronomers, including Copernicus, had searched for parallax, none had found
it, which meant that no one could say how distant individual stars were
or how big the Milky Way was.

Parallax eluded astronomers for so long because the stars are so far
from Earth. Indeed, even the nearest star to the Sun has a tiny parallax,
less than an arcsecond (one arcsecond is 1/3600 of a degree)—too small
for any astronomer, including Herschel, to detect. Efforts to record
parallax almost always focused on the brightest stars, since astronomers
figured the bright stars were nearby and had large parallaxes that were
easy to observe: Sirius, Arcturus, Vega, Capella, Procyon, Altair, Aldebaran, Deneb, and Polaris were all targets, but astronomers either reported no parallax or found only spurious ones. For example, in 1699
British astronomer John Flamsteed published a parallax for Polaris that
gave the faithful North Star a distance of only 0.08 light-years, which

*Figure 5. As the Earth circles the Sun, the positions of the stars appear to shift
slightly. The closer the star to the Sun, the greater is this shift, or parallax;
therefore, the measurement of parallax reveals the star's distance from the
Sun.*

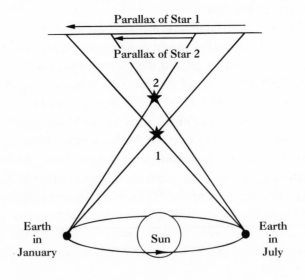

would have placed the star inside the solar system. The star's actual distance is over 4000 times greater.

Ironically, when astronomers finally detected true stellar parallax, in 1838, the winning star was not a leading luminary like Sirius or Vega but a dark horse named 61 Cygni. Just barely visible to the naked eye, 61 Cygni lies in the constellation Cygnus and consists of two orange dwarfs. In the Earth's sky, over a thousand other star systems look brighter. Yet 61 Cygni opened the pathway to the Galaxy.

Despite its dimness, the star attracted the attention of perceptive astronomers because of its large proper motion, a sign that the star might be nearby. Proper motion is the apparent motion of a star, year after year, across the sky; it reflects both the star's true velocity across the line of sight (the greater the star's velocity, the greater the proper motion) and the star's distance from Earth (the smaller the distance, the greater the proper motion). Anyone riding down a highway has seen the same phenomenon: look to the side and nearby objects, like road signs, whiz by fast, having what astronomers would call a large proper motion. But distant objects, such as hills and clouds, appear to move slowly, having a small proper motion. Proper motion itself had first been discovered in 1718 by British astronomer Edmond Halley, of comet fame, who noticed that the positions of Arcturus and Sirius differed from those on ancient star maps. He also found other evidence that the star Aldebaran had moved. This motion was called "proper" to distinguish it from various "improper" motions that have nothing to do with the star itself, such as the apparent movement a star shows as the Earth rotates. Of all the bright stars visible from England, Arcturus has the greatest proper motion, moving the equivalent of one lunar diameter—half a degree—every 820 years. This works out to a proper motion of 2.3 arcseconds per year.

In 1804, Sicilian astronomer Giuseppe Piazzi was constructing a star catalogue when he discovered that the otherwise unspectacular star 61 Cygni had a huge proper motion of 5.2 arcseconds per year, over twice that of Arcturus. Like the blur of a nearby road sign rushing past, the large proper motion indicated that the star might be nearby, and Piazzi suggested that 61 Cygni would make a good parallax target. But no one followed his lead, perhaps because astronomers were not ready to believe that such a faint star could be close to the Earth.

In 1812, however, Prussian astronomer Friedrich Wilhelm Bessel published the same result for 61 Cygni's proper motion, and in 1838 he finally succeeded in determining the star's parallax, finding a distance within 10 percent of the modern value of 11.4 light-years. The heavens were at last unlocked; the Milky Way was fathomable. "Esteemed Sir," wrote Bessel in a letter to John Herschel, the only son of William. "Having succeeded in obtaining a long-looked-for result, and presuming that it will interest so great and zealous an explorer of the heavens as yourself, I take the liberty of making a communication to you thereupon. Should you consider this communication of sufficient importance to lay before other friends of Astronomy, I not only have no objection, but request you to do so." Almost immediately, Wilhelm Struve in Estonia determined the parallax of Vega, and Thomas Henderson in South Africa measured the parallax of Alpha Centauri, the nearest star system to the Sun.

Despite these successes, the determination of parallax remained so difficult that even by the end of the nineteenth century only a few dozen reliable parallaxes had been measured. Meanwhile, astronomers were gathering clues to the Milky Way from what were then called "nebulae," objects now known to be other galaxies beyond our own. In 1828 and 1830, John Herschel observed a nebula south of the Big Dipper's handle named M51 and saw it as a bright nucleus surrounded by a ring. In 1833, he asked, "Can it, then, be that we have here a brother-system bearing a real physical resemblance and strong analogy of structure to our own?" The following decade, M51 yielded more of its secrets after William Parsons, the third earl of Rosse, built a mighty 72-inch telescope at his castle in Ireland. The telescope was the world's largest, and in April 1845 Lord Rosse turned it to M51 and discovered that the nebula had a spiral pattern. He later discovered dozens of other spiral nebulae. But the structure of what is now known to be the nearest spiral—mighty Andromeda—eluded astronomers until 1887, when Isaac Roberts in Wales used photographic plates to capture Andromeda's spiral shape. From these pioneering discoveries, however, many astronomers incorrectly deduced that the spiral nebulae were not systems similar to and outside the Milky Way, as John Herschel had suggested, but were instead much smaller and nearer, lying within the Milky Way. To these astronomers, the spiral nebulae were newborn solar systems: the bright center was a young star, and the spirals surrounding it were developing planets.

THE SMALL UNIVERSE OF
JACOBUS KAPTEYN

The invention of photography gave astronomers a new tool in their study of the Milky Way. After all, if photographic plates could reveal the spiral shape of Andromeda, they might also unravel the nature of the Milky Way. A single photographic plate could capture thousands of stars of different brightnesses, distances, and proper motions. With photography, astronomers could redo what William Herschel had done by eye, counting the number of stars in different directions to map the Milky Way.

One of the first to seize this opportunity was Jacobus Kapteyn, a Dutch astronomer who devoted four decades of his life to the tedious task of counting stars. As one commentator said, "Never did he shrink from any labour, however great, to elucidate a doubtful point, perhaps uninteresting in itself, but necessary as a part of the whole." During the eighteenth century the study of the Milky Way had centered on the musings of a philosopher and the observations of a musician; but during the late nineteenth and early twentieth centuries Kapteyn placed the subject at the forefront of astronomical research.

Yet to no other astronomer was the Galaxy more cruel. Despite his hard work, the model of the Milky Way that Kapteyn derived from his star counts was wrong, for like William Herschel's it mistakenly placed the Sun near the center of a small disk of stars. In admiration for Kapteyn's work, British astronomer James Jeans dubbed the model the "Kapteyn universe," but the name ultimately did Kapteyn more harm than good, for when the Kapteyn universe collapsed, a few years after his death, it took Kapteyn's name with it. In his day Kapteyn was one of the world's greatest astronomers, but today even books on the Milky Way often fail to mention him.

Born in a small village in 1851, Jacobus Kapteyn was one of fifteen children. In 1875, he received a doctorate in physics from the University of Utrecht, and at the age of twenty-seven became a professor of astronomy at the University of Groningen. Even then, the structure of the Milky Way was on his mind, for his opening address to the university was "The Parallax of the Fixed Stars." But Groningen had no observatory, and despite repeated attempts, Kapteyn failed to find funds to build one.

"Kapteyn presented the unique figure of an astronomer without a telescope," wrote Frederick Seares many years later. "More accurately, all the telescopes of the world were his." In his ambition to fathom the Milky Way, Kapteyn established links with astronomers around the world who supplied him with the data he needed. His first major project was a collaboration with David Gill, director of the observatory at the Cape of Good Hope in South Africa. Gill was about to photograph most of the southern sky, and Kapteyn knew that these photographs would capture hundreds of thousands of stars, so in 1885 he volunteered to measure Gill's plates in Holland and determine the positions and brightnesses of all the stars. The work took over a decade; at one point, in need of money, Kapteyn asked for, and received, convicts from a Holland prison to help with the computations. The resulting star catalogue, the *Cape photographic Durchmusterung*, appeared in three volumes—in 1896, 1897, and 1900—and contained positions and magnitudes for 454,875 stars.

The massive star catalogue held a wealth of data for an investigation of the Milky Way's structure. But Kapteyn did not know most of the stars' distances, because only a few dozen stars had reliable parallaxes. Instead, Kapteyn relied on proper motion, which was much easier to determine than parallax. Thousands of stars had accurate proper motions, and Kapteyn used these as distance indicators. On average, the smaller a star's proper motion, the greater its distance, so proper motion served as a surrogate for parallax. From the few nearby stars that had both known parallaxes and proper motions, Kapteyn found a relation between a star's mean parallax, proper motion, and apparent brightness. He then extrapolated to stars that had known proper motions and apparent brightnesses but not parallaxes. And he then extrapolated still further—to stars so distant that their proper motions were too small to be detected—by using these stars' apparent brightnesses.

Proper motion, in fact, led Kapteyn to discover a faint star in the southern constellation Pictor that now bears his name. In 1897 Kapteyn compared the star's position on Gill's plates with the position previous astronomers had measured, and discovered that the star had a proper motion of 8.7 arcseconds per year, displacing the previous record holder, Groombridge 1830, whose proper motion of 7.0 arcseconds per year had been found in 1841. Even today, the proper motion of Kapteyn's Star is second only to that of Barnard's Star (discovered in 1916), whose proper motion is 10.3 arcseconds per year. As astronomers now know, Kapteyn's Star has a large proper motion because it is the nearest star that orbits the Milky Way backward; the star revolves in the opposite direction from the Sun and so has a large velocity relative to us.

In 1906, Kapteyn conceived an ambitious program to probe the Milky Way more thoroughly. With advances in photography, astronomers could capture fainter and fainter stars, but a survey of the entire sky would produce far more data than even he could analyze. Kapteyn therefore marked off 206 small areas in different parts of the sky and called on astronomers around the world to obtain data on the stars within them. Just as holes drilled at various points through a mountain could reveal the mountain's structure, so the study of each of Kapteyn's selected areas might unveil the structure of the Milky Way.

Although astronomers were familiar with the results of Kapteyn's labor before they were published, the culmination of his work came in 1920 and 1922, when he laid out in detail his model of the Milky Way. Kapteyn's picture was similar to William Herschel's, describing the Milky Way as a disk of stars with a diameter 5.1 times greater than its thickness. But unlike Herschel, Kapteyn had used parallax and proper motion to estimate the Milky Way's size. His 1922 paper said the Milky Way's diameter was 55,000 light-years, its thickness was 11,000 light-years, and it contained 47.4 billion stars.

The Sun, according to Kapteyn, was near the center of this disk. In his 1920 paper, written with his student Pieter van Rhijn, Kapteyn had simply assumed this, but his 1922 paper expressed discomfort with the idea. "It is of course infinitely probable that the Sun must be at a certain distance from the center of the system," he wrote. Later in the paper, Kapteyn nevertheless concluded that the Sun was close to the Milky Way's center, lying between 2000 and 2300 light-years from it, and that the center was in the northern constellation Cassiopeia.

Kapteyn well knew that his entire model rested on the same assumption Herschel had made: the space between the stars was free of material that absorbed starlight. He recognized that any interstellar fog would incorrectly yield a Galaxy too small, with the Sun near its center. Although observers did see individual clouds of interstellar gas, such as the Orion Nebula, few astronomers thought that a general absorbing medium existed throughout space.

In 1904, however, German astronomer Johannes Hartmann had reported interstellar gas along the line of sight to Mintaka, one of the stars in Orion's belt. Mintaka is really three stars, two of which orbit each other every 5.7 days. As these two stars revolve around each other, one star approaches an observer while the other recedes. The approach of the first star scrunches up its light waves, reducing their wavelength. This is called a blueshift, because blue light has a short wavelength. Likewise, light waves from the receding star get stretched out to longer —or redder—wavelengths, the phenomenon that astronomers call a redshift. However, Hartmann noticed one line in Mintaka's spectrum that did not exhibit blueshifts and redshifts but instead remained stationary as the two stars circled each other. He correctly deduced that the spectral line did not arise from either star but was produced by stationary gas that lay somewhere between Earth and Mintaka and absorbed Mintaka's light.

Kapteyn himself also began to study interstellar absorption. In a 1909 paper he wrote, "[In 1904] I tried to show how fundamentally our results for the arrangement of the stars in space are changed by admitting even an absorption of light considered as small by some astronomers. Now there can be little doubt, in my opinion, about the existence of absorption in space." In another paper that year, as well as one in 1914, Kapteyn again argued for the possibility of interstellar absorption, noting that distant stars looked redder than nearby ones. This reddening of distant starlight could arise because gas and dust scatter and absorb blue light, leaving only red light to reach an observer. For the same reason, the Sun at sunset appears orange or red, because its light passes through large amounts of gas and dust on the Earth's horizon.

But before publishing his 1922 magnum opus on the Milky Way, Kapteyn changed his mind and claimed that no interstellar absorption existed. To support this view, on which his entire model depended, Kapteyn cited the work of a young astronomer in California. This astronomer had observed distant clusters of stars, and the stars showed the full range of color from blue to red, indicating that they suffered no reddening or absorption of their light. Ironically, the young astronomer whom Kapteyn cited was Kapteyn's nemesis, a man who was pushing a radically different model that would ultimately demolish Kapteyn's life work: Harlow Shapley.

3

BIG GALAXY,
BIG UNIVERSE

THE UNIVERSE OF Jacobus Kapteyn was a cozy place, a small, safe Milky Way with the Sun near the center and most stars revolving around it at the modest speed of twenty kilometers per second, less than the speed with which the Earth circled the Sun. Into this comfortable world charged Harlow Shapley, who shattered the Kapteyn universe and replaced it with a far grander vision. Just as Copernicus had removed the Earth from the center of the solar system, so Shapley would yank the Sun from the center of the Milky Way and put it in the celestial equivalent of a suburb. Shapley's Galaxy was also enormous: if the small Kapteyn model were a village, Shapley's model would be a great metropolis that embraced the entire universe.

Born in Missouri, Harlow Shapley discovered astronomy by accident. He had worked as a newspaper reporter and wanted to study journalism, but when he arrived at college the journalism school had not yet opened, so he had to pick something else. He chose astronomy, he said later, because it was near the front of the college course catalogue; it started with A.

In 1914, after completing graduate school at Princeton, Shapley went to Mount Wilson Observatory in Pasadena, California. Mount Wilson owned the world's largest telescope: built in 1908, the telescope had a mirror that was 60 inches in diameter. (By then, Lord Rosse's 72-

inch telescope had fallen into disuse.) There Shapley turned his attention to the globulars, dense star clusters that typically packed hundreds of thousands of stars into just a hundred or so light-years of space. Using the globular clusters, Shapley would construct his radical new model for the Milky Way.

CELESTIAL FIREFLIES: TRACKING THE CEPHEIDS

Shapley began his work by using remarkable stars called Cepheids, yellow supergiants that pulsate like a human heart, first growing and then shrinking. As the star expands and contracts, it brightens and dims, so the pulsations can be detected on Earth. Most Cepheids have pulsation periods between one and fifty days and pulsate with the pre-

Figure 6. The Cepheid period-luminosity relation: the longer a Cepheid's pulsation period, the greater its intrinsic brightness, or luminosity. A comparison of the intrinsic brightness with the apparent brightness will yield the Cepheid's distance. This graph is based on data from the 1990s.

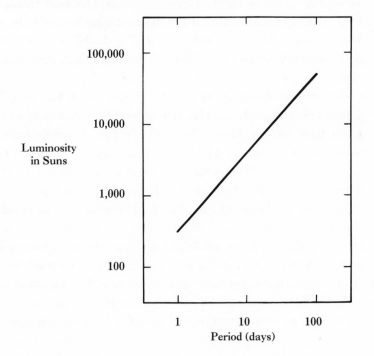

cision of a Swiss watch. For example, the first Cepheid discovered, Eta Aquilae, pulsates exactly once every 7.176779 days.

The Cepheids' most astonishing property, though, is that they obey a period-luminosity relation, which allows an astronomer to determine their distances. The longer a Cepheid's period, the larger the star and the greater the Cepheid's mean intrinsic brightness. Measuring the period, which is easy, therefore determines the Cepheid's intrinsic brightness; and comparing the intrinsic brightness with the Cepheid's apparent brightness determines the star's distance, since the fainter the star looks, the farther away it is. If the Cepheid belongs to a star cluster, the distance to the Cepheid yields the distance to the entire cluster.

The first two Cepheids had been discovered in 1784, a century and a half before Shapley started to use them. At that time, few variable stars of any kind were known. Indeed, that stars can vary in brightness had been anathema to the classical view, espoused by Aristotle, who believed the heavens were perfect and changeless. The first variable star to be discovered was surely Algol, an otherwise ordinary star in the constellation Perseus. Ancient people must have known that the star's light flickered, for the name means the Demon's Head, and medieval astrologers considered it the most dangerous star in the sky.

Algol is not a Cepheid, however; it is a double star that dims whenever one member passes in front of the other. The first to suggest this explanation for Algol's variability was a wealthy English amateur astronomer named Edward Pigott, who in the 1780s had begun to search for variable stars. Like other astronomers, he knew only a handful of variables, such as Algol, Mira, and Chi Cygni, but he correctly suspected that many others might exist. In his work, Pigott served as mentor to young John Goodricke, who was deaf and mute. Goodricke began observing with Pigott at the age of seventeen, and Pigott, eleven years older, guided Goodricke in the proper conduct of astronomical observations.

On the night of September 10, 1784, Pigott detected fluctuations in the light of the first Cepheid, Eta Aquilae. He monitored the star and by year's end had determined the period to within 1 percent of its correct value. Meanwhile, on October 20, Goodricke had noticed that one of the stars in the northern constellation Cepheus was variable, and three nights later he became "almost convinced" that the culprit was Delta Cephei.

Eta Aquilae lay in the summer constellation Aquila, so it could not be observed during the coming winter. In contrast, Delta Cephei was so far north that it never set and therefore could be seen every clear night. Goodricke observed the star over a hundred times in late 1784 and 1785, finding a period close to the modern value of 5.3663 days. But in 1786, Goodricke contracted pneumonia, possibly from exposure to the cold nights, and died at the age of twenty-one. As Pigott later wrote, "I had the misfortune to lose the best of friends . . . which took away the pleasure I ever had in astronomical pursuits."

Eta Aquilae, Delta Cephei, and other stars like them came to be called Cepheids (after Delta Cephei), a word apparently coined by the great British astronomy writer Agnes Mary Clerke. In 1907, Harvard astronomer Henrietta Leavitt discovered the remarkable Cepheid period-luminosity relation. She was studying the Large and Small Magellanic Clouds, which astronomers now know are galaxies orbiting the Milky Way. The Magellanic Clouds lie in the southern sky, and Harvard owned a telescope in Peru, where photographs of the Magellanic Clouds had been taken. As Leavitt examined the photographic plates, she discovered over a thousand variable stars blinking on and off like a nest of fireflies. In her 1907 paper, she listed 969 variables in the Small Magellanic Cloud and 808 variables in the Large Magellanic Cloud.

For 16 Cepheids in the Small Magellanic Cloud, Leavitt had enough data to determine periods. "It is worthy of notice that . . . the brighter variables have the longer periods," she wrote. Unfortunately, that one sentence was her only mention of the period-luminosity relation, and it appeared toward the paper's end, after the long tables of variable stars. As a result, no one took notice of it. But since all the stars in the Small Magellanic Cloud were about the same distance from the Earth, the relation Leavitt had found meant that Cepheids with longer periods were not only apparently brighter but intrinsically brighter as well.

In 1912, Leavitt published a three-page paper devoted solely to the Cepheid period-luminosity relation, this time based on 25 Cepheids of known period in the Small Magellanic Cloud. The following year, Danish astronomer Ejnar Hertzsprung recognized the implications of Leavitt's work and used the period-luminosity relation to measure distances.

THE BIG GALAXY

Harlow Shapley also realized that the Cepheids were powerful weapons in his scientific arsenal. Because he was studying the globular clusters, he wanted to know their distances, so he used what he thought were Cepheids that belonged to the clusters. The Cepheids were faint and therefore distant, and the distances that Shapley calculated were so large that he at first thought the globulars lay outside the Milky Way. They were certainly outside the small Kapteyn universe.

In 1917, however, in a bold move, Shapley reversed himself, bravely proclaiming that the globulars not only belonged to the Milky Way but also traced its structure the way streetlights outline a city. Kapteyn had used stars to trace the Milky Way, but Shapley used the globulars. After measuring their distances, he knew the size of the Milky Way; after determining how far the center of the globular cluster system lay from the Sun, he knew how far the Sun was from the Galactic center.

Shapley's method was elaborate, but it allowed him to deduce the distances of 69 different globular clusters. He began with three clusters in which he used Cepheids to determine the clusters' distances. Unfortunately, the other clusters had no known Cepheids with well-determined periods. But many globulars, including the three that Shapley started with, abounded with a fainter type of variable star. Astronomers now call these RR Lyrae stars, after a star in the constellation Lyra whose variability was reported in 1901. RR Lyrae stars are white or yellow-white giants that pulsate the way Cepheids do, but with shorter periods and lower luminosities. They are even easier to use than Cepheids, since all RR Lyraes have nearly the same intrinsic brightness, independent of their period. As a consequence, measuring the apparent brightness of an RR Lyrae star reveals the star's distance.

Shapley used the distances of the first three globular clusters, those containing Cepheids, to find the intrinsic brightness of the RR Lyrae stars there. Assuming that the RR Lyrae stars in four other globulars had the same intrinsic brightness, he determined the distances to those clusters. At that point, he had seven globulars with known distances, and he noticed that in all seven, the brightest stars—which outshone the Cepheids and the RR Lyrae stars—had similar intrinsic brightnesses. He assumed that the brightest stars in other globulars had this intrinsic brightness as well. From this, and the apparent brightnesses of these stars, he derived the distances of 21 other globulars, giving him a total of 28 clusters with known distances.

He went still further. For the 28 globulars whose distances he knew, Shapley found that the more distant the cluster, the smaller it looked. From this and the apparent sizes of 41 additional clusters, he determined the distances of those clusters. Altogether, then, Shapley had measured distances to 69 globulars—3 from Cepheids, 4 from RR Lyraes, 21 from the brightest stars, and 41 from the cluster's apparent sizes. By assuming these clusters traced the Milky Way, Shapley estimated its size and the Sun's place within it. In 1918, his work gave the Milky Way the enormous diameter of 330,000 light-years, far larger than that of the Kapteyn universe. Furthermore, he yanked the Sun from the Milky Way's center and put it 65,000 light-years out. Shapley also determined the location of the Galactic center, placing it in the southern sky, near the border of the constellations Scorpius and Ophiuchus, just west of Sagittarius, where astronomers now know the Galactic center actually lies. This deduction followed from the globulars, which congregate in this part of the sky, indicating that the Galactic center lies there and the Sun far outside it. In the same way, a suburbanite atop a hill could determine the location of downtown by seeing in which direction streetlights were most numerous.

Shapley's model was basically correct, although astronomers now know that it is about two times too large, for the actual diameter of the Milky Way's disk is some 130,000 light-years, and the Sun lies about 27,000 light-years from the Galactic center. Shapley's overestimate of the Galaxy's size had two causes. First, the "Cepheids" in the first three globular clusters that Shapley used are actually W Virginis stars. Like Cepheids, W Virginis stars pulsate and obey a period-luminosity relation, but at any given period they are intrinsically fainter than Cepheids, so the stars are closer than Shapley thought. This error made him overestimate the first three clusters' distances, which led him to overestimate all the other distances. The second source of error was his neglect of interstellar absorption. If gas and dust partially block the view of a cluster, it will look fainter and farther away than it really is.

Nevertheless, Shapley would fare far better than Kapteyn, because Shapley had chosen globular clusters, rather than stars, to trace the Milky Way's structure. Globular clusters do indeed trace the Milky Way; they are luminous and so can be seen at great distances; and most of them lie outside the plane of the Milky Way, which contains the thickest gas and dust.

"I have always admired the way in which Shapley finished this whole problem in a very short time, ending up with a picture of the Galaxy that just about smashed up all the old school's ideas about [G]alactic dimensions," said astronomer Walter Baade many years later. "It was a very exciting time, for these distances seemed to be fantastically large, and the 'old boys' did not take them sitting down."

One of the first confrontations occurred in 1920, when Shapley and Heber Curtis of Lick Observatory held a discussion in Washington, D.C., concerning "The Scale of the Universe." Often incorrectly called "The Great Debate," the discussion let each side present its view, but neither side could respond to the other, so the event was hardly a true debate. The first question the participants "debated" was the Milky Way's size. Shapley thought it was big, Curtis small, even smaller than the Kapteyn universe.

Shapley had accepted the invitation to debate with reluctance, for he was angling to become director of Harvard College Observatory, whose director had died in 1919. Shapley correctly suspected that Harvard men would be in the audience to judge his performance, and he knew that his opponent, Curtis, was an outstanding public speaker. Shapley unsuccessfully tried to get another, less articulate astronomer to oppose him, and he also tried to reduce the amount of time each person would speak.

Shapley went first. Playing to an audience that included two Harvard men who would help decide his fate, Shapley made his presentation elementary, defining "light-year" only a third of the way through his talk, so that much of what he said was noncontroversial. Listening to Shapley, Curtis was appalled; he had prepared a much more technical argument, and as he waited for Shapley to finish, he thought of scrapping it entirely. But in the end, Curtis proceeded with his original plan, and though he was wrong about the Milky Way's size, his presentation was excellent and he won the debate.

The Harvard men were not impressed with Shapley. Wrote one, "Shapely [sic] lacks maturity and force, and does not give the impression of being a big enough personality for the position." Harvard offered the position that Shapley had sought to another astronomer. He declined, and Harvard then offered a lesser position to Shapley, who left Mount Wilson Observatory for Harvard in early 1921. Later that year, despite his performance in the debate, Shapley became Harvard College Observatory's new director.

In his quest to become director, Shapley had had an odd ally: Jacobus Kapteyn. In a letter to Harvard, Kapteyn wrote, "Shapley is a brilliant man and personally I, who know him mainly through his scientific work, would think him best fitted for the position." Nevertheless, Shapley's model for the Milky Way threatened to topple Kapteyn's life work, and in early 1922 Kapteyn and Pieter van Rhijn attacked Shapley's model. They tried to discredit the intrinsic brightnesses Shapley had derived for the RR Lyrae stars, which directly or indirectly had given him the distances to 66 of his 69 clusters. The two Dutch astronomers looked at the proper motions of 14 RR Lyrae stars and found them high, which suggested to Kapteyn and van Rhijn that the stars were nearby and therefore intrinsically fainter than Shapley had claimed. This in turn meant that the globular clusters containing RR Lyrae stars were much nearer than he had said, which led to smaller distances for most of the other globulars Shapley had used. When Kapteyn and van Rhijn worked out the numbers, they found that the Milky Way was 7.6 times smaller than Shapley's figure.

Shapley challenged the results. True, he said, the 14 RR Lyrae stars have high proper motions, but that does not necessarily mean the stars are nearby, since the proper motions depend not only on the stars' distances but also on their velocities. If the RR Lyrae stars have high velocities, they would still have high proper motions, even if the stars were distant. As Shapley pointed out, and as Kapteyn apparently forgot, the RR Lyrae stars do indeed have high velocities, because their spectra show large redshifts or blueshifts.

Kapteyn died later in 1922, but his model lived on. The two competing versions of the Milky Way both seemed based on secure observations and careful analysis, yet one was clearly wrong. It would be years before the verdict came in, in Shapley's favor, and years more before astronomers knew the reason that Kapteyn had erred: because space was full of obscuring gas and dust.

THE DISCOVERY OF OTHER GALAXIES

Meanwhile, an even larger question was looming in astronomy, one that concerned not just the Milky Way but the entire universe. This too had been a topic during the so-called Great Debate: what exactly were the spiral "nebulae," like Andromeda and M51? So big was Shapley's Milky Way that he believed the spiral nebulae to be small systems located within it. In fact, to Shapley, the Milky Way *was* the universe, a view once held by many astronomers. By the time of the debate, however, Shapley was in the minority, and Curtis argued, correctly, that the spiral nebulae were similar to and well outside the Milky Way. Shapley therefore got only half the debate correct; he was right about the Galaxy but wrong about the universe. As Dimitri Mihalas and James Binney wrote many years later, "Shapley was as conservative (and wrong) on the second question as he was radical (and correct) on the first."

Proof that Shapley was wrong about the spiral nebulae came just three years after the debate. In late 1923, while working at Mount Wilson Observatory's new 100-inch telescope, astronomer Edwin Hubble detected a dim Cepheid in the spiral nebula Andromeda. Shapley did not like Hubble: though also from Missouri, Hubble spoke with an affected Oxford accent that annoyed Shapley. Now, using the Cepheid relation that Shapley had developed, Hubble could estimate Andromeda's distance. The Cepheid that Hubble had found in Andromeda was faint, which meant that Andromeda was distant, contrary to Shapley's belief that it lay within the Milky Way. Hubble's discovery proved that the universe was immense and that other Milky Ways—or "galaxies," after the Greek for "milky ways"—existed throughout it.

Shapley was slow to appreciate Hubble's result. In early 1924 he wrote, "Your letter telling of the . . . variable stars in the direction of the Andromeda nebula is the most entertaining piece of literature I have seen for a long time." But he could hardly argue with a Cepheid, and with chagrin he eventually accepted that there was much more to the universe than the "Big Galaxy" he had constructed during his years at Mount Wilson Observatory. Just as Shapley had challenged Kapteyn's universe, so had Hubble demolished Shapley's. Furthermore, when Shapley had abandoned Mount Wilson, site of the world's largest telescopes, he probably figured that, having deciphered the Milky Way, he had also deciphered the universe. But the Cepheid in Andromeda meant that the universe was vast and uncharted, promising great discoveries that Shapley could have made if only he had remained at Mount Wilson. Worst of all, the discoveries Shapley had forfeited would likely be made by the astronomer he so disliked. Indeed, in 1929 Hubble went on to announce the greatest astronomical discovery of the twentieth century, the expansion of the universe.

In 1941, Harvard astronomer Cecilia Payne-Gaposchkin spoke before a crowd commemorating twenty years of Shapley's directorship. "There are not many of you who can remember the time when there were, so to speak, no galaxies," she told the gathering. "I remember it very well. I remember the day Dr. Shapley said: 'My universe has been shattered.' "

4

GIVING THE GALAXY
A SPIN

TO ASTRONOMERS THROUGHOUT the world, Edwin Hubble's discovery of Cepheids in Andromeda brought two major revelations. First, by proving the existence of other galaxies, Hubble diminished the role of our own. The Milky Way was simply one galaxy among countless others, scattered throughout an unimaginably large universe, and these galaxies were just as deserving of study as the Milky Way. Yet at the same time that Hubble undermined the unique stature of the Galaxy, he blazed a new trail to unraveling its secrets. Armed with Hubble's discovery, astronomers could now attempt to deduce the nature of our Galaxy by examining others.

The most obvious deduction that followed from Hubble's discovery of other galaxies was that our own was rotating. From their appearance alone, the spiral galaxies almost demanded rotation, for they resembled huge pinwheels in space. So strong is this impression that two spiral galaxies even have nicknames suggesting rotation: M51 is the Whirlpool Galaxy, and M33, the third largest member of the Local Group, is sometimes called the Pinwheel Galaxy. If the Milky Way resembles these, and if they rotate, then so should it.

Firm proof of the Galaxy's rotation came in 1927, just four years after Hubble's discovery, and finally settled the Kapteyn-Shapley dispute over the size of the Milky Way. By measuring the velocities of the

Milky Way's stars, astronomers not only discovered that the Galaxy rotates but also found that it does so quickly. This in turn implied that the Milky Way was far larger than Kapteyn had said, because a small Galaxy would not have enough gravitational force to retain such fast-flying stars. Instead, both the speed and direction of the rotation matched Shapley's model of a large Galaxy whose center lay far from the Sun in the direction near Sagittarius.

THE ROTATION OF OTHER GALAXIES

Over a decade before the discovery of the Milky Way's rotation, astronomers confirmed that other galaxies rotate. The evidence came in 1913 from American astronomer Vesto Slipher, who worked at Arizona's Lowell Observatory, a place one astronomer called "the traditional 'bad boy' in astronomy." Lowell Observatory had been founded in 1894 by wealthy Bostonian Percival Lowell, a flamboyant astronomer who observed the solar system and believed that the planet Mars had intelligent beings. The public loved Lowell's theory of a dying Martian civilization that had dug canals to carry precious water from the polar caps to the equator, but most other astronomers ostracized him. As part of Lowell's work, he had Slipher examine the spiral nebulae, the objects now known to be spiral galaxies. Like most astronomers of his time, however, Lowell believed that the spiral nebulae were newborn solar systems, and he wanted Slipher to see whether they were rotating.

Slipher's instrument for studying the spiral nebulae was the spectroscope, a device that had recently ushered in a revolution. Just as water droplets split sunlight into different colors and produce a rainbow, so a spectroscope splits starlight into its individual colors. The resulting rainbow of starlight—or spectrum—contains dark lines where different elements absorb certain wavelengths. These spectra, which could be recorded on photographic plates, would thereby help reveal the composition of distant planets and stars. Moreover, the spectrum also determines an object's radial velocity, the speed at which the object travels toward or away from the observer, because light waves from an approaching object get compressed—or blueshifted, since blue light has a short wavelength—whereas light waves from a receding source get elongated—or redshifted. This so-called Doppler shift reveals the star's radial velocity: the greater the blueshift or redshift, the faster the object is moving toward or away from Earth.

In 1912, Slipher began to measure the spectra of spiral nebulae. The work was arduous, because the nebulae were faint, and Slipher often had to observe the same nebula for several nights in order for the photographic plate to build up a good image of the spectrum. But Slipher's hard work paid off with three major discoveries. First, the spectra revealed that most spiral nebulae had enormous radial velocities, larger than the velocity of any star in the Milky Way, which convinced Percival Lowell and many other astronomers that the spirals were not part of the Milky Way but were instead other galaxies beyond it. Second, most spiral nebulae were rushing away from the Galaxy. Although Slipher did not interpret it as such, this was the first evidence of the expansion of the universe.

Third, in 1913 Slipher discovered the rotation of a galaxy. He observed what is now called the Sombrero Galaxy, a spiral whose disk we view nearly edge-on in the constellation Virgo. Although the entire galaxy is moving away from Earth and so is redshifted, Slipher found that one edge of the galaxy had a larger redshift than the other. Thus, relative to the galaxy's center, one edge was receding from Earth and the other approaching; that is, the galaxy was rotating. But Slipher was as cautious as Lowell was flamboyant and did not announce the discov-

Figure 7. As the edge-on spiral galaxy rotates, the left side recedes and the right side approaches; therefore, as viewed from Earth, the left side shows a greater redshift than the right. By observing this phenomenon, Vesto Slipher discovered the rotation of the Sombrero Galaxy.

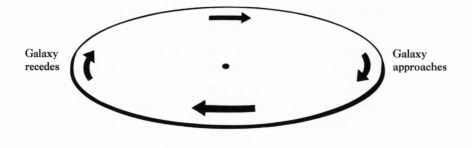

Galaxy recedes

Galaxy approaches

Earth

ery until 1914, after he had obtained a second spectrum of the Sombrero Galaxy that confirmed the first.

KINEMATICS OF THE MILKY WAY'S STARS

In the 1920s, when astronomers discovered the rotation of the Galaxy, they did so by observing the motions of the Milky Way's individual stars. To determine a star's total velocity through space, an astronomer must know the star's distance, proper motion, and radial velocity. Proper motion is the apparent movement of a star across the sky rather than a real velocity. To translate a star's proper motion into its velocity across our line of sight—the so-called tangential velocity—an astronomer must also know the star's distance. A star with a large proper motion could be nearby and have a low tangential velocity, or it could be distant and have a high tangential velocity. For example, the stars Alpha Centauri and Mu Cassiopeiae both have nearly identical, and large, proper motions, but Mu Cassiopeiae is six times farther away than Alpha Centauri and so has a tangential velocity six times higher. Thus, a proper motion is of limited value if astronomers do not know the star's distance, and most stars are so far from Earth that their distances are poorly determined.

In contrast, no such problem arises with a radial velocity, the star's velocity *along* the observer's line of sight. The Doppler shift of a star's spectrum gives the star's radial velocity in real velocity units—miles per hour or, in the units astronomers prefer, kilometers per second (one kilometer per second is 2237 miles per hour). To obtain the radial velocity, an astronomer need not know anything about the star's distance.

Although the concept of proper motion dated back to Edmond Halley's discovery of it in 1718, radial velocities entered the astronomer's arsenal much later. The first attempts to measure radial velocities came in 1868, when English astronomer William Huggins tried to determine the Doppler shift of Sirius, the brightest star in the night sky. Unfortunately, the work was so difficult that Huggins got a radial velocity six times the modern value, and he also said that the star was receding when it is actually approaching. Subsequent attempts by other astronomers also produced radial velocities that were wide of the mark. Furthermore, many astronomers were skeptical of such efforts, because certain physicists had argued that the Doppler shift—which had been proven for sound waves—did not apply to light.

The first reliable stellar radial velocities came only in 1888, when German astronomers Hermann Vogel and Julius Scheiner applied photography to spectroscopy and captured the Doppler shifts of bright stars on photographic plates that could be examined later. The future, however, lay in California, where that same year Lick Observatory began operation atop Mount Hamilton. During the first three decades of the twentieth century, astronomers at this observatory measured the radial velocities of most of the stars visible to the naked eye, in large part because of the efforts of William Campbell. Like Jacobus Kapteyn, Campbell appreciated vast quantities of data, realizing that a thorough investigation of the Milky Way would require radial velocities for as many stars as possible. Campbell had a colleague in this work, William Wright, who journeyed to Chile and set up an observatory to measure the radial velocities of stars that lay too far south to be seen from California.

This work advanced astronomers' knowledge of the Galaxy tremendously, for radial velocities and proper motions allowed astronomers to paint a new portrait of the Milky Way. In the past, when they mapped the Milky Way, astronomers had been studying only the Galaxy's structure. Now, with the new data, they could begin to investigate the Milky Way's kinematics, or the motion of its stars. The mating of the two—structure and kinematics—would produce new insights into the Milky Way, the way a lyric and a melody combine to make both richer and more memorable than either alone. For example, in 1918 Harlow Shapley published a diagram comparing the positions of the Galaxy's Cepheids with those of its RR Lyrae stars. Both types of stars pulsate, but the Cepheids, which are more luminous, cling tightly to the Milky Way's plane, whereas the RR Lyrae stars spread around it. Shapley correctly deduced that the different structures of the Cepheid and RR Lyrae systems arose because of their different kinematics: Cepheids have low velocities and therefore remain close to the Galaxy's plane, whereas RR Lyrae stars have high velocities that carry them above and below the Galactic plane. The structure follows from the kinematics, and the kinematics follow from the structure.

STAR STREAMS

In 1904, Jacobus Kapteyn announced a startling result concerning the Galaxy's kinematics. He was in St. Louis for the World Exposition, where Simon Newcomb was having astronomers give talks. "I wanted Kapteyn to speak of Durchmusterungs [sky surveys]," said Newcomb at the time, "but he wants to talk about something else." The "something else" proved to be what British astronomer Arthur Eddington later said was "Kapteyn's greatest discovery": stars tend to move in one of two opposite directions, forming what Kapteyn called star streams. One star stream carried roughly half of all stars toward what Shapley later identified as the Galactic center, while the other stream carried the remaining stars away from the Galactic center. Kapteyn's star streams were not rigid, for all stars in a stream did not move exactly the same way. On average, though, the stars did follow the stream they belonged to, just as commuters living north of a large city tend to head south during the early morning.

Kapteyn had actually discovered the star streams two years earlier, in 1902, through his analysis of proper motions. But he found the result so surprising that he did not publish it then, preferring instead to confirm it through the radial velocities that Campbell was obtaining at Lick. Campbell, however, guarded his data so jealously that Kapteyn could test his discovery only with the few stars whose radial velocities had been published in scientific papers. Though meager, these data supported the concept of two parallel star streams, and Kapteyn felt confident enough to announce the discovery.

Kapteyn's discovery of the star streams motivated other scientists to explain them. In 1907 and 1908, German astronomer Karl Schwarzschild published an interpretation of the star streams that astronomers still use today. Brilliant and versatile, Schwarzschild had published his first scientific papers at the age of sixteen and held a broad view of science. "Mathematics, physics, chemistry, and astronomy march in one front," he said before the Berlin Academy of Science in 1913. "Whichever lags behind is drawn after. Whichever hastens ahead helps

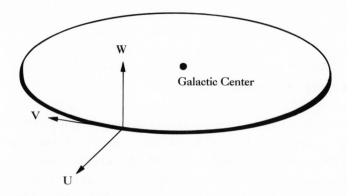

Figure 8. U, V, W velocities: As used by astronomers today, U is a star's velocity away from the Galactic center; V is the star's velocity in the direction of Galactic rotation; and W is the star's vertical velocity, perpendicular to the Galactic plane.

on the others." Today Schwarzschild is best known for a 1916 paper in which he used Einstein's general theory of relativity to describe what are now called black holes.

Schwarzschild's interpretation of Kapteyn's work was also crucial, for he showed that it was not necessary to think in terms of two star streams. He set up a coordinate system in which a star's velocity through space was split into three components. Today astronomers call these U, V, and W, each of which is perpendicular to the other two. A star's U velocity is the component of its velocity directed away from what astronomers now recognize is the Galactic center; the star's V velocity is the component of its velocity in the direction of what astronomers now recognize is the Galaxy's rotation; and the star's W velocity is the speed at which the star travels toward or away from the Galactic plane. For example, the Sun has a U velocity of −9 kilometers per second, which means that it moves toward the Galactic center with this speed; the Sun has a V velocity of +12 kilometers per second, which means that it revolves around the Galaxy this much faster than the average star; and the Sun has a W velocity of +7 kilometers per second, which means that it moves up through the Galactic plane at this speed.

Every other star also has its own U, V, and W velocities. Schwarzschild showed that the U velocities of stars tend to be greater than the V velocities, which in turn tend to be greater than the W velocities. In other words, stars usually move fastest along the U axis. This behavior was manifested as Kapteyn's two star streams: stars with positive U velocities moved away from the Galactic center, while stars with negative U velocities moved toward it. The V and W velocities, which were smaller, simply perturbed these two main flows. But in Schwarzschild's view, the stars did not divide into two discrete groups. Kapteyn's view is like saying that all people are either rich or poor, whereas Schwarzschild's view is like saying that a continuous distribution of income exists, with the rich and poor at either extreme.

The relatively high U velocities of the stars turned out to be a consequence of the Galaxy's rotation, but this is hardly obvious, and the realization eluded astronomers for two decades after Schwarzschild did his work. Even Kapteyn interpreted the result incorrectly, because he thought that stars revolved around the Galaxy in the U direction. In Kapteyn's view, stars in one stream orbited the Galaxy clockwise and stars in the other counterclockwise.

THE HIGH-VELOCITY STARS: HARBINGERS OF GALACTIC ROTATION

Clearer clues to the Galaxy's rotation actually lay in the rare stars that have high speeds relative to the Sun and do not belong to either of Kapteyn's star streams. Most stars have small radial velocities, less than 30 kilometers per second. Indeed, back in 1888, when the first radial velocities were measured, the fastest known star was Aldebaran, which moves away from the Sun at 54 kilometers per second. But as astronomers measured more spectra, they uncovered a few stars with much higher radial velocities. For example, in 1908 astronomers discovered that Kapteyn's Star had a huge radial velocity of 244 kilometers per second, larger than that of any other star whose radial velocity had been measured.

These rare high-velocity stars held the clues to a profound discovery. In 1918, American astronomer Benjamin Boss, editor of *The Astronomical Journal,* analyzed the few stars that had radial velocities exceeding 75 kilometers per second. He found that these stars showed a striking asymmetry. On one half of the sky, nearly all the high-velocity stars were redshifted and moving away from us, whereas on the other half of the sky, nearly all the high-velocity stars were blueshifted and moving toward us. Boss also found that the high-velocity stars moved fastest in the V direction, perpendicular to Kapteyn's star streams.

In 1923, Mount Wilson astronomer Gustaf Strömberg analyzed a larger set of data and reached the same conclusions. He went further, though, and suggested a possible explanation that is now known to be correct: the Sun and most other stars revolve around the Galaxy at high speed—not in the U direction, as Kapteyn had thought, but in the V direction, orbiting the Galaxy's center in Sagittarius. However, a few stars do not orbit the Galaxy rapidly. As the Sun passes them by, astronomers see that the stars have high velocities *relative to the Sun* in the V direction. On the side of the sky the Sun is approaching, these stars appear to be rushing toward us. On the side of the sky the Sun is leaving behind, these stars appear to be speeding away. But it is really the Sun, not those stars, which does most of the moving. Ironically, then, the Galaxy's rotation was first betrayed by the very stars that do not partake in it.

Strömberg also discovered that the stars that lagged behind the Sun and its neighbors had higher velocity dispersions than do ordinary stars. A velocity dispersion is the spread around the mean velocity: it indicates how stars move relative to one another. A group of stars that have nearly the same velocity will have a small velocity dispersion, whereas a group of stars whose velocities are wildly different from one another will have a large velocity dispersion. Strömberg found that the more the stars lagged behind the Sun, the greater was their velocity dispersion. Astronomers now understand Strömberg's result as follows. A star that revolves slowly around the Galaxy (in the V direction) must have a high speed in the U and/or W direction to compensate, or else it would succumb to the Galaxy's gravity and fall into the Galactic center—just as a satellite that does not circle the Earth fast enough will fall into the atmosphere. Such stars will therefore have a large velocity dispersion, since their velocities will be large and in different directions.

PROOF OF THE GALAXY'S ROTATION

In 1925, Bertil Lindblad in Sweden seized on Strömberg's work and published a paper in which he claimed that the Galaxy was indeed rotating. He imagined that the stars in the Milky Way formed a number of subsystems. Some subsystems were flat and rapidly rotating, and most stars, including the Sun, belonged to them. But other subsystems were more spherical and slowly rotating. The rare stars in these slowly rotating subsystems got overtaken by the Sun, so the stars had high velocities relative to it and were the "high-velocity" stars that Strömberg had studied.

Furthermore, in 1927 Lindblad also explained Kapteyn's star streams as a consequence of the Milky Way's rotation, albeit in a way different from what Kapteyn had envisioned. Lindblad said that most stars, including the Sun, revolved fast around the Galaxy's center in Sagittarius. The stars moved in the V direction, at a speed of hundreds of kilometers per second, revolving clockwise as viewed from above the Galactic plane. But because the stars revolved together, their V velocities relative to one another tended to be small, so the stars had a small velocity dispersion. It was similar to cars racing down the highway. Though they may be traveling in the same direction at around sixty miles per hour, relative to one another they may have a velocity dispersion of only five miles per hour.

Just as each car follows its own path down the highway, each star follows its own orbit around the Galaxy. Each orbit is somewhat elliptical, so as the star revolves around the Galaxy, the star moves toward and away from the Galactic center, a motion that gives the star a velocity in the U direction. After analyzing the consequences, Lindblad found that the stars tend to have greater U velocities than V or W velocities—the very phenomenon that Schwarzschild had noted twenty years earlier in his interpretation of Kapteyn's star streams. Now, though, Lindblad explained it as a consequence of the Galaxy's rotation.

Earlier in 1927 came the first proof of the Galaxy's rotation, this time from Dutch astronomer Jan Oort. Born in 1900, Oort was one of the greatest astronomers of the twentieth century. "Few can doubt," wrote Harvard astronomer Charles Whitney in 1981, "that he has, in fact, contributed more to our understanding of galaxies than any astronomer in history." Oort's proof of the Galaxy's rotation was simple and elegant. He began by assuming that the Galaxy rotated differentially—that is, different parts rotated at different speeds. In particular, the closer a star was to the Galactic center, the faster it revolved around the Galaxy. This pattern, Oort knew, appeared in the solar system, where planets near the Sun, such as Mercury, moved faster than distant planets, such as Pluto. From this assumption Oort calculated the radial velocities that stars lying in different parts of the sky should have. In some parts of the sky, the Sun was catching up to the stars, so they should show blueshifts. In other parts of the sky, the Sun was speeding away from the stars, so they should show redshifts. As Oort analyzed the radial velocities of stars in different areas, he found just this behavior.

Oort had admired Kapteyn—Kapteyn's portrait hung in Oort's office—yet Oort's work spelled the end of the Kapteyn universe. Because the Sun and most other nearby stars were revolving at hundreds of kilometers per second, the small Kapteyn universe had too weak a gravitational force to hold on to them. Furthermore, Oort confirmed that the Sun was moving in the direction it should if it were revolving around a Galactic center that lay in the southern constellation Sagittarius, near where Shapley had placed it. In contrast, in 1922 Kapteyn had put the Galactic center in the opposite direction, in the northern constellation Cassiopeia.

AN INTERSTELLAR FOG

Where, then, had Kapteyn gone astray? In his 1927 paper, Oort suggested the answer: interstellar gas and dust blocked distant starlight and fooled Kapteyn into believing that the Milky Way was much smaller than it really is. Shapley never believed in interstellar absorption, and Kapteyn, who had earlier believed in it, finally came not to, largely because of Shapley's work. But Oort pointed out that interstellar absorption could reconcile the models of Shapley and Kapteyn.

Confirmation of this idea came in 1930, just three years after Oort's work, with a paper by Lick Observatory astronomer Robert Trumpler. Even after the work of Strömberg, Lindblad, and Oort, Trumpler still believed in the Kapteyn universe and rejected Shapley's larger model of the Galaxy. Like Shapley, Trumpler studied star clusters, but star clusters of a different type. Whereas Shapley had investigated globular clusters, which are densely packed with hundreds of thousands of stars, Trumpler observed open clusters, looser objects that typically contain only a few hundred stars. The nearest open clusters are the Hyades and the Pleiades, both in the constellation Taurus.

Open clusters cling to the Galactic plane, whereas the globular clusters spread around it.

Using the colors of the open clusters' brightest stars as guides to their intrinsic brightness, Trumpler estimated the distances of a hundred open clusters. From the clusters' distances and apparent sizes, he determined their true diameters. But Trumpler noticed a peculiarity: the farther away the clusters were, the larger they seemed to be. He concluded that this was impossible. Trumpler realized that he had overestimated the distances of the farthest clusters, which in turn had led him to overestimate those clusters' sizes as well.

Trumpler then deduced that the cause of his overestimates was interstellar gas and dust, which absorbed starlight and made the farther clusters look fainter and even farther than they actually were. To support this view, he noted that the distant clusters were redder, which indicated that their blue light was being removed by large amounts of gas and dust. Shapley thought such absorption did not exist, for he had seen all colors of stars in distant globular clusters, which meant that their light suffered no absorption or reddening on the way to Earth. Trumpler correctly explained Shapley's result by postulating that the absorbing layer was only in the Galaxy's disk and therefore affected only the open clusters, which lay in the disk. In contrast, because most globulars lay above or below the disk, they were little obscured by intervening gas and dust.

With that, the Kapteyn universe was finally dead, killed in part by Trumpler's work but more by the previous studies of Strömberg, Lindblad, and Oort, all of which had revealed the Galaxy's rapid rotation. Thus, by 1930, astronomers had, in broad terms, deduced the structure and kinematics of the Milky Way: the Galaxy was a huge, rapidly spinning disk of stars containing so much gas and dust that astronomers could see only a minute fraction of it. Further progress would come only after they understood the nature and evolution of the Galaxy's citizens, its countless stars.

5

CITIZENS OF THE GALAXY

STARS ARE TO galaxies what people are to nations. Just as one can understand the United States by studying Americans, astronomers can probe the Milky Way Galaxy by exploring its stars. Stars generate nearly all the Galaxy's light and have transformed the Milky Way from what was once a lifeless entity into a system that supports at least one thriving civilization. In addition, each star preserves a record of the Galactic conditions that prevailed when and where the star was born. Consequently, by comparing stars of different ages, astronomers can try to piece together how the entire Milky Way has changed with time, and from this, attempt to deduce the evolution of galaxies throughout the universe.

THE H-R DIAGRAM:
LUMINOSITY AND COLOR

The cornerstone of stellar astronomy is the Hertzsprung-Russell diagram, which is as central to the field as the periodic table is to chemistry. The periodic table arranges the chemical elements by mass in a way that reveals new relations among them. For example, the elements in the first column—such as lithium, sodium, and potassium—react violently with water, while the elements in the eighth and final column —such as helium, neon, and argon—react with little or nothing.

The H-R diagram does for stars what the periodic table does for the elements, arranging stars in a way that segregates one type from another. The "H" in the H-R diagram comes from Danish astronomer Ejnar Hertzsprung. His father had a degree in astronomy but was not an astronomer himself, and Hertzsprung never took a course in the subject. In fact, when Hertzsprung was twenty and his father died, the latter's astronomy books were sold, because, as Hertzsprung later noted, "nobody imagined that I should become an astronomer."

But he did, and in 1905 and 1907 Hertzsprung published papers that linked stellar luminosities and colors. Stars come in a variety of colors from blue to red, and Hertzsprung found that all the blue and white stars were intrinsically bright, whereas the orange and red stars split into two groups, one bright, the other faint. But Hertzsprung reported this finding in a photography journal, so few astronomers heard of his result. In 1911, he plotted what are now called H-R diagrams when he drew graphs illustrating the relation between luminosity and color. But he again published his work in an obscure journal, albeit this time one devoted to astronomy.

Meanwhile, Harlow Shapley's graduate advisor at Princeton University, Henry Norris Russell, was noticing the same relation. Russell had become interested in astronomy when he was five years old, after his parents had him watch Venus cross the face of the Sun in 1882. By 1913, when he first plotted his own diagram, he was one of America's foremost astronomers, and his colleagues soon recognized the importance of what they called the Russell diagram. After realizing that Hertzsprung had already drawn a similar plot, astronomers renamed it the Hertzsprung-Russell diagram.

Like many great discoveries, the H-R diagram is simple. It plots luminosity against color, two basic properties of any star. Luminosity is a star's true, or intrinsic, brightness, the amount of energy the star radiates into space. Stellar luminosities span an enormous range, with the most luminous stars outshining the least a trillion times over. The Sun lies in the middle of this range, which causes careless writers to call it average; but it isn't, not by a long shot, for 95 percent of all stars are less luminous than the Sun.

To determine a star's luminosity, an astronomer must first know the star's apparent brightness—that is, how bright the star looks at night. To quantify a star's apparent brightness, astronomers use a number called apparent magnitude. The smaller the apparent magnitude, the brighter the star looks, just as a lower golf score indicates a better player. For example, a star with an apparent magnitude of 1 looks brighter than a star with an apparent magnitude of 2. Each successive magnitude corresponds to a factor of 2.5, so the magnitude 1 star is 2.5 times brighter than the magnitude 2 star. A few stars are so bright that their apparent magnitudes are actually negative. On a clear dark night, the average unaided eye can see stars as faint as apparent magnitude 6, and binoculars can detect stars as faint as apparent magnitudes 9 or 10. Telescopes can detect still fainter stars, with the largest discerning objects down to apparent magnitude 30—four billion times fainter than the naked eye can see.

Although apparent magnitude indicates how bright a star looks, it does not reveal the star's luminosity. To determine that, astronomers must also know the star's distance, because a star that looks bright could be intrinsically faint and nearby or intrinsically bright and far away. To quantify a star's intrinsic brightness, astronomers use what they call absolute magnitude, which is the apparent magnitude a star would have if it were at a standard distance of 32.6 light-years (10 parsecs) from Earth. For example, the Sun has an absolute magnitude of +4.83, so if it were seen from a distance of 32.6 light-years, it would have an apparent magnitude of +4.83—barely visible to the naked eye. Absolute magnitudes of stars run the gamut from −10 (a million times brighter than the Sun) to +20 (a million times fainter than the Sun), but the faint stars far outnumber the bright ones.

As important as a star's luminosity is its color. Stars range in color from blue and white to yellow, orange, and red. These colors reflect the star's temperature. Blue and white stars are hot, yellow stars are warm, and orange and red stars are cool, just as a white-hot metal rod is hotter than a red-hot one.

A star's temperature also affects the spectrum of its light, because different elements respond differently to different temperatures. These elements absorb light at certain wavelengths and imprint themselves on the star's spectrum. Thus, hot stars have different spectral lines than cool ones do, so astronomers can determine a star's temperature from its spectrum. The present system of spectral types was developed around 1890 at Harvard. Edward Pickering led the effort, but most of the work was done by his female assistants, especially Williamina Fleming and Annie Jump Cannon. The Harvard astronomers determined the spectral types of over 200,000 stars and published them between 1918 and 1924 in the nine-volume *Henry Draper Catalogue*, named for a wealthy amateur astronomer whose widow had given money to the university.

The Harvard astronomers devised seven main spectral types and designated each with a letter. From blue to red, the spectral types are O, B, A, F, G, K, and M (mnemonic: "Oh, Be A Fine Guy/Girl, Kiss Me!"). O and B stars are blue; A stars white; F stars yellow-white; G stars yellow; K stars orange; and M stars, the coolest of all, red. The Sun is a yellow G star.

TABLE 5-1: SPECTRAL TYPES

Spectral Type	Color	Examples
O	Blue	Alnitak
B	Blue	Rigel, Regulus, Spica
A	White	Sirius, Vega, Altair, Deneb
F	Yellow-white	Procyon, Canopus
G	Yellow	Sun, Capella, Alpha Centauri A
K	Orange	Arcturus, Aldebaran, 61 Cygni
M	Red	Betelgeuse, Antares, Wolf 359

Hertzsprung and Russell both noticed that, of the stars whose distances they either knew or could infer, the blue B and white A stars were more luminous than the Sun, whereas the orange K and red M stars divided into two branches, one brighter than the Sun, the other fainter. A modern version of the H-R diagram shows the same thing. Luminosity appears along the vertical axis, so intrinsically bright stars lie at the top of the diagram and intrinsically faint stars at the bottom. Color, or spectral type, appears along the horizontal axis, so the blue O and B stars lie on the left, the white A, yellow-white F, and yellow G stars in the middle, and the orange K and red M stars on the right.

As Hertzsprung and Russell discovered, the most remarkable result of such a plot is that the stars do not scatter over it randomly. Instead, some regions of the H-R diagram abound with stars and others have none. Moreover, three main groups appear on the H-R diagram, each containing a different kind of star. The first and most prominent group is a sequence that runs diagonally across the H-R diagram from upper left (bright and blue) to lower right (faint and red). This diagonal band includes stars of all spectral types—O, B, A, F, G, K, and M. It embraces 90 percent of all stars, including the Sun, and is therefore called the main sequence. The second group of stars appears in the upper right of the H-R diagram. They are bright and cool, with spectral types G (yellow), K (orange), or M (red), and are called yellow giants, orange giants, and red giants. Often astronomers refer to all three simply as red giants. The very brightest stars, at the top of the H-R diagram, are called supergiants—blue supergiants on the left, yellow supergiants in the middle, and red supergiants on the right. The third and final group of stars on the H-R diagram appears below and parallel to the main sequence. These small, faint stars are the white dwarfs.

Figure 9. The H-R diagram plots stellar luminosity against stellar temperature, or color. Intrinsically bright stars appear at the top, intrinsically faint stars at the bottom. Hot blue stars are on the left, cool red stars on the right.

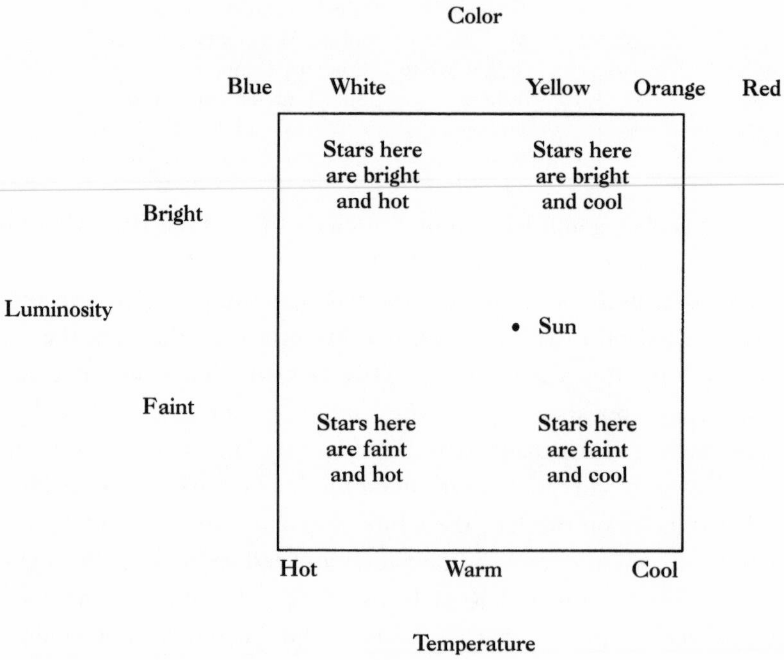

Color

	Blue	White		Yellow	Orange	Red
Bright		Stars here are bright and hot		Stars here are bright and cool		
Luminosity				• Sun		
Faint		Stars here are faint and hot		Stars here are faint and cool		
	Hot		Warm		Cool	

Temperature

THE STELLAR MAJORITY:
THE MAIN SEQUENCE

Because the main sequence includes 90 percent of all stars, it is the most important group on the H-R diagram. But the first impression one gets is of its enormous diversity, for main-sequence stars seem to have nothing to do with one another. The main sequence includes highly luminous blue O and B stars, very bright white A and yellow-white F stars, bright yellow G stars, modest orange K stars, and dim red M stars.

Figure 10. The H-R diagram splits stars into three groups: 1) THE MAIN SEQUENCE, which runs diagonally from upper left (bright and blue) to lower right (faint and red) and includes the five stars marked here: Regulus (spectral type B), Vega (type A), the Sun (type G), 61 Cygni (type K), and Wolf 359 (type M). 2) GIANTS AND SUPERGIANTS, the former represented here by the orange K giant Arcturus and the red M giant Mira, the latter by the red M supergiant Betelgeuse. The blue supergiant Rigel is also shown. 3) WHITE DWARFS, the sequence below and parallel to the main sequence. One white dwarf is marked here, Sirius B.

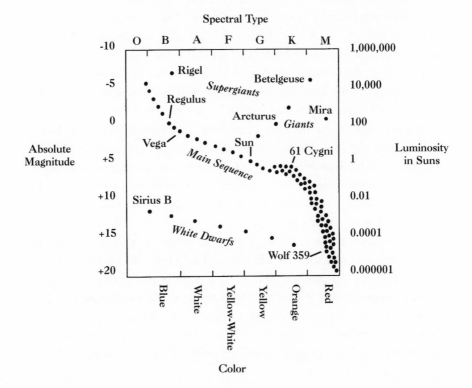

Yet the main sequence is the H-R diagram's greatest revelation, for it connects stars that would otherwise seem unrelated. As astronomers now realize, all main-sequence stars share one fundamental property. Every one, bright or dim, generates energy the same way, by fusing hydrogen nuclei into helium nuclei at its center.

TABLE 5-2: MAIN-SEQUENCE MASSES

Spectral Type	Main-Sequence Mass (Sun = 1.0)
O	16 to 100
B	2.5 to 16
A	1.6 to 2.5
F	1.1 to 1.6
G	0.9 to 1.1
K	0.6 to 0.9
M	0.08 to 0.6

Main-sequence stars differ so greatly from one another in luminosity and color because they differ in mass. The more mass a main-sequence star has, the hotter its center is and the faster it fuses hydrogen into helium; so the brighter and hotter the star burns. Thus, the blue O and B main-sequence stars outshine the Sun and have more mass than it does, while the K and M main-sequence stars are fainter than the Sun and have less mass. Each star is born with a different mass, which determines the spectral type it will have on the main sequence. The most massive stars have a hundred times the Sun's mass, while the least massive main-sequence stars have only 8 percent the mass of the Sun. If a star is born with less mass than this, it never gets hot enough to ignite its hydrogen and thus does not join the main sequence. Such a star, called a brown dwarf, glows faint red when young and slowly fades as it ages.

In addition to determining a main-sequence star's luminosity and color, mass dictates the star's lifetime. The more mass a star has, the faster it burns up its fuel and dies. For example, Regulus, a blue B-type main-sequence star with four times the Sun's mass, will live about 100 million years; the Sun, a yellow G-type main-sequence star, has a total lifetime of about 10 billion years; and Proxima Centauri, a red M-type main-sequence star with a tenth of the Sun's mass, will live for over a trillion years.

Along the main sequence, the most massive stars are the rarest. The brightest main-sequence stars (spectral types O, B, A, and F) account for only 1 percent of all stars in the Galaxy. About 4 percent of stars are yellow G-type main-sequence stars like the Sun, and 15 percent are orange K-type main-sequence stars, or orange dwarfs. By far the most common main-sequence stars are the faint M stars, which are called red dwarfs. Though faint, cool, and small, red dwarfs make up 70 percent of the stars in the Milky Way. They outnumber all other stars put together, and many of the nearest stars, such as Proxima Centauri and Wolf 359, are red dwarfs. But due to their intrinsic faintness, not one of these stars is visible to the naked eye.

In short, five rules govern the lives of main-sequence stars. The more mass a main-sequence star has (1) the brighter it is (2) the hotter it is (3) the bluer it is (4) the shorter-lived it is and (5) the rarer it is.

BIG AND BRIGHT: GIANTS AND SUPERGIANTS

Red giants and supergiants form the second group of stars on the H-R diagram. As their names imply, they are much larger than main-sequence stars, which are sometimes called dwarfs. Aldebaran, an orange K-type giant, is forty times the size of the Sun. If it replaced the Sun, Aldebaran would extend halfway to Mercury. The red M-type supergiant Betelgeuse is even larger, with a diameter about a thousand times the Sun's. If Betelgeuse were at the center of the solar system, the star would engulf Mercury, Venus, Earth, Mars, and possibly Jupiter.

Because they are so luminous, giants and supergiants make a splash in the sky, and many well-known stars are members of this class. But giants and supergiants are actually rare, accounting for less than 1 percent of all stars in the Galaxy. Hertzsprung likened the sky to the sea, which has many fish but few whales. Main-sequence stars are the fish, giants and supergiants the whales.

Although they outshine the Sun, red giants and supergiants are not necessarily more massive, but they are in a more advanced stage of evolution. Giants and supergiants started their lives as main-sequence stars, and how an individual star evolves from the main sequence to the giant or supergiant stage depends critically on the star's mass. Stars born with masses between 0.5 and 8 Suns evolve in a similar way. This range includes most B main-sequence stars, all white A, yellow-white F, yellow G, and orange K main-sequence stars, and a few red dwarfs. The Sun lies in this mass range, so it will follow this evolution, too. As a main-sequence star, the Sun presently converts hydrogen into helium at its core. When the Sun's core fills up with helium, the Sun will begin burning hydrogen in a shell surrounding the core, at which time it will no longer be a main-sequence star. It will expand and brighten, first becoming a so-called subgiant, which is intermediate between a main-sequence star and a giant. Then, as the Sun expands further, it will cool and brighten, becoming a red giant hundreds of times brighter than it is now. Stars with more than about 3 solar masses, which were bright blue B stars while on the main sequence, do not brighten much but simply maintain their high luminosities as they expand and cool. On the H-R diagram, these stars move to the right, whereas less massive stars, like the Sun, move both rightward and upward. Thus, gianthood is the great equalizer, for giants of all masses shine brightly.

When a star first becomes a red giant, it burns only hydrogen into helium, but eventually the helium core ignites. Helium fuses into carbon and oxygen, so these elements begin to fill the core. When the core runs out of helium, the star burns helium in a layer surrounding the core. During this time, the star may cast huge amounts of material into space until its outer atmosphere of hydrogen and helium is ejected. Like a butterfly emerging from a cocoon, the small carbon-oxygen core at the star's center now appears. The core is extremely hot, even hotter than an O star, so it is blue and appears to the left of the conventional H-R diagram. Because of its extreme heat, the star gives off so much ultraviolet radiation that the ejected atmosphere glows and forms a planetary nebula around the star.

Within a few tens of thousands of years, the planetary nebula expands into space and disappears from view, and only the star's core, hot and blue, remains. By now the star is dead, for it no longer generates energy by nuclear fusion. It shines simply because it is hot. As it radiates energy into space, it cools and fades. Once a mighty giant, the star is now a tiny white dwarf, little larger than Earth.

SMALL AND FAINT: WHITE DWARFS

White dwarfs constitute the third and final group of stars on the H-R diagram and make up 10 percent of all stars in the Galaxy. When Hertzsprung began his work, however, white dwarfs were unknown, because they are so dim that none is visible to the naked eye. Indeed, Hertzsprung thought that all intrinsically faint stars were of spectral type M—red dwarfs. The first *white* dwarf, Omicron^2 Eridani B, was discovered in 1910, when astronomers realized that this intrinsically faint star was of spectral type A rather than M. The discovery occurred when Henry Norris Russell was visiting Edward Pickering, the Harvard astronomer behind the effort to classify stellar spectra. The men were discussing the importance of determining spectral types for nearby stars with reliable distances.

As Russell later recalled it, the conversation went like this: "Pickering said, 'Well, name one of these stars.' Well, said I, for example, the faint component of Omicron Eridani. So Pickering said, 'Well, we make rather a specialty of being able to answer questions like that.' And so we telephoned down to the office of Mrs. Fleming and Mrs. Fleming said, 'Yes,' she'd look it up. In half an hour she came up and said 'I've got it here, unquestionably spectral type A.' I knew enough, even then, to know what that meant. I was flabbergasted. I was really baffled trying to make out what it meant. Then Pickering thought for a moment and then said with a kindly smile, 'I wouldn't worry. It's just these things which we can't explain that lead to advances in our knowledge.' Well, at that moment, Pickering, Mrs. Fleming and I were the only people in the world who knew of the existence of white dwarfs."

Five years later, in 1915, astronomers discovered the second white dwarf, Sirius B, another dim star they had thought was a red dwarf. Sirius B is the nearest white dwarf to Earth, lying only 8.6 light years away.

As the core of a former sun, a white dwarf is remarkably dense. The average white dwarf has 60 percent of the Sun's mass packed into a sphere little larger than Earth, and a spoonful of white dwarf matter would weigh over a ton. Despite their name, white dwarfs are various colors, for the stars cool and redden as they age. This is why they form a sequence parallel to the main sequence. The hottest, brightest, and bluest white dwarfs, such as Sirius B, are the stars that became white dwarfs most recently, whereas white dwarfs that are yellow or orange have been in this stage for billions of years.

SUPERNOVAE

White dwarfdom represents the humble destiny of any star, like the Sun, that is born with less than eight solar masses. In contrast, more massive stars end their lives with pizzazz. While on the main sequence, these high-mass stars are hot and blue, with spectral types of O or B. They quickly exhaust the hydrogen at their cores, and most evolve into red supergiants, like Betelgeuse and Antares, and ignite helium.

After igniting helium, the red supergiant may contract and heat up, becoming a blue supergiant like Rigel or a white supergiant like Deneb. Once it has exhausted the helium at its core, the star may move back toward the red. The star soon burns heavier fuels, such as carbon, neon, oxygen, and silicon. But none of these fuels does the trick, for none creates as much energy as hydrogen did, and the star desperately needs energy to balance its inward-pulling weight. The final fuel, silicon, spells death, for silicon burns into iron, and iron does not burn at all. The star now faces an energy crisis of immense proportions. When it runs out of energy, it collapses and explodes, becoming a supernova that can shine with the light of a billion Suns.

What remains of the star collapses into a sphere even denser than a white dwarf, either a neutron star or a black hole. A neutron star consists almost entirely of neutrons, for the collapse rams the star's positively charged protons into its negatively charged electrons, creating the neutral neutrons. When they are born, neutron stars spin rapidly, and, like lighthouses, many emit a pulse of radiation toward Earth every time they spin. The first of these so-called pulsars was discovered in 1967, and today over six hundred are known.

If the remnant star has too much mass, however, the star collapses into a black hole, an object with such intense gravity that not even light can escape from it. Because of their nature, black holes are obviously difficult to discover. The first black hole candidate, Cygnus X-1, was found in 1971, and only a few more have been found since. The black hole of Cygnus X-1 gave away its presence because another star circles it and dumps material into it. Before this material falls into the black hole, it gets heated and emits x-rays, which astronomers can detect.

Although most supernovae arise from massive stars, white dwarfs can also explode. The most mass a white dwarf can have is 1.4 solar masses; anything more, and the star explodes. This boundary is called the Chandrasekhar limit, after Subrahmanyan Chandrasekhar, who discovered it during the 1930s. If a white dwarf receives material from a companion star and exceeds this limit, the carbon in the white dwarf begins to fuse in a runaway nuclear reaction that annihilates the star and produces what astronomers call a type Ia supernova. Although type Ia supernovae arise from small stars, they are actually the most luminous, outshining the supernovae that occur when high-mass stars explode. (Massive-star supernovae are labeled type Ib, Ic, or II, depending on various details.)

The exact evolution of a star, then, can be quite complex. But the broad outline is simple:

main-sequence star $\Rightarrow$ red giant or supergiant $\Rightarrow$ white dwarf, neutron star, or black hole.

Those who study the Milky Way, however, are interested not only in individual stars but also in how they relate to one another and to the Galaxy. The most powerful example of this interrelatedness was discovered in 1943, when Walter Baade, an astronomer at Mount Wilson Observatory, introduced a concept that has stimulated research ever since. Baade found that stars in one part of the Galaxy share certain characteristics with one another but differ from stars in other parts of the Galaxy. Baade called these Galaxy-wide groups "stellar populations." Astronomers then discovered that different stellar populations had formed at different times in the Galaxy's life, so each population preserves information about a different era in the Milky Way's life. By studying differences among these populations, astronomers could at last begin to see how the entire Milky Way Galaxy had changed since its formation. Indeed, Baade's breakthrough provided the key to deciphering the very origin of the Milky Way—and of galaxies throughout the cosmos.

6

THE DEMOGRAPHICS OF THE MILKY WAY

"THE IDEA OF stellar populations," wrote astronomer Allan Sandage in 1986, "has become one of those grand unifying themes in science, appearing every century or so, that ties together a number of intricate details and diverse parts of a particular field." Indeed, so powerful is the concept that most work on the Milky Way during the last half century has in one way or another involved studying the Galaxy's different stellar populations and how they relate to one another, for they hold the clues to the origin and evolution of the entire Milky Way Galaxy.

STELLAR POPULATIONS

As presently defined, a stellar population is a Galaxy-wide group of stars of all types that have similar ages, locations, kinematics, and compositions. Nevertheless, each stellar population is diverse, so all stars in the population do not have exactly the same age, location, kinematics, and composition. Furthermore, each stellar population encompasses billions of stars and a plethora of different types. For example, the stellar population to which the Sun belongs includes not only yellow

G-type main-sequence stars but also blue supergiants, red giants, white dwarfs, yellow giants, red dwarfs, and more.

Of the four properties that mark a stellar population, the first—age —is the most important, for it is what astronomers use to trace the Galaxy's evolution. Each population preserves information about the state of the Galaxy when that population was born and marks an epoch in the Galaxy's life.

The second characteristic, location, is the simplest. Stars from different stellar populations tend to reside in different parts of the Galaxy. Specifically, stars in one stellar population may cling to the Galactic plane, while stars in another may spread around the plane. Harlow Shapley had found one example of this. When he plotted the positions of Cepheids and RR Lyrae stars, he found that the former tend to lie close to the Galactic plane and the latter away from it.

The third characteristic of a stellar population, kinematics, is linked to the second but is more complicated. Kinematics describe how stars orbit the Galaxy. Most stars near the Sun take about 230 million years to make one orbit of the Galaxy. To determine a star's exact orbit, astronomers must know the star's U, V, and W velocities. (Recall that U is the component of the star's velocity directed away from the Galactic center, V is the component of the star's velocity in the direction of Galactic rotation, and W is the component of the star's velocity in the direction of Galactic north; see Figure 8 in Chapter 4.) Astronomers measure U, V, W velocities relative to the local standard of rest, an imaginary point located at the Sun's distance from the Galactic center that revolves clockwise around the Galaxy on a circular orbit at about 220 kilometers per second. Thus, in general, the greater a star's U, V, W velocities, the more the star deviates from the local standard of rest and the more elliptical the star's orbit is. In the Sun's case, the U, V, W velocities are small and its orbit is nearly circular. Astronomers express an orbit's shape in terms of what they call eccentricity, a number that ranges from 0 percent for circular orbits to nearly 100 percent for extremely elliptical orbits. The Sun has an orbital eccentricity of only 6 percent, which means that the Sun's greatest distance from the Galactic center is 6 percent greater than its mean distance and its smallest distance from the center is 6 percent less than this mean distance. Right now the Sun is near perigalacticon, the point closest to the Galactic center, and lies about 27,000 light-years from the

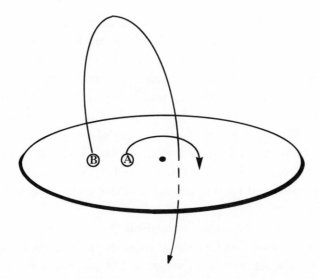

Figure 11. Stars with small U, V, W velocities—such as star A here—have nearly circular orbits around the Galaxy and remain close to the Galactic plane. In contrast, stars with large U, V, W velocities—such as star B here—usually have extremely elliptical orbits and often travel far from the Galactic plane.

center. When the Sun reaches its farthest point, apogalacticon, it will be about 30,000 light-years from the Galactic center.

Of the three velocities—U, V, and W—the V velocity reflects how rapidly a star revolves around the Galaxy. Stars that revolve rapidly, as the Sun does, have V velocities close to zero (the Sun's is +12 kilometers per second), whereas stars that revolve slowly have very negative V velocities. Some stars even revolve around the Galaxy backward, counterclockwise rather than clockwise; these stars have V velocities more negative than −220 kilometers per second. The W velocity is also important. As a star orbits the Galaxy, it moves up and down, bobbing like a horse on a merry-go-round, and the W velocity that the star has when it crosses the Galactic plane determines how far the star will climb from the plane. The Sun has a modest W velocity of +7 kilometers per second, so it will always lie within about 250 light-years of the plane, whereas stars with larger W velocities travel to greater heights.

The fourth and final characteristic of a stellar population is chemical composition. This measures a star's abundance of different elements, especially those heavier than hydrogen and helium, the two lightest and most common elements. Astronomers call these heavy elements "metals," a practice that annoys chemists, for to an astronomer not only are iron and copper metals but so are oxygen and neon. A star's total abundance of metals is its "metallicity," which astronomers measure by studying the star's spectrum. A metal-rich star has stronger metal lines in its spectrum than a metal-poor star of the same temperature. Astronomers express stellar metallicities in terms of the Sun's. Hydrogen and helium account for 98 percent of the Sun's mass, so only 2 percent of its mass is in heavy elements. By Galactic standards, though, the Sun is metal-rich. Different stellar populations have different metallicities, and for one population the metallicity is less than a tenth of the Sun's.

As astronomers presently understand the Milky Way, every star falls into one of four different stellar populations. The brightest is the thin-disk population, to which the Sun and 96 percent of its neighbors belong. Sirius, Vega, Rigel, Betelgeuse, and Alpha Centauri are all members. Stars in the thin disk come in a wide variety of ages, from newborn objects to stars that are 10 billion years old. As its name implies, the thin-disk population clings to the Galactic plane, with a typical member lying within a thousand light-years of it. Kinematically, the stars revolve around the Galaxy fast, having fairly circular orbits and small U, V, W velocities. Thin-disk stars are also metal-rich, like the Sun.

The second stellar population in the Galaxy is called the thick disk, which accounts for about 4 percent of all stars near the Sun. Arcturus is a likely member. The thick disk is old and forms a more distended system around the Galactic plane, with a typical star lying several thousand light-years above or below it. The stars have more elliptical orbits, higher U, V, W velocities, and metallicities around 25 percent of the Sun's.

The third stellar population is known as the halo. Halo stars are old and rare, accounting for only 0.1 to 0.2 percent of the stars near the Sun. Kapteyn's Star is the closest halo star to Earth. These stars make up a somewhat spherical system, so most members of the halo lie far above or far below the Galactic plane. Kinematically, halo stars as a group show little if any net rotation around the Galaxy, and a typical member therefore has a very negative V velocity. The stars often have extremely elliptical orbits: a halo star may lie 100,000 light-years from the Galactic center at apogalacticon but venture within a few thousand at perigalacticon. Metallicities are even lower than in the thick disk, usually between 1 and 10 percent of the Sun's.

The fourth and final stellar population is the bulge, which lies at the center of the Galaxy and can be seen in other edge-on spiral galaxies as the bump that extends above and below the galaxy's plane at the center (see Figure 2 in Chapter 1). The Galactic bulge is old and metal-rich. Most of its stars lie within a few thousand light-years of the Galactic center, so few if any exist near the Sun. Consequently, the bulge is the least explored stellar population in the Milky Way.

THE DISCOVERY OF STELLAR POPULATIONS

Today, the idea of stellar populations is such a useful probe of the Milky Way's formation and evolution that even the concept's originator, if he were still alive, might be astonished at its vast scope and power. The idea arose in 1943, when German-born astronomer Walter Baade synthesized a wide array of astronomical data and split the Galaxy into two broad populations. "Hubble was the outstanding astronomer of the twentieth century," said Lick Observatory's Donald Osterbrock, who has studied Baade's life. "Baade was the second outstanding astronomer of the twentieth century. He was a fantastically smart and technically skilled astronomer and also had a wonderful personality—he got along very well with others. He was very interested in the physical meaning of what he was doing. All of Baade's work was aimed at understanding what was going on, including finding the two stellar populations."

Baade developed the concept of stellar populations while he was at Mount Wilson Observatory, which he had joined in 1931. At the time, Mount Wilson's great astronomer was Edwin Hubble, who approached the universe in an opposite and complementary way to Baade. Hubble, a cosmologist, was after the universe as a whole—big questions such as its origin and destiny, for which he studied the countless galaxies strewn throughout space. In contrast, Baade pre-

ferred to examine each individual galaxy, scrutinizing its stars and nebulae. If Hubble surveyed the forest, it was Baade who studied the
trees.

Baade's interests steered him toward the galaxies in the Milky
Way's backyard. "The Local Group is our main field of exploration if
we want to do detailed work on galaxies," he said during lectures at
Harvard University in 1958. "In fact, if our own Galaxy were not a
member of such a group, we should be at a very great disadvantage."
Indeed, then there would have been no nearby galaxies for Baade to
dissect. Today the Local Group has about thirty known members, but
during the 1930s astronomers knew less than a dozen. The two most
important were the Andromeda Galaxy and the Milky Way. Each had
smaller companion galaxies that orbited it the way moons circle a
planet. The Milky Way, for example, was then known to have two, the
Large and Small Magellanic Clouds.

The galaxies in the Local Group were of various types, and differences among them provided Baade with the first clues to the stellar
population concept. In 1926 Hubble had published a classification system that divided galaxies into three main types. Hubble's first type was
the elliptical galaxies, featureless and rather boring objects that look
like eggs or tennis balls. Examples included M32 and NGC 205, which
circle Andromeda. Hubble's second type was the more attractive spirals, disk-shaped galaxies containing beautiful spiral arms. The nearest
spirals were the Andromeda Galaxy and M33, another Local Group
member. (The Milky Way is also a spiral, but this was then unknown.)
And Hubble's third type was the irregular galaxies—"the wastebasket
in Hubble's system," as Baade called them—which were neither ellipticals nor spirals. The nearest irregulars were the Magellanic Clouds.

But there was a problem that perplexed Baade for decades and
would be resolved only after he developed the stellar population concept. It concerned the Milky Way's relation to the other galaxies of the
Local Group, especially Andromeda. After Hubble detected Cepheids
in Andromeda, he estimated that the galaxy was just under a million
light-years away. By using this distance and Andromeda's apparent
size, he calculated the galaxy's true size and found that it was smaller
than the Milky Way. This in itself was not troubling, but disturbing
implications followed. When the intrinsic brightnesses of Andromeda's
objects were compared with their counterparts in the Milky Way, it

turned out that the Milky Way's were much more luminous—odd, since similar objects should have similar luminosities, no matter which galaxy they inhabit. Yet the Galaxy's brightest globular clusters were four times more luminous than Andromeda's brightest globulars, and novae in the Milky Way also outshone their peers in Andromeda. Worst of all, not only was the Milky Way larger than Andromeda, but it was also larger than every other galaxy in the universe.

"These discrepancies interested me exceedingly," said Baade, "and since I wanted to stay out of Hubble's field, cosmology (in which I was anyhow not interested), I tried to see how one could at least clear up some of these discrepancies." The problem, thought Baade, required studying individual stars in the Local Group galaxies. To do that, he had to "resolve" the galaxies—that is, break them into stars. If a galaxy is too far, or is viewed through too small a telescope, it looks like a blur of light. But when a galaxy is resolved, some of its stars can be seen and studied. Imagine a large garden that contains a myriad of flowers. From a distance, the garden looks like a swirl of color, and you cannot see the individual flowers. If you walk toward it, though, you eventually get close enough to resolve the garden, seeing individual red roses, orange nasturtiums, and yellow marigolds. During the 1920s, Hubble had used Mount Wilson's 100-inch telescope and successfully resolved the spiral arms of the Andromeda Galaxy and M33, where he detected red supergiants, yellow Cepheids, and blue supergiants.

"In the course of this work on the members of the Local Group," said Baade, "one remarkable thing had shown up: in the Andromeda [Galaxy], Hubble had found it impossible to resolve the central part. Also—even more significant—he had been unable, in spite of many attempts, to resolve the companions, which were the closest representatives, and apparently normal representatives, of the [elliptical] galaxies. It was quite clear that here was a problem that had to be solved, and a place where one might expect to find an answer to the other problems that he had encountered." Especially striking was Hubble's failure to resolve the Local Group's elliptical galaxies, two of which—M32 and NGC 205—lay next to Andromeda. So great was the problem that James Jeans in England had even suggested that ellipticals were balls of gas that had no stars.

In addition to the contrast in resolution, spirals and ellipticals differed in color, which would prove to be another key clue as Baade formulated the population concept. The arms of spiral galaxies were blue, whereas the bulges of the spirals were yellow-orange. Elliptical galaxies were also yellow-orange. The galaxies therefore mirrored the Milky Way's star clusters. In open clusters, which line the Galaxy's plane, the brightest stars are blue or white, but in globular clusters, which typically lie away from the Galactic plane, the brightest stars are orange or red. This was another clue to the presence of two different stellar populations—one in the open clusters, the other in the globulars—but at the time no one connected cluster colors with galactic colors.

Yet another clue emerged in 1938. Harlow Shapley reported two peculiar objects that are now known to be dwarf elliptical galaxies orbiting the Milky Way. Both lay in the southern sky and appeared on photographic plates taken by astronomers in South Africa. On one of the plates, a faint smudge surfaced in the constellation Sculptor. Shapley at first thought the smudge might be only a fingerprint or defect in the plate, but other plates of the same region showed the same smudge and convinced Shapley that the object in Sculptor was real. Later, the plates revealed a similar object in the neighboring constellation of Fornax.

The Sculptor and Fornax objects stumped astronomers: no one had ever seen anything like them. They were several thousand light-years across, and each consisted of stars. But the stars were so widely separated that Sculptor and Fornax were barely noticeable. They had what astronomers call a low surface brightness, meaning that their light is spread over a large area, and even Shapley did not recognize that they were galaxies.

In the summer and fall of 1938, Baade and Hubble turned the 100-inch telescope on the new objects. Because Sculptor and Fornax lie far to the south and do not rise much above Mount Wilson's horizon, Baade and Hubble could observe them for only a couple of hours each night. Despite this handicap, the astronomers discovered dozens of RR Lyrae stars in Sculptor and two globular clusters in Fornax. In both Sculptor and Fornax, the brightest stars were orange—similar in color to the elliptical galaxies, the bulges of spirals, and the brightest stars in the Milky Way's globular clusters. But again, no one made the connection.

These diverse clues would not converge until after Baade attempted to do what Hubble had failed to, resolve the bulge of the

Andromeda Galaxy and scrutinize its stars. Baade began the task in 1942. Because of World War II, most astronomers had deserted Mount Wilson to help in the war effort. Baade had to stay, however, for as a German citizen and therefore an enemy alien he was not even permitted to leave Los Angeles County. Not that he wanted to: with most of his colleagues gone, Baade had nearly free run of the mighty 100-inch telescope. Furthermore, the Mount Wilson sky was darker than Baade had ever seen it. After Japan bombed Pearl Harbor, West Coast cities dimmed their lights to thwart similar attacks, so light pollution from nearby Los Angeles was minimal. Under these excellent astronomical conditions, Baade used the 100-inch telescope to take aim at Andromeda's bulge. He photographed the bulge using the same type of plates Hubble used, which were sensitive to blue light.

But like Hubble, Baade still failed to see any stars in Andromeda's bulge, achieving only incipient resolution. "What is meant by incipient resolution," said Baade, "can best be understood from the reaction of everybody who looks at this photograph. The plate is very irritating to the eye; a definite structure is emerging all over, but one does not yet see any stars."

Nevertheless, said Baade, "The resolution of [Andromeda] was now clearly within our grasp; the question was, how could it be attained?" The blue plates had failed to resolve the bulge. "The only hope was to try something else," he said, "and I decided to see whether red plates would do the trick."

The red plates were new. Astronomers traditionally used blue plates because they were more sensitive to light than red plates and captured the faintest stars the fastest. But improved red plates had recently been developed, and though they built up star images more slowly than the blue, they offered two advantages. First, because Andromeda's bulge was yellow-orange, and because the blue plates had failed to reveal stars, Baade suspected that the brightest stars in the bulge were not blue. They might be orange or red, in which case red plates would pick them up better. Second, the longest exposure Baade could take with the blue plates was only ninety minutes, because the sky emits blue light that fogs the plates and eventually overwhelms the stars. In contrast, the sky emitted less light at the red wavelengths Baade chose, so he could expose the red plates for hours.

"No success could be expected by simply relying on these

guesses, snapping a red plate into the plateholder of the 100-inch, making the exposure, developing, and hoping to see something," said Baade. "It was quite clear that the stars would be very faint, and also very likely that they would be extremely crowded. It would tax the resolution of the 100-inch, and evidently one would have to be very careful and use every trick of the game." Baade chose nights with excellent seeing, when the atmosphere was stable and stars appeared sharp and clear. He also worried about the telescope's mirrors, whose shapes changed as the temperature did. During the day, the dome housing the telescope was closed, but sunlight heated the dome and the air and mirrors inside. At night, when the temperature dropped, the telescope mirrors cooled unevenly and distorted the images of stars. To circumvent this problem, Baade had the dome opened during the early afternoon to air it out and cool the mirror before he began his nightly observations. During the observations, he also constantly examined the guide star and quickly responded to any changes in the focus.

"These preparations lasted from the fall of 1942 to the fall of 1943, when I was finally ready to make the real test," said Baade. "Indeed, after taking all these precautions, the resolution was a very simple matter. In August and September [of 1943] I resolved the central part of [Andromeda] and the two nearby [elliptical] companions [M32 and NGC 205] in rapid succession." Baade then went further. A pair of fainter elliptical galaxies, NGC 147 and NGC 185, lay near Andromeda, and Baade resolved them as well, proving that they were nearby and probably satellites of Andromeda. Thus, Baade had resolved the bulge of Andromeda and four elliptical galaxies.

The stars that Baade saw—by the tens of thousands—differed greatly from the stars that lit the spiral arms. In the arms the brightest stars were blue supergiants, but in Andromeda's bulge and the four elliptical galaxies the brightest stars were orange giants. Furthermore, Baade realized that the orange giants were much brighter than ordinary giants, because otherwise he would not have seen them. Ordinary giants appear in open clusters, but the bright orange giants Baade saw resembled the brightest stars in globular clusters, whose giant stars are more luminous. The difference in luminosity, astronomers now know, is due to a difference in metallicity: metal-poor giants, like those in globular clusters, outshine the metal-rich giants in open clusters.

Baade announced the discovery in two landmark papers, published back to back in *The Astrophysical Journal* in 1944. "There can be no doubt," he wrote, "that, in dealing with galaxies, we have to distinguish two types of stellar populations." The first was population I, which ruled the spiral arms and the open star clusters. The brightest population I stars were blue supergiants, although the population also included fainter stars of other colors, such as the Sun, which could not be seen in galaxies outside the Milky Way. The second population, population II, filled the spiral bulges, elliptical galaxies, and globular clusters. Here the brightest stars were more modest, being orange or red giants; there were no blue or white supergiants.

TABLE 6-1: WALTER BAADE'S STELLAR POPULATIONS

Population I	Population II
In the Milky Way:	
O and B stars	RR Lyrae stars
Open star clusters	Globular star clusters
Low-velocity stars	High-velocity stars
Galactic disk	Galactic halo and bulge

Beyond the Milky Way:	
Disks and spiral arms of spiral galaxies	Elliptical galaxies Bulges of spiral galaxies

Baade applied his population idea to the Milky Way. Population I, he said, dominated the stars near the Sun and, by extension, the entire Galactic disk, for near the Sun the most luminous stars were blue or white supergiants like Rigel and Deneb. These stars had low velocities with respect to the Sun, which meant they circled the Galaxy fast and remained near the Galactic plane. Population II stars had large velocities with respect to the Sun, revolved slowly around the Galaxy, and spread above and below the Galactic plane. Their most prominent representatives in the Milky Way were the globular clusters. This population included the RR Lyrae stars, which had elliptical orbits around the Galaxy and also abounded in the globulars.

In addition, Baade recognized that the mysterious Sculptor and Fornax systems were actually elliptical galaxies. They were full of population II stars. The brightest stars were orange, and Sculptor had RR

Lyraes and Fornax had globular clusters. Baade came to this crucial realization after resolving NGC 147 and NGC 185. Like M32 and NGC 205, these elliptical galaxies orbit Andromeda but have lower surface brightnesses, between those of normal ellipticals and the extremely low surface brightnesses of Sculptor and Fornax. Thus, NGC 147 and NGC 185 linked Sculptor and Fornax with the greater ellipticals, all of which had the same population of stars, population II. Sculptor and Fornax then linked the elliptical galaxies with the globular clusters, which are even smaller than Sculptor and Fornax.

But Baade kicked himself for not having seen the connection five years earlier, when he and Hubble had observed the Sculptor and Fornax systems. "The ridiculous part of the situation is that, after all these efforts, it appears that the whole difficulty should have been unnecessary. When Shapley discovered the Sculptor and Fornax systems he described them as a new type of stellar system because they looked so peculiar. Hubble and I were of course immediately interested, set out to make a rapid survey, and discovered the [RR Lyrae] variables. We treated the Sculptor system as if it were a normal globular cluster . . . and never realized what we had got hold of. We described the systems, even though they were in some ways different, as exhibiting the well-known features of globular clusters. The paper was soon published and was discussed by everybody, but nobody had the wits at the time to see that the whole problem was solved and that we knew what the [elliptical] galaxies consisted of." Baade added, "As Einstein said, mankind is very stupid and progress is very slow."

But Baade's discovery of the two populations was a major breakthrough. "It became evident that in this division we had found something very much deeper," he said. By the early 1950s, astronomers realized that the basic difference between populations I and II was age, for population I was young and population II old. Population I stars appeared along the spiral arms, which contain the clouds of gas and dust that create stars, whereas population II systems possessed little gas and dust, having used it up long ago to form stars that are now old.

Age, in fact, explains why the brightest population I stars are blue and white and the brightest population II stars are orange or red. An extremely young group of stars has main-sequence members of all spectral types—O, B, A, F, G, K, and M. The brightest and most massive stars are the blue O and B stars. When they exhaust the hydrogen at their cores, these stars evolve into red and blue supergiants and explode. But the blue stars outnumber the red ones, because the stars spend more time blue than red. Thus, in a young, population I system, most of the brightest stars are blue. As the group ages, though, all the O stars die, then the B stars die, and so on. Eventually, after many billions of years, the brightest main-sequence stars will be type F or G, which are yellowish. As these stars leave the main sequence, they evolve into orange and red giants that are much brighter. So, in an old population, the brightest stars will be orange or red giants, as is the case for population II systems.

If this idea was right, the globular clusters were old, and their brightest main-sequence stars should be yellow, too faint for the 100-inch to see because of the globulars' great distances. But in 1948, the mammoth 200-inch telescope went into operation atop Palomar Mountain in California, and Baade assigned two of his graduate students, Halton Arp and Allan Sandage, the task of finding the main sequence in the globulars. Arp and Sandage succeeded, and the brightest main-sequence stars were indeed yellow.

In the population concept also lay the source of the discrepancies that had long troubled Baade—the large size of the Milky Way and the overluminosity of its globulars and novae. Hubble had used Cepheids to estimate that the Andromeda Galaxy lay just under a million light-years away. But Cepheids actually occurred in both the spiral arms (population I) and the globular clusters (population II). There was no guarantee that the population I Cepheids obeyed the same period-luminosity relation as the population II Cepheids, yet Hubble's estimate rested on this very assumption.

Using the 200-inch telescope, Baade examined Andromeda. If the galaxy were really as near as Hubble had claimed, then the new telescope should have revealed the galaxy's RR Lyrae stars. But it did not, proving that Andromeda was farther. Hubble's error resulted from confusing the population I Cepheids (which today are simply called Cepheids) with the population II Cepheids (which today are called W Virginis stars, after the prototype star W Virginis). A W Virginis star is intrinsically fainter than a Cepheid of the same period, so if one mistakes a Cepheid for a W Virginis star, one will underestimate the star's distance. When Baade worked out the numbers, in 1952, he found that Andromeda was twice as far as Hubble had claimed.

By placing Andromeda farther away, Baade increased the estimate of the galaxy's true size. Andromeda was now the largest galaxy in the Local Group; the Milky Way came in at number two. The greater distance also removed all the other discrepancies that had troubled Baade, because it meant that Andromeda's globulars and novae were as luminous as those in the Milky Way. Moreover, Baade's revision of the Cepheid distance scale doubled the size of the entire universe, because estimates of galaxy distances depended on the distances to Andromeda and the few other nearby galaxies in which astronomers had detected Cepheids.

But Baade's population concept had equally great implications for research on the Milky Way. By studying and comparing the stellar populations, astronomers could begin to decode the rich fossil record written in the Galaxy's stars. Indeed, so influential was Baade's discovery that every subsequent decade has enriched and complicated it, providing new insights into the origin and evolution of the Milky Way that Baade himself never even suspected.

One of the first revelations was both simple and profound. When Hubble had classified galaxies, no one knew which type of galaxy we inhabit, whether it was elliptical, spiral, or irregular. But in 1951, after a century of speculation by others, astronomers at Yerkes Observatory in Wisconsin applied Baade's ideas and proved that the Galaxy is the most beautiful type in the universe, a spiral.

7

A SPIRAL GALAXY

ALL GALAXIES ARE not created equal. Of the three main types of galaxies, one stands out from the rest for its stunning beauty: the spiral. In contrast, an elliptical galaxy looks like a mere ball of haze, and an irregular galaxy is just a random assortment of stars and gas. But a giant spiral galaxy is the most magnificent type of galaxy in the universe, boasting the spiral pattern of a seashell on a scale spanning over 100,000 light-years. Lit by bright blue stars and lined with clouds of gas and dust, the spiral arms wrap around such a galaxy as if it were a huge vortex in space. Since the first spiral had been found, in 1845, astronomers had discovered thousands of others, strewn across the universe like flowers in a vast garden. But only in 1951 did astronomers prove that the Milky Way itself belongs to this spectacular class.

THE MILKY WAY'S SPIRAL SHAPE

The idea that the Milky Way was a spiral galaxy had been around a long time, but concerted attempts to prove this had failed: far easier to see the spiral arms of a galaxy we view from the outside than to penetrate the thicket of stars in which we lie entangled. In 1852, seven years after Lord Rosse glimpsed the spiral nature of M51, American astronomer Stephen Alexander first suggested that our Galaxy was also a spiral. Later, in 1869, British astronomy writer Richard Proctor made a similar proposal. In 1900, and again in 1913, an astronomy writer in Holland named Cornelis Easton published beautiful maps that purported to show a spiral form for the Galaxy, and Easton compared the Milky Way with what are now known to be other spiral galaxies. Although Easton's claims turned out to be true, his maps were incorrect, for they delineated spiral arms in the wrong direction and placed the Galaxy's center in the constellation Cygnus rather than in Sagittarius.

The next to take up the challenge of uncovering the Galaxy's spiral arms was Dutch astronomer Bart Bok, who later, with his wife, wrote a popular book on the Milky Way that went through five editions. After studying in Holland under Jan Oort and Pieter van Rhijn, Bok had inherited Jacobus Kapteyn's love for star counts, which Bok considered the perfect technique for detecting the Milky Way's spiral structure. In 1929, he came to Harvard and attempted to put the star-counting plan to work.

It failed. "In the late 1930's," Bok wrote, "it seemed an almost hopeless task to trace the spiral structure of our Galaxy. In public lectures during that period I often said it was unlikely the problem would be solved in my lifetime." Bok and his colleagues had spent over a decade counting stars in different directions. He hoped and expected that the number of stars he counted would shoot up when he reached a spiral arm and then plummet when he passed beyond it. At the time, this seemed a reasonable assumption, because the arms of a spiral galaxy were brighter than the rest of its disk, implying that the arms had more stars. But in his work on the Milky Way, Bok discerned little more than a hint of the arm we inhabit.

Fortunately, Walter Baade's work on the Andromeda Galaxy provided an alternative to the tedium of star counts, for Baade had found that the blue O and B supergiants lined only Andromeda's spiral arms. Ordinary stars, which Bok had been counting, may not trace the Galaxy's spiral arms, but these bright blue beacons do—just as streetlights outline city streets better than do porch lights. Baade's idea to determine the Milky Way's shape required that one measure the distances and therefore the absolute magnitudes of the Galaxy's OB stars, and Baade never pursued this approach himself. But his suggestion caught the attention of William Morgan, an astronomer at Yerkes Observatory in Williams Bay, Wisconsin. Morgan was in an excellent position to exploit Baade's idea, for during the 1930s and 1940s he and his colleagues had developed a careful scheme for classifying stars according to their spectral type and absolute magnitude.

The son of a physically abusive minister, Morgan had become interested in astronomy as a child. "My father beat me up frequently," said Morgan. "The stars gave me something that I felt I could stay alive with. The stars and the constellations were with me, in the sense that on walks in the evening, I was a part of a landscape which was the stars themselves. It helped me to survive."

As a child, Morgan looked forward to the changing seasons and the different stars they brought. "Certain stars attracted me—their color and also the time of year when they appeared," he said, calling Deneb in Cygnus one of his favorites. "There were certain groupings which for some reason attracted me deeply, because—this seems strange to you, perhaps—but I felt comfortable with them, and they helped me get through the problems I was having."

In 1926, at the age of twenty and against his father's wishes, Morgan landed a job at Yerkes Observatory. There he classified stellar spectra more precisely than had the Harvard astronomers who compiled the *Henry Draper Catalogue*. Years later, after he had begun his search for the Milky Way's spiral arms, Morgan enlisted two young graduate students at Yerkes, Stewart Sharpless and Donald Osterbrock. Sharpless has since left the profession, but Osterbrock is now at California's Lick Observatory and has written several books on astronomy.

Osterbrock's background was quite different from Morgan's, for Osterbrock's parents had encouraged him in science—his father was a professor of electrical engineering and his mother had been a lab technician before she married. Osterbrock's advisor at Yerkes was Subrahmanyan Chandrasekhar, the astronomer who had earlier done theoretical calculations to derive the maximum mass that a white dwarf can have.

"I was doing mostly theoretical research with Chandra," said Osterbrock. "But I was very interested in observational work, too. I took a graduate course from Morgan in the spectroscopic classification of stars. I was convinced that what he was doing was very important, and I made it clear to him that I'd like to do some observational research."

To map the arms as Baade had suggested, Morgan needed to determine the distances of the high-luminosity O and B stars that lit the arms. But most of the stars lay over a thousand light-years away, so their parallaxes were too tiny to measure. As a result, Morgan had to deduce the stars' distances from their absolute magnitudes. Fortunately, stars of different absolute magnitudes have slightly different spectra. For example, B-type supergiants have narrower hydrogen lines than less luminous B-type stars, and Morgan had earlier embarked on a massive project to refine the spectroscopic classification of stars. By scrutinizing a spectrum, he could determine the star's absolute magnitude and, from that, its distance from Earth.

"Morgan got into spectral classification long, long before Sharpless and I came," said Osterbrock. "Then he gradually shifted his interest to high-luminosity stars, partly because the Harvard classification was so bad. Before Morgan, people were using spectral types out of the *Henry Draper Catalogue* that were not very good. If you take the spectral types as published in the HD and try to use them today, they're just terrible." The poor quality of the Harvard data resulted from the nature of the blue stars themselves. Classifying the spectral types of stars requires seeing the lines in the stars' spectra, but O and B stars have weak spectral lines. Furthermore, distant O and B stars are faint, which made their lines impossible for the Harvard astronomers to see. Consequently, the very stars that were vital for discerning the Galaxy's spiral structure had the least accurate spectral types.

During the 1930s, however, Morgan gathered better data and devised a precise system for classifying stellar spectra. He still used the Harvard spectral types—O, B, A, F, G, K, and M—but he segregated the stars by luminosity, marking different luminosity classes with roman numerals. For example, he placed supergiants like Rigel into luminosity class I, giant stars like Arcturus into luminosity class III, and main-sequence stars like the Sun into luminosity class V. Soon after, spectroscopist Philip Keenan joined Morgan, and in 1943, with Edith Kellman, they published a thorough atlas that showed characteristic spectra of stars with different spectral types and luminosity classes. The MK (Morgan-Keenan) classification system is still the standard today.

In 1947, Morgan began to exploit this system in an attempt to map the Galaxy's spiral arms. From 1947 to 1949, Morgan and Jason Nassau, of Warner and Swasey Observatory in East Cleveland, discovered and classified hundreds of O and B stars. To find the stars, Morgan and Nassau used a 24-inch telescope in East Cleveland; to classify the stars by spectral type and luminosity, the astronomers used Yerkes's 40-inch telescope. From these observations, Morgan and Nassau determined the absolute magnitudes and distances for some of the stars. In 1950, the two presented their results during a meeting at the University of Michigan. They had the distances for 49 individual OB stars as well as for three groups of OB stars. Unfortunately, when the astronomers plotted the stars' positions, the resulting map did not show a clear spiral pattern. Although a possible spiral arm ran from Cygnus through the Sun and into the constellation Carina, Nassau cautioned that the Galaxy might not have any spiral arms at all.

To achieve greater success required a more powerful tactic, and once again it came from Baade's work on the Andromeda Galaxy. At the same 1950 conference, Baade displayed a spectacular photograph that caught Morgan's attention. The photograph showed Andromeda's H II regions—glowing areas where hot O and B stars had ionized the interstellar hydrogen near them. Hydrogen atoms come in two varieties: neutral hydrogen, or H I ("H one"), consisting of a proton bound to an electron; and ionized hydrogen, or H II ("H two"), when the proton has lost its electron. Neutral hydrogen pervades interstellar space, but a single O or B star can ionize all the hydrogen for dozens or even hundreds of light-years around. All O and the hottest B stars are so hot they emit extreme ultraviolet light that rips electrons off protons, ionizing the hydrogen and creating an H II region. In the Galaxy, the most famous H II region is the Orion Nebula, which harbors several O and B stars and glows so brightly that it is visible to the naked eye, even though it lies 1500 light-years away.

Baade's photograph of Andromeda's H II regions showed a remarkable feature. "The large H II regions defined the spiral arms," said Osterbrock. "You see that in Andromeda." The H II regions marked the arms even better than individual OB stars did. "So Morgan's idea was to identify large H II regions in our Galaxy," said Osterbrock, "and then find the distances to them by the spectroscopic classification of the stars within them. Each of these H II regions has not just one O star but a lot of O and B stars in it." If individual O and B stars were like Galactic streetlights lining and lighting the spiral arms, large H II regions were akin to well-lit shopping centers boasting hordes of streetlights. Because all the stars creating a particular H II region lay at nearly the same distance from us, determining all the stars' distances in an H II region gave a much more accurate distance to the region than a single O or B star could.

"Morgan always believed in getting the best data and forgetting about lousy data," said Osterbrock. "This was opposite to the Harvard approach, which was to use any distances you could get, average them all together, and hope for the best."

The next step was to find the H II regions in the Galaxy. In 1950 and 1951, Sharpless and Osterbrock equipped their camera with a filter that transmitted red light and took dozens of photographs of the Milky Way and aurorae. Most H II regions are red, even though the O and B stars that create them are blue; indeed, Baade had photographed Andromeda's H II regions on red-sensitive plates. The red glow comes in part from ionized hydrogen that has recaptured an electron and become neutral. As the electron settles into position, it can emit light with a wavelength of 6563 angstroms, which is red. The red light also comes from singly ionized nitrogen—nitrogen with one of its electrons torn away—which radiates at 6548 and 6583 angstroms.

The red filter emphasized the H II regions on the resulting photograph and made them easier to find. After searching the plates for H II regions, Sharpless and Osterbrock turned the plates over to Morgan, who inspected them thoroughly and discovered even more H II regions than his younger colleagues had. From the spectra of the O and B stars in each H II region, Morgan determined the stars' absolute magnitudes and the distance to each H II region.

The final step was to plot the positions of the H II regions on a map of the Galaxy. "For me," said Morgan, "it was a jewel all the way. It was absolutely perfect. In some respects, it's the most interesting thing that happened in my life." Morgan's map, which he plotted in late 1951, revealed two spiral arms. The Sun lay along the inner edge of one, now called the Orion arm, which stretched from Cygnus to Monoceros and contained H II regions such as the North American Nebula in Cygnus and the Orion Nebula. A second spiral arm, the Perseus arm, ran parallel to the Orion arm and lay some seven thousand light-years farther from the Galactic center. Its two most prominent members were a double star cluster in Perseus named h and Chi Persei.

Morgan also saw hints of a third arm, interior to the Orion arm, that ran through Sagittarius. "Those are more southern stars," said Morgan. "They're harder to work with than ones in the northern hemisphere." On his first map, Morgan plotted only one H II region in the Sagittarius arm, but two years later he added southern OB stars to the map and confirmed the third arm's existence.

Morgan announced the discovery of the first two spiral arms at the next meeting of the American Astronomical Society, which took place in Cleveland in December 1951. Jan Oort chaired the session and introduced Morgan to the large audience. During his fifteen-minute talk, Morgan described the search for the H II regions, explained how he

Figure 12. The spiral arms near the Sun include the Orion arm, which contains the Sun and all the bright stars in the night sky; the Perseus arm, which lies farther from the Galactic center than the Orion arm; and the Sagittarius arm, which lies closer to the Galactic center. The arms are bright because they contain extremely luminous stars, such as Rigel and Deneb, whereas the regions between the arms are dark because they have no very luminous stars. On this diagram, the Galactic center lies well below the Sagittarius arm.

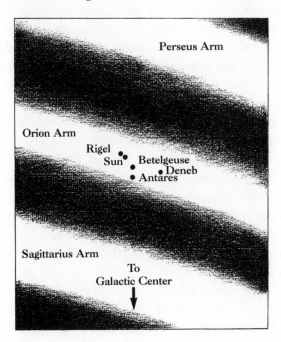

had measured their distances, and then played his trump card, the Galactic map that revealed the spiral arms. The audience went wild: Morgan received thunderous applause and, by some accounts, a standing ovation.

"It was very moving, deeply moving," said Morgan. "I wasn't any big shot, just an ordinary type. I was surprised as to how it was accepted by the audience. It's the sort of thing that people live for."

Morgan listed Sharpless and Osterbrock as co-authors of the paper. "I got much more credit than I deserved," said Osterbrock. "That paper—the concept and everything about it—was Morgan's, and he was very, very generous to include us as authors. The fact that my name was on it was probably the best-known thing about me for the first ten years of my astronomical career and undoubtedly opened all kinds of doors for me."

Morgan's team had succeeded where Bart Bok and his years of tedious star counting had failed. Osterbrock explained why. "Stars don't define the spiral arms," he said. "H II regions and OB stars do. Bok didn't realize that." Though O and B stars light the spiral arms and make them bright, ordinary stars are no more common in the arms than between them, so counting ordinary stars cannot reveal spiral structure. "Anyhow," said Osterbrock, "star counts don't give you any decent distances at all, because they're a mishmash of all kinds of stars. You have to make lots of simplifying assumptions to get anything out of them. In my time, 'star counts' was a bad word at Yerkes Observatory."

But no paper reporting the momentous discovery ever appeared. All that surfaced in the technical literature was a brief abstract, five paragraphs long, in the April 1952 issue of *The Astronomical Journal.* Written by Morgan, Sharpless, and Osterbrock, the abstract had no map, and it bore the understated title "Some Features of Galactic Structure in the Neighborhood of the Sun." A more thorough account, complete with a detailed map, actually ran in the popular magazine *Sky and Telescope,* but even that report could not substitute for a thorough technical paper.

The full paper never appeared because in 1952 Morgan suffered a mental breakdown. "I was in the hospital—helpless, completely helpless," said Morgan. "Some people thought I would never come out of it." Morgan attributed his recovery to the same thing that had helped him survive his troubled childhood: astronomy. "It brought me out of the deepest problem in my whole life, which was my collapse," he said. By the time he was better, later in the year, other astronomers had raced ahead and mapped more of the Milky Way, so Morgan never wrote the paper.

THE RADIO GALAXY

Morgan, Sharpless, and Osterbrock had done their work at optical wavelength, relying on the study of visible light, but hot on their trail was the new science of radio astronomy. Unlike light, which dust absorbs, radio waves zip through dust and reach astronomers from all parts of the Galaxy. Because Morgan and his colleagues had observed visible light, they could map only the spiral structure that lay within a few thousand light-years of the Sun. In contrast, radio astronomers could see the entire Milky Way, even regions on the opposite side of the Galaxy.

However, astronomers were slow to appreciate the advantage. In 1932, two decades before Morgan's team discovered the spiral arms, Karl Jansky detected the first radio waves from space. Jansky, a radio engineer at Bell Telephone Laboratories in New Jersey, was trying to eliminate static in radio communications. He found that some of the radio noise came from Sagittarius, where the Galactic center lay.

But Jansky's work attracted little interest from astronomers, and Jansky himself never followed it up. That was left for an amateur radio operator outside Chicago. In 1937, Grote Reber built a radio antenna in his backyard and for years was the world's sole radio astronomer. With this primitive radio telescope, Reber captured the Milky Way's radio waves, but most American astronomers still ignored the new field.

However, in Holland, then under Nazi occupation, Jan Oort read Reber's work and recognized the implications for astronomy. In 1944, Oort asked his student Hendrik van de Hulst to see whether there was some way to put radio astronomy to use, and van de Hulst found that indeed there was. He concentrated on hydrogen, the most common element in the Galaxy. Studying hydrogen in its neutral form—H I, as opposed to the H II that Morgan's team had investigated—van de Hulst calculated that neutral hydrogen should emit radio waves. In a neutral hydrogen atom, both the proton and the electron can be thought of as spinning particles. The spins of the proton and electron are said to be parallel if they spin in the same direction and antiparallel if they do not. About once every 11 million years, a neutral interstellar hydrogen atom in the parallel state spontaneously switches to the antiparallel state, emitting energy by producing a radio wave that is 21 centimeters long. Although 11 million years is a long time, van de Hulst calculated that there were so many neutral hydrogen atoms in space that interstellar clouds should constantly buzz with 21-centimeter radiation—just as a large city always has a few traffic lights that are yellow at any one time, even though each light's yellow color lasts just a few seconds.

After the war, Oort and his colleagues attempted to detect this 21-centimeter radiation, but two physicists at Harvard, Harold Ewen and Edward Purcell, beat them to it. On March 25, 1951, they picked up signals from neutral hydrogen, broadcasting at 21 centimeters. The Dutch team succeeded on May 11, and an Australian team followed in July. Unfortunately, the young radio engineer who had created the new field did not live to see these successes. Karl Jansky had died in 1950, only forty-four years old, a victim of kidney failure and heart disease.

The successful detection of 21-centimeter radiation meant that radio astronomers could try to map the spiral arms, since the arms contain neutral hydrogen gas. In fact, within this gas new stars are born, which is why all the short-lived O and B stars lie in the spiral arms. By the time the teams in Holland and Australia set out to map the Milky Way, Morgan, Sharpless, and Osterbrock had already detected the spiral arms. Nevertheless, the radio astronomers could confirm and extend Morgan's work, because unlike light, 21-centimeter radiation passed unimpeded across the Galaxy.

This advantage, though, was weakened because astronomers could not directly measure the distance to any particular cloud of gas they saw. Instead, they had to infer the cloud's distance from its velocity relative to us. As Oort had discovered in 1927, the Galaxy rotates differentially—the inner parts spin somewhat faster than the outer parts—and the radio astronomers exploited this in mapping the Milky Way, because the blueshift or redshift of the 21-centimeter line indicated how fast a gas cloud was moving toward or away from us. Using the cloud's velocity and a model of the Milky Way's rotation, the astrono-

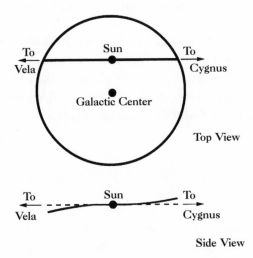

Figure 13. Although the Galactic disk looks flat from above (top diagram), a slice through the Sun's position shows that the outer disk is warped, sloping down on the left and up on the right (bottom diagram).

mers could then determine the cloud's approximate distance. By combining the observations from Holland and from Australia, the astronomers successfully constructed a complete map of the Milky Way. The first radio map showing part of the Galaxy's spiral structure was published in 1953, and a better map appeared in 1958. The second map showed the Perseus, Orion, and Sagittarius arms, which Morgan's team had discovered, and revealed spiral arms wrapped all around the Galaxy.

But the work also held a surprise. In 1956, before the final map was completed, it became apparent that the outer part of the Galaxy's disk was not flat. Rather, the region was warped like a record that had been left out in the sunlight. Some parts of the outer disk curved above the Galactic plane and others below it. Imagine the Sun within a giant clock, with the Galactic center in the direction of six o'clock. In the direction of three and four o'clock, the outer disk lies north of the Galactic plane; in the direction of nine and ten o'clock, it curves south of the plane.

Still, the far greater discovery was that of the Galaxy's spiral shape, a discovery that had resulted from Walter Baade's work on stellar populations. By the early 1950s, however, astronomers began to realize that the Galaxy was much more complex than Baade had thought. In fact, even as some astronomers were unraveling the Milky Way's spiral structure, others were learning that the Galaxy's stars differed from one another in a fundamental way. Two rare stars in Baade's population II were found to harbor far fewer heavy elements than the Sun—a result that astonished astronomers and would add a striking new twist to their understanding of the Galaxy's origin and evolution.

8

THE RISE OF THE SUBDWARFS

THE MILKY WAY Galaxy sustains life: it harbors large amounts of the heavy elements that life needs, such as the oxygen in the air, the calcium in bones, and the iron in blood. Indeed, of the four properties that mark a stellar population, the one that living beings carry within them is metallicity, the abundance of elements heavier than hydrogen and helium. Life emerged in the Milky Way's disk rather than in its halo because the former is metal-rich and the latter is not. However, when Walter Baade divided the Galaxy's stars into two populations, he recognized differences only in their locations and kinematics. He did not know that the population II stars were older than the population I stars. Nor did he know that the two differed in metallicity.

"At that time," said Lawrence Aller of the University of California at Los Angeles, "there was a general belief among a lot of astronomers that all stars had exactly the same chemical composition." This belief seemed to have a secure basis, for with few exceptions the spectra of two stars with the same temperature were identical: if one star showed a strong iron line at a particular wavelength, so did the other star, and the line was equally strong in both, indicating that the stars had equal iron-to-hydrogen ratios. This ratio, in turn, matched the Sun's. A few stars had cropped up with abnormal spectra, but the experts usually explained them away as arising from anomalies in the stars' atmospheres rather than from metallicity differences.

INTERLOPERS FROM THE GALACTIC HALO

But in 1951, Aller and a graduate student published a paper that Allan Sandage later called "unbelievably revolutionary." The paper claimed that two stars in the Galaxy's halo population, HD 19445 and HD 140283, had much lower iron and calcium abundances than the Sun, a finding that many of Aller's peers disputed. "I won't mention any names," said Aller, "because these people were very fine spectroscopists and astronomers, and we all make fools of ourselves at one time or another. They were just plain wrong." Still, because of this prejudice, Aller had to fight to get the 1951 paper published.

In fact, long before that, Aller had had to fight simply to become an astronomer. "My parents were dead set against it," said Aller, who grew up in the western United States. "They couldn't see why anyone would want to fool with this kind of stuff. They said, 'Oh, you should be out enjoying yourself, playing basketball or something.' More than that, my old man had a crazy idea that he was going to find gold. His father had found gold in 1849, so he figured he could do it in 1929. That the gold had all been mined out never occurred to him." As a consequence, Aller never graduated from high school. "I'm not a high school dropout," he said. "I was a high school pullout. My worthless old man dragged me out to work on his crackpot mining scheme."

Nevertheless, with the help of astronomers at Lick Observatory, Aller managed to escape from his father, and in 1943 received a doctorate from Harvard. There he had analyzed Wolf-Rayet stars—hot blue stars with abnormal spectra. Earlier, a few astronomers had gone against convention by claiming that certain other stars had peculiar abundances, and as Aller studied the Wolf-Rayet stars he concluded that some had large quantities of nitrogen and others large amounts of carbon and oxygen. These abundances arise, astronomers know today, because the stars have blown off their outer layers, exposing the carbon, nitrogen, and oxygen they have manufactured beneath.

"I took a beating for it," said Aller. "The paper was all but declared anathema in certain quarters. I was a young fellow, in those days

I had no permanent job, and I got my nose bloodied on all that kind of stuff."

Years later, Aller moved to the University of Michigan, where he performed even more controversial work. From the observations of other astronomers, Aller had spectra of HD 19445 and HD 140283, two stars named for their places in the *Henry Draper Catalogue*. The stars lay on opposite sides of the sky—HD 19445 was in Aries, HD 140283 in Libra—but they had long and related histories. In 1911, before the *Henry Draper Catalogue* was published, Walter Adams and Jennie Lasby at Mount Wilson reported that the stars were among the fastest then known. According to modern data, HD 19445 and HD 140283 are rushing toward the Sun at 140 and 171 kilometers per second, respectively.

Two years later, Adams reported that both stars had peculiar spectra, and in 1922 he and Alfred Joy confirmed it. Adams and Joy were attempting to deduce stellar luminosities from subtle differences in the stars' spectral lines. The 1922 paper dealt with the white stars of spectral type A, the class to which the astronomers had assigned HD 19445 and HD 140283. Like other A-type stars, these two had weak iron lines. At wavelengths that are visible to the eye, the most prominent iron lines arise from neutral and singly ionized iron atoms. But the high temperatures of A stars strip several electrons off each iron atom, producing more highly ionized iron atoms whose spectral lines are at wavelengths too short for the eye to see, in the ultraviolet. So, in optical spectra, A-type stars have weak iron lines. Although HD 19445 and HD 140283 shared this trait, the stars also had weaker lines of *hydrogen* than other A stars. In the 1922 paper, Adams and Joy listed three of these odd A-type stars.

By 1934, Adams, Joy, and their colleagues had found three more A-type stars with weak hydrogen lines, boosting the total to six. In addition to their puzzling spectra, these stars had luminosities fainter than those of main-sequence stars but brighter than those of white dwarfs, so Adams's team called them "intermediate white dwarfs." In 1939, Gerard Kuiper, at McDonald Observatory in Texas, coined the term astronomers use today. Main-sequence stars are sometimes called dwarfs, so Kuiper named the peculiar stars "subdwarfs," because they were less luminous than dwarfs. Subdwarfs are indeed main-sequence stars, but for reasons then unknown, their luminosity is about a magnitude fainter than main-sequence stars.

Like HD 19445 and HD 140283, most subdwarfs had high velocities with respect to the Sun and therefore followed eccentric, or highly
elliptical, orbits around the Galaxy. In 1943, American astronomer
Daniel Popper discovered that a subdwarf in the constellation Co-
lumba, which lies south of Orion, was flying away from the Sun at
record speed. The star's name was CD −29°2277, and the modern
value for its radial velocity is 546 kilometers per second; it remained
the radial-velocity champion until the 1980s, when astronomers found
an even faster star, also a subdwarf.

Aller first encountered the subdwarfs several years after Popper
did his work. "One time when I was out at Mount Wilson," said Aller,
"Roscoe Sanford showed me these spectra of HD 19445 and HD
140283. They looked pretty interesting, and he wanted me to do some
work on them." Aller took the spectra back to Michigan, where he and
a graduate student, Joseph W. Chamberlain, analyzed them. As Aller
and Chamberlain began a thorough investigation, they found that the
stars had been misclassified all along and were actually cooler than
spectral type A.

"They had lines of neutral iron," said Aller. "In a star as hot as an
A-type star, you'd expect the singly ionized iron lines to be more prominent, and they weren't. This indicated a lower temperature." All the
iron lines were weak, but the strength of the neutral lines relative to
the ionized ones convinced the astronomers that the stars really belonged to spectral class F, which meant the stars were yellow-white
instead of white. Although no one had yet measured the stars' colors,
spectral type F better fit the weak hydrogen lines, since F stars have
weaker hydrogen lines than A stars.

But the shift from A to F had a major implication, for the stars'
iron lines were also weak. If the stars were truly type A, the weak iron
lines would have been normal, a consequence of the high stellar temperature. But if HD 19445 and HD 140283 were cooler stars of spectral
type F, the weak iron lines could mean only one thing: the stars were
iron-poor, and by no small amount. Aller and Chamberlain estimated
that the stars' iron-to-hydrogen ratios were down by a whopping factor
of 100, meaning their iron abundances were only 1/100 that of the Sun.

Trouble was imminent. "We got this result," said Aller, "and there
was no way we could get off the hook. There was no way that we could

finagle it to make the abundances come out the same as in the Sun."
Aller and Chamberlain sent their work to *The Astrophysical Journal; The
Astrophysical Journal* sent it back.

"The referee didn't believe any of this stuff," said Aller. "He
screamed and hollered and made us push all the parameters as far as
we could until we reached a point and said, THIS 'FUR' AND NO FURTHER."
To accommodate the referee, Aller and Chamberlain altered their models and cut the depletion of iron. "We finally had him backed into a
corner," said Aller. "There was no way that he could get out of the
thing. We lowered our ratio from a hundred to ten and showed him that
by pushing everything—all the parameters, temperatures, everything
you could think of—as far as we could, we couldn't do any better than
that. The depletion had to be at least ten."

The paper was finally accepted, but its publication did not quiet
the critics. Although Aller and Chamberlain had moderated their
claims, an iron depletion of ten was still radical, and the leading author-
ity on stellar atmospheres, Albrecht Unsöld, did not believe it. But a
simple test could prove or disprove their result, because Aller and
Chamberlain had said the stars were spectral type F.

"The crucial thing was the colors of these stars," said Aller. "If
they were as cool as we said they were, they would be yellowish." In
December 1951, at the same meeting of the American Astronomical
Society where William Morgan announced the discovery of the Galaxy's spiral arms, a team of astronomers from Warner and Swasey Observatory reported on the color of HD 19445. The star was yellow.

"This was presented at the meeting, and I was sitting there, quite
satisfied," said Aller. "[Princeton astronomer] Martin Schwarzschild
was sitting next to me and Martin says, 'Oh, that's wonderful.' Oh,
yeah. So this was the beginning—the breaking down of this prejudice
that all stars had the same composition." Indeed, at about the same
time, Schwarzschild and his wife Barbara were uncovering milder examples of metal deficiency in stars having less extreme velocities than
the subdwarfs.

Ironically, the value of 100 that Aller and Chamberlain had originally derived for the subdwarfs' iron depletions, before the referee's
interference, was closer to the truth. According to modern estimates,
both stars are even more metal-poor than Aller and Chamberlain had

first found. The iron depletions in HD 19445 and HD 140283 are 130 and 600, and both stars are even cooler, being spectral type G rather than F.

Astronomers soon realized that the subdwarfs' low metallicities explained why the stars were less luminous than ordinary main-sequence stars. Metals make up only a small fraction of even a metal-rich star's mass, but these few atoms absorb light of certain wavelengths and alter the star's color and luminosity. According to detailed calculations, low metallicity changes a star's color and luminosity in such a way that metal-poor main-sequence stars form a sequence parallel to but fainter than the normal, metal-rich main sequence. Nearby subdwarfs like HD 19445 and HD 140283 were thus rare stars, metal-poor interlopers from the Galactic halo that were diving through the metal-rich disk.

AN ULTRAVIOLET EXCESS

As Aller and Chamberlain were deciphering the subdwarfs, an astronomer at Yerkes Observatory in Wisconsin was finding other peculiarities in the supposedly normal stars she was observing. The Yerkes astronomers were determining precise spectral types for bright stars; different astronomers worked on stars of different spectral types, obtaining spectra to classify their assigned stars. To Nancy Roman fell the yellow-white F stars, the yellow G stars, and the orange K stars as well as some of the white A stars. But while determining the stars' spectral types, she made two important discoveries.

"It was very serendipitous," said Roman. "We needed good spectral types, and all I was trying to do originally was to classify the bright stars." But like garnets hidden in a large stone, a few subdwarfs lay in her sample, and she discovered that they glowed brightly at ultraviolet wavelengths, a property that other astronomers would use to uncover more of the rare but important stars. She also found that even ordinary population I stars—which were not subdwarfs—differed in metallicity, and that the more metal-poor a star was, the higher its velocity and the more elliptical its orbit around the Galaxy tended to be.

Roman's interest in astronomy dated back to childhood. Her father, a geophysicist, supported her interest, but her mother worried that science might not be right for a woman. "I didn't get any opposition at home," said Roman. "I got *plenty* of opposition everywhere else." She recalled an incident in high school when she had lobbied for an extra year of algebra in place of Latin. "I had already had Latin, from seventh grade on," she said. "And the guidance counselor, who in my memory was seven feet tall, drew herself up to her full height and looked down her nose at me and said, 'What *lady* would take mathematics instead of Latin?' "

In 1943, Roman entered Swarthmore College in Pennsylvania. "When the dean of women had a woman who was interested in science," said Roman, "she tried to talk her out of it. And if she couldn't, that was the last she had anything to do with her." Roman was not dissuaded, however. "If you can call it encouragement," she said, "the first encouragement I got at all was from the head of the physics department. He came up to me in lab my junior year and said, 'You know, I usually try to talk women out of going into physics. But I think, *maybe*, you might make it.' "

After college, Roman went to Yerkes, where she earned her doctorate under Morgan in 1949. She then remained at Yerkes and worked on the classification project. As part of her assignment, she examined stars that were supposedly spectral type A but had strong lines from singly ionized calcium, a feature of cooler stars.

"In looking at those," said Roman, "I realized that they weren't A stars at all; they were F stars with very, very weak metal lines—subdwarfs. Several of them were near enough to have fairly decent parallaxes, and the parallaxes told the same story: they were definitely F-type main-sequence stars and not A stars at all." The parallaxes revealed the stars' distances and hence their luminosities, and the stars proved to be fainter than genuine A-type main-sequence stars should be. Like other subdwarfs, the stars had large U, V, W velocities.

Roman then measured the stars' output of ultraviolet, blue, and yellow light. "This showed very clearly that the stars with the weaker lines were also brighter in the ultraviolet," she said. Roman reported her discovery of this so-called ultraviolet excess in a 1954 paper, where she presented data on seventeen subdwarfs. "This whole business that the 'normal stars' were not all alike was a new one," she said.

Roman's discovery of the ultraviolet excess would later help other investigators formulate theories for the Galaxy's origin, but the effect itself is simple and signals a low metallicity. In general, the bluer and hotter a star, the greater the percentage of its light that emerges in the ultraviolet. But Roman's discovery meant that two stars having identical visual colors but different metallicities emitted different percentages of ultraviolet radiation. For example, a subdwarf that was precisely as yellow as the Sun actually gave off a greater percentage of its light in the ultraviolet. In fact, Roman found that her most metal-poor stars—the subdwarfs—were 25 percent brighter in the ultraviolet than they would have been if they were normal, metal-rich stars.

This ultraviolet excess results from the subdwarfs' low metallicity. Metal atoms absorb ultraviolet light and block part of a metal-rich star's radiation at these wavelengths. But in a metal-poor star, such as a subdwarf, the ultraviolet radiation emerges unimpeded, so the star is especially bright in the ultraviolet. Roman's discovery of the ultraviolet excess had a practical consequence for the study of the Galaxy's oldest stars. Determining a star's output of ultraviolet light was easier than obtaining the star's spectrum, so astronomers could now pick out the rare subdwarfs from the vast sea of disk stars by looking for stars with large ultraviolet excesses.

Even before she uncovered the ultraviolet excess, Roman had made another significant discovery, one that involved not the subdwarfs, which were members of the Galactic halo, but ordinary disk stars like the Sun. "I noticed that the [disk] stars differed in the intensity of their lines," she said, "and that the stars with weaker lines tended to have higher velocities. They were not high-velocity stars in the conventional sense, but they were higher on the average than the velocities of the stars with the stronger lines." These faster stars, astronomers know today, have lower metallicities than the Sun but are not nearly as metal-deficient as the subdwarfs.

When Roman published her first results, she listed 94 F and G stars. She segregated the stars into two groups based on the strengths of their metal lines. When she examined the stars' velocities, she noticed that the two groups had different kinematics. The strong-line (metal-rich) stars had a mean speed with respect to the Sun of 28 kilometers per second, whereas the weak-line (less metal-rich) stars had a mean speed of 44 kilometers per second. Neither velocity was extreme, meaning that the stars belonged to the Galaxy's disk rather than to its halo, but the discovery suggested that even within the disk, stars had somewhat different metallicities. In 1955, Roman published a catalogue of 571 high-velocity stars, including both subdwarfs and disk stars. For most of the stars, she computed U, V, W velocities and orbital eccentricities, which measured how elliptical the stars' orbits were. She found that the larger a star's ultraviolet excess and therefore the lower its metallicity, the higher its U, V, W velocities tended to be. This in turn meant that metal-poor stars had more eccentric—that is, more elliptical—orbits around the Galaxy, a fundamental correlation that other astronomers would later exploit to probe the Milky Way's very origin.

With that, however, Roman's work at Yerkes came to an end, for she knew that the observatory would never grant tenure to a woman. Furthermore, although her colleagues respected her work, Yerkes paid her less than it paid a man. Said Roman, "The chairman of the department said to me, 'We don't discriminate against women; we can just get them for less.'" Roman left the Wisconsin observatory for a position in Washington, D.C., where she worked first for the Naval Research Laboratory and then for NASA.

THE ARCHITECTURE OF THE GALAXY'S DISK

Roman's work suggested that the stars of the Galactic disk had different metallicities and velocities. The different velocities, in turn, meant the stars had different locations with respect to the Galactic plane. The strong-line stars, which had low velocities, could not climb far from the plane and therefore remained closer to it than the weak-line stars, which had somewhat higher velocities.

That different types of stars have different locations about the Galactic plane had been known since the 1800s, when astronomers noted that most of the brightest blue stars lay near the Milky Way. Around 1890, as Jacobus Kapteyn and David Gill were working on the mammoth star catalogue of the southern sky, they noticed that white stars lay near the Galactic plane, and spectra from Harvard's Edward Pickering confirmed that the stars were indeed of the Sirius type—that is, spectral type A. In 1904, Pickering himself used his spectra to report that blue B-type stars cling even more tightly to the Galactic plane than the A stars do.

Modern data verify and extend this trend. Blue and white stars do live closer to the Galactic plane than yellow, orange, and red stars. To quantify these differences, astronomers measure how far stars of different spectral types lie above or below the Galactic plane and then compute the stars' mean distance from the plane, which is called their "scale height." The blue O and B stars have the smallest scale height, only about 300 light-years, so they lie the closest to the Galactic plane. The white A-type stars, whose concentration to the Galactic plane Kapteyn and Gill first noted, have a scale height of 350 light-years, and the yellow-white F stars have a scale height of 600 light-years. Yellow G-type main-sequence stars have a still greater scale height, 1000 light-years, which means that they typically lie farther from the Galactic plane. Orange and red dwarfs also have a scale height of about 1000 light-years.

The scale height correlates with stellar age: young stars tend to lie closer to the Galactic plane than older stars do. This follows from considering the lifetimes of stars of different spectral types. Because the blue O and B stars have high masses and short life spans, the average OB star is only tens of millions of years old, whereas less massive main-sequence stars—such as G, K, and M stars—live longer and are, on average, billions of years old.

The stars' velocities also reflect this distribution. Younger stars in the Galaxy's disk, which lie closer to the plane, tend to have lower velocities than older stars. This result first emerged in early 1910, after Kapteyn and William Campbell independently analyzed radial velocity data and discovered that blue stars had lower velocities than yellow stars, which in turn had lower velocities than red stars. Modern data illustrate the same progression: B stars, which are young, have a W velocity dispersion of 6 kilometers per second, which means that about two thirds of all B stars have a W velocity between −6 and +6 kilometers per second. A typical B star thus has only a small velocity perpendicular to the Galactic plane. In contrast, G, K, and M main-sequence stars, which are generally older, have a W velocity dispersion of 16 kilometers per second, and white dwarfs, most of which are older still, have a W velocity dispersion of 25 kilometers per second.

"We may well regard this relation of age and velocity as one of the most startling results of modern astronomy," wrote British astronomer Arthur Eddington in 1914. Indeed, this elegant relation held an important clue to the Galaxy's evolution, but decades passed before astronomers recognized what it was. Some assumed that the velocity difference arose from what physicists call the equipartition of energy, a form of socialism that forced all stars to have about the same kinetic energy. This principle applies to molecules in the atmosphere. The kinetic energy of a molecule depends on both its mass and its velocity, and the larger either one is, the larger the molecule's kinetic energy. To maintain equipartition of energy, therefore, heavy molecules move more slowly than light ones, giving all molecules about the same kinetic energy. Almost as soon as Kapteyn and Campbell discovered that stars of different colors moved at different speeds, astronomers thought that the same concept might apply to the Galaxy. The more massive stars were slower than the lighter ones, so they all had roughly the same kinetic energy.

But this explanation for the stars' varying velocities ran into trouble. Molecules in a gas maintain equipartition because they constantly collide with one another and exchange energy. If a molecule moves faster than it should, it bumps into other molecules, which slow it down; if a molecule is too slow, it steals energy from a faster one and speeds up. As a consequence, no molecule moves too fast or too slow, and all have about the same kinetic energy. But in the Galaxy, stars lie so far from one another that they hardly ever exchange much energy. In fact, meaningful stellar encounters occur only in a dense environment, such as a star cluster.

"In a cluster of stars," said Lyman Spitzer, Jr., an astronomer at Princeton University and an expert on stellar dynamics, "encounters between stars *do* have a dramatic effect. But in the Galaxy, where the stars are farther apart, the encounters between stars are much less frequent and have a much smaller effect." The reason is that the Milky Way is mostly empty space. Shrink the Galaxy so that the Earth is only an inch from the Sun, and on the same scale Alpha Centauri will still lie over four miles away. Therefore, outside of star clusters and the dense Galactic center, stars do not interact with each other. Thus, the equipartition principle does not apply and cannot explain why heavy stars move more slowly than light ones.

In the early 1950s, Spitzer explored a different theory for the disk stars' velocities. He traced his interest in theoretical physics and astronomy to high school, when he and other students had persuaded a physics teacher to offer a course in astronomy. "He took us into Boston one evening to hear a lecture by Henry Norris Russell, who was my predecessor here at Princeton," said Spitzer. "I was very much impressed by his talk. Of course, at the time, I had no idea that I would succeed Russell."

While Spitzer was developing a theory to explain the disk stars' velocities, his colleague was German-born astronomer Martin Schwarzschild, whom Spitzer had convinced to come to Princeton. Astronomy ran in Schwarzschild's family, for his father was Karl Schwarzschild, the astronomer who had explained Kapteyn's star streams.

"My mother maintained," said Martin Schwarzschild, "that from the moment I could say the word *sterne*—that is, in German, 'stars'—I wanted to become an astronomer." Karl Schwarzschild, who was Jewish, had died in World War I just before Schwarzschild turned four. But otherwise, said Schwarzschild, "I had a very, very happy childhood, indeed a very happy growing period—until the Nazis came."

Schwarzschild received his doctorate from Göttingen University, where Walter Baade had studied, in December 1935. "Less than a month later," said Schwarzschild, "I wasn't even permitted to enter a university building." In 1936 he fled to Norway and then to the United States. In 1947 he joined Spitzer at Princeton and taught a graduate course that dealt in part with stellar dynamics.

"Lyman and I were discussing these dynamical processes over lunch," said Schwarzschild, "just so that in the afternoon I could teach them. And I remember that I had quite a hard time really understanding the phenomenon of star-star encounters." Such encounters occur when two stars pass each other, and Schwarzschild wondered how these encounters depended on stellar mass. The Sun, for example, feels a tiny gravitational pull from Alpha Centauri, 4.35 light-years away, which slightly bends the Sun's path through the Galaxy and barely changes its velocity. But suppose the Sun passed the same distance from a more massive star, like Rigel. Then the force of attraction would be greater, and the Sun's velocity would be more greatly affected.

Here, thought Spitzer and Schwarzschild, might lie the explanation for the different scale heights and velocities of different disk stars.

Suppose all disk stars were born near the Galactic plane and originally revolved around the Galaxy on circular orbits. The blue O and B stars, which are young, would have completed just a fraction of an orbit around the Galaxy since their birth and would therefore have encountered few if any massive stars during their lives. As a result, they would stay close to the Galactic plane, and their orbits would remain circular. But the white A stars are, on average, somewhat older and would have suffered a greater number of encounters with massive stars. These encounters would scatter the stars away from the Galactic plane, increasing their scale heights and velocities. And the G, K, and M main-sequence stars, which on average are older still, would have encountered many massive stars that had perturbed them. They therefore have greater scale heights and velocity dispersions than the O, B, and A stars.

Spitzer and Schwarzschild envisioned that this scattering mechanism might account for the disk stars' different scale heights and velocity dispersions. Unfortunately, when Spitzer actually computed the numbers, the idea did not pan out. The most massive stars have a hundred times the Sun's mass, which was not enough to perturb other stars to their observed speeds. But the proposal was not in vain, for the Galaxy harbored objects far heavier than its most massive stars.

"In the course of these discussions," said Spitzer, "we realized that the interactions between stars and interstellar clouds could in principle accelerate the stars and give the older stars in the disk a greater velocity. An interstellar cloud with a mass of a million Suns will produce an enormously different deflection than a single star would."

In two papers, published in 1951 and 1953, Spitzer and Schwarzschild formulated the necessary equations to see whether collisions between stars and interstellar clouds of gas and dust could explain the different velocities of disk stars. Assume, they said, that all stars in the disk were born on circular orbits in the Galactic plane. Such stars, when young, would cling to the Galactic plane and have small U, V, W velocities. But as the stars orbited the Galaxy, they would meet different interstellar clouds, which would kick the stars away from the Galactic plane, increase their U, V, W velocities, and make their orbits more elliptical. Old stars, which had circled the Galaxy many times, would on average have suffered many such encounters and would have a larger scale height and velocity dispersion than younger stars. This, suggested Spitzer and Schwarzschild, might explain the increase in scale height and velocity dispersion along the main sequence as one progressed from the young blue stars to the older yellow and red ones.

Schwarzschild recalled the reaction to their theory. "It was friendly skepticism," he said. "Other astronomers considered Lyman and me serious contributors but also a bit on the speculative and theoretical side. Historically, this country became great in astronomy by incredible pioneering—building large telescopes at excellent astronomical sites. And in the fifties, there was a difficult transition from dominance by observational astronomers toward a balance between observers and theoreticians."

To illustrate how many observers regarded theorists, Schwarzschild recounted an incident with Edwin Hubble. "Hubble once gave me a very, very harsh sermon," said Schwarzschild. "He told me, 'If Spitzer and you and [Leo] Goldberg and [Lawrence] Aller dominate American astronomy in the coming decades, we will never again discover anything, because you will persuade everybody to make observations that will confirm your theories.' "

There was, however, a less ideological reason for astronomers to question the Spitzer-Schwarzschild scattering mechanism. To perturb stars sufficiently and produce the observed increase in scale height and velocity dispersion from the young disk stars to the old, the interstellar clouds would have to contain a large amount of mass. In fact, Spitzer and Schwarzschild's equations demanded that the interstellar clouds have a million solar masses, which was ten times heavier than the most massive clouds then known.

"We ourselves were a little bothered by that large number," admitted Schwarzschild. "At the time, it was quite a jump of faith." Said Spitzer, "Yes, it was considered large, but not hopelessly so. There were clouds of appreciable size and mass, within an order of magnitude of what we needed. So we were close, and we felt that the theory was worth considering."

Vindication for the Spitzer-Schwarzschild scattering mechanism came two decades later, in the 1970s, when astronomers discovered that interstellar clouds were indeed more massive than previously thought. Astronomers can measure the mass of an interstellar cloud by detecting the radio waves it emits: the stronger the radiation, the more massive the cloud. The largest clouds are now called giant molecular clouds, because they consist almost entirely of molecular hydrogen. But the hydrogen molecule does not radiate at radio wavelengths, so astronomers cannot use it to deduce the mass of an interstellar cloud. In 1970, however, radio astronomers detected a molecule that does radiate: carbon monoxide. Though much less abundant than molecular hydrogen, carbon monoxide allowed astronomers to map interstellar clouds and estimate their masses. The largest clouds had roughly a million times the mass of the Sun, just as Spitzer and Schwarzschild had predicted.

Still, the Spitzer-Schwarzschild scattering mechanism could not explain the extremely high velocities of the halo stars, such as the subdwarfs. "We could not touch those stars," said Schwarzschild, "so we believed from the outset that we were talking only about what we now call disk stars, which at that time were called population I stars." The halo stars were so fast that interstellar clouds could not have accelerated the stars to their present velocities, a fact that provided insight into the Galaxy's origin and evolution. Spitzer and Schwarzschild deduced that the ancient halo stars were born with their high velocities, possibly from turbulent interstellar material—an idea that most astronomers accept today. The interstellar material then collapsed into a plane, where the disk stars subsequently arose.

In the disk, though, Spitzer and Schwarzschild painted a different picture from the one in the halo. Here different stars had different scale heights and velocities not because they had been born with them but because they had acquired them during their lives, by being scattered off interstellar clouds. With this knowledge, astronomers could understand the different locations and kinematics of different disk stars. Nancy Roman had found that the disk stars with weaker metal lines tended to have higher velocities. On average, therefore, disk stars with lower metallicities must be older than disk stars with higher metallicities, for only by orbiting the Galaxy for billions of years could a star encounter enough interstellar clouds to alter its velocity significantly. In addition, since low-metallicity disk stars are older than metal-rich disk stars, *and* since a star's metallicity does not change but rather reflects the Galactic conditions in which that star was born, the Galaxy's disk has become more metal-rich with time. But the strongest proof of the Galaxy's metal enrichment came from the work of Lawrence Aller and Joseph Chamberlain, who showed that the most metal-poor stars of all were the ancient subdwarfs in the Galactic halo, high-velocity stars that had formed before the Galaxy became metal-rich.

To astronomers who accepted these revolutionary results, the implication was clear. Over its life, the Milky Way Galaxy has grown more and more metal-rich. The Galaxy's stars have forged heavy elements like oxygen, calcium, and iron. When these stars died, they ejected the heavy elements into space, where the elements drifted for millions or billions of years. Some of these elements encountered and enriched the interstellar clouds of gas and dust that gave birth to new stars and planets. In at least one case, on one water-covered planet circling one yellow star, these elements then gave rise to life. We are therefore made of elements forged in the Milky Way's stars billions of years ago: the oxygen we breathe, the calcium in our bones, and the iron flowing through our veins all came from stars that died long ago and cast their remains into the Galaxy. We are heirs to these ancient stars. We are stardust.

9

THE ALCHEMY OF THE HEAVENS

Burbidge, Burbidge, Fowler, Hoyle
Took the stars and made them toil:
Carbon, copper, gold, and lead
Formed in stars, is what they said.

THE 1957 RESEARCH paper opens with words from Shakespeare, runs a hundred and four pages long, and reveals where gold, carbon, platinum, silver, oxygen, iron, and nearly every other element came from. It will tell you, for example, that the gold in the world's treasuries and the iodine added to salt were forged in the fiery explosion of a supernova, while most barium and zirconium on Earth slowly blossomed within the heart of a red giant star. By setting an audacious goal—to explain how and where in the universe every element from hydrogen to uranium had originated—and largely achieving that goal, the four authors of the paper created one of the greatest monuments in the history of science.

"Originally," said British astronomer Fred Hoyle, "we didn't intend anything more than a normal paper. But it grew and grew as the months slipped by, and eventually it was of such enormous length that only the *Reviews of Modern Physics* could handle it." In the 1940s, when

Hoyle began advocating the idea that the stars had created the elements, most scientists thought the idea preposterous, but by the mid-1950s the discovery that old stars had lower metallicities than young stars had convinced many astronomers that he was right—that the Milky Way, through its stars, was the true creator of the elements. For most astronomers, however, the epic 1957 paper proved Hoyle's concept beyond a doubt, for the paper described in exquisite detail exactly how each element had arisen and explained why some elements, such as oxygen, were common, and others, such as gold, were rare. The classic paper takes its nickname—B^2FH—from its authors, who listed their names alphabetically: E. Margaret Burbidge, Geoffrey Burbidge, William Fowler, and Hoyle.

THE ORIGIN OF THE ELEMENTS

Although the alphabet put him last, Fred Hoyle was actually the first of the four to suggest that the heavy elements had formed from nuclear reactions in stars. Hoyle could easily investigate such matters, because he had studied nuclear physics in graduate school. But in the late 1930s renowned physicist Paul Dirac swayed him away from the field. "Dirac said to me," Hoyle recalled, "that when he started, even people who were not very good could solve important problems. But now people who are very good can't find important problems to solve. And I was not so stupid as to think that if Dirac was telling me that people of his class couldn't find important problems, I was going to find one."

Hoyle searched for more fertile fields. "In retrospect," he said, "I realize it could have been either biology or astronomy that I went into. I sensed that there were a lot of interesting things to come in biology. It just happened that astronomy threw up an interesting problem first." During the early 1940s, Hoyle investigated the accretion of material onto stars and then, after a brief visit to Mount Wilson Observatory in late 1944, began studying the origin of the elements.

While at Mount Wilson, Hoyle had met Walter Baade, shortly after Baade had discerned the two stellar populations. By then astronomers knew that the Sun and most other stars fused hydrogen into helium, the nuclear reaction that began a main-sequence star's life. But Baade convinced Hoyle that the real action came at the end of a star's life, in the supernovae that Baade could see exploding in other galaxies.

After returning to England, Hoyle began to investigate what elements a dying star might manufacture. In 1946, even before astronomers had discovered metallicity differences among the stars, Hoyle published a 41-page article in which he proposed that after a massive star's core used up its hydrogen, the core would collapse, heat up, and forge heavier elements, including iron (atomic number 26). Physicists regard iron as the most stable element, because nuclear reactions do not easily destroy it. As the stellar core collapsed, said Hoyle, it would rotate faster, just as a spinning ice skater does when he draws in his arms. Because the core spun fast, centrifugal force would fling the newly created heavy elements into space and produce a supernova. In fact, thought Hoyle, supernovae might explain why iron was a relatively common element in the cosmos.

Hoyle was not the first to suggest that the elements had arisen in stars, but other astronomers believed that nucleosynthesis had occurred in the big bang. During the 1940s, Russian-born American astronomer George Gamow, the big bang's leading proponent, suggested that all the elements had originated then, in the universe's fiery birth. As later developed with Ralph Alpher and Robert Herman, Gamow's theory started with hydrogen, the lightest element, from which the nuclear fury of the big bang built up all the heavier elements.

When he first proposed it, Gamow's idea certainly seemed plausible, for it explained why the two lightest and simplest elements—hydrogen and helium—were the most abundant: because the universe had started with these and had not progressed much beyond them. But several problems soon emerged. First, as astronomers found in the early 1950s, old stars had lower metallicities than young ones. If the big bang had created all metals, every star in the Galaxy should have equal metallicity, because the material emerging from the big bang was uniform in composition. Second, the big bang itself would have had trouble making nuclei heavier than helium, because no stable nuclei have mass numbers of five or eight. (The mass number is the number of nucleons—protons and neutrons—in an atom.) The big bang scheme started with hydrogen-1, the simplest nucleus, which got hit by a neutron to form hydrogen-2, which then got hit by a proton to make helium-3, and so on. But nucleosynthesis could not get much past helium-4, because no stable nucleus had a mass of five or eight; as soon as such a nucleus arose, it split back into lighter nuclei.

This latter problem, of course, also confronted Hoyle's theory that the elements were created by stars during their lives and deaths. But Hoyle recognized that his theory had three advantages. First, stars live so long that rare reactions have plenty of time to occur. In contrast, the early universe remained hot enough for nucleosynthesis to take place for only a few minutes. Second, the great diversity of the elements means that different elements formed under extremely different temperatures and pressures. During its life, a high-mass star's core achieves a variety of temperatures and pressures and so can cook a wide variety of elements, whereas the more uniform conditions of the big bang could not. Third, the cores of stars are dense. Density promotes nuclear reactions, because the greater the density, the more frequently nuclei collide—just as busy streets have more traffic accidents than quiet ones. In contrast with stars, and contrary to popular belief, when the universe was only a few minutes old it was already no denser than the Earth's atmosphere. During the first minute, when the universe was denser, it was so hot that hydrogen-2, the first step in the nuclear chain, got torn apart as soon as it formed, so no nucleosynthesis occurred until the universe was a few minutes old.

Density is crucial, because if the density is low, each nuclear reaction involves only two nuclei: one nucleus strikes another. The problem with these so-called two-body reactions is that they cannot create elements much heavier than helium-4, because no nucleus of mass number five or eight is stable. For example, if two helium-4 nuclei hit each other, the product has mass number eight and immediately decays instead of forming heavier elements.

If the density is high, however, then *three* nuclei can strike one another simultaneously—a near-impossibility in the low-density conditions of the early universe, just as the chance of three cars colliding in a desert is almost zero. In 1951, Ernst Öpik and Edwin Salpeter independently used such a three-body reaction to catapult across the mass eight chasm and form heavier elements. They said that if three helium-4 nuclei hit one another nearly simultaneously, they would create carbon-12, which is stable:

$$^4He + {}^4He + {}^4He \rightarrow {}^{12}C.$$

Carbon is the second most common metal in the universe, after oxygen, and this carbon-creating reaction could occur in the dense cores of red giants.

But Hoyle saw a problem. "Carbon had to be made much more prolifically," he said. "Otherwise, the amount of carbon in the world would be minuscule." That would have a major consequence: we, being carbon-based life, would not exist. The problem that Hoyle identified was that the small amount of carbon formed by the triple-helium reaction quickly slipped into oxygen, via another reaction with helium:

$$^{12}C + {}^{4}He \rightarrow {}^{16}O.$$

It was like feeding a few dog biscuits to a hungry dog: the dog biscuits don't last long. Hoyle therefore endeavored to make more carbon-12 by speeding up the three-body reaction that converted helium into carbon, so that the "dog biscuits" could be fed to the dog faster than the dog could eat them. Hoyle would be able to do this if carbon-12 had what physicists call a resonant state. A resonance fosters nuclear reactions, in the same way that pushing a child's swing at the right frequency gives the swing a large motion.

In a nucleus, a resonance occurs only when the nuclei have a specific energy, analogous to the ideal frequency of pushing a swing. "I knew approximately what the energies were inside the stars," said Hoyle, "so I was able to calculate where to put the resonance to have optimal effect." He then predicted that the carbon-12 nucleus would exhibit a resonance at this energy.

In January 1953, with this problem in mind, Hoyle visited the office of William Fowler—the F in $B^{2}FH$—at Caltech's Kellogg Radiation Laboratory. Raised in Ohio, Fowler had studied engineering and physics as an undergraduate and after college went to the Kellogg Radiation Laboratory. It was here that a test could now determine whether carbon-12 had the resonance Hoyle predicted it must. Many of Fowler's colleagues were skeptical, arguing that if the resonant state existed, they would already have detected it. But Fowler thought it worth a try.

"Sure enough," said Fowler, "they proved Fred was right." The experiment showed that carbon-12 had the resonance, and exactly where Hoyle had said it would be. "That really impressed me about

Fred, that he was able to predict this thing," said Fowler. "He was one smart guy." Hoyle's success convinced Fowler and many other physicists that perhaps the stars, and not the big bang, had indeed made the elements. Said Hoyle, "It was the most rapid turnaround of people's views I've ever experienced, my going from a voice in the wilderness to suddenly a very large number of people believing the idea."

From 1954 to 1955, Fowler was in Cambridge, England, where Hoyle was. Also in Cambridge were Margaret and Geoffrey Burbidge— the B^2 of B^2FH—two British astronomers who were studying stars with peculiar compositions. Margaret Burbidge had discovered astronomy when she was four, as her family sailed across the English Channel one summer evening. "I began to feel a little bit queasy, from the motion of the boat," she said, "and I was told to look out of the porthole and look at the stars. I'd never noticed the stars in the sky in London, because the sky is bright and you don't see the stars well. And I suddenly saw in this dark clear sky a lot of bright stars, spread like diamonds in the sky. I thought how beautiful they were."

Geoffrey Burbidge, whom Margaret married in 1948, was originally a theoretical physicist. "My life is a series of accidents," he said. "I got into astronomy by marrying an astronomer." Another accident occurred in the fall of 1954, following a talk Burbidge gave on the peculiar stars that he and Margaret were studying. "After the meeting," said Burbidge, "this man came up and talked to me. He was an American with a rather bald head and a very cheerful demeanor. He was very interested and intrigued by all this, and he said he was an experimental nuclear physicist. So that's how we met Willy Fowler."

DISCOVERING THE S-PROCESS

Fowler and the Burbidges began discussing the abnormal abundances that the Burbidges had been finding in their stars. In particular, some elements heavier than iron were especially abundant, which suggested that the stars themselves were manufacturing the heavy elements. How the stars could make elements heavier than iron, though, was a mystery, because iron is the most stable element. When elements lighter than iron fuse into somewhat heavier ones, they produce the energy that sustains a star. For example, the fusion of hydrogen into helium releases energy, as does the fusion of helium into carbon and oxygen, and so on, up to iron. But iron is the end of the line. If iron were to fuse into a heavier element, the reaction would steal energy from the star. Thus, a star cannot easily engage in such energy-depleting reactions, just as a company cannot afford continually to lose money.

For that reason, Fowler and the Burbidges had to search for a nuclear reaction that would create the heavy elements. The scientists suspected that some of these elements might arise from collisions with neutrons. If neutrons struck the iron nuclei that the star had been born with, they would stick and form heavier nuclei. Neutrons can work in this way because they have no electric charge and thus are not repulsed by the nuclei, which have positive charge and repel any protons that try to strike. Fowler and the Burbidges named this sequence of neutron-capture reactions the s-process. The s stood for "slow," because free neutrons are rare and hit the nuclei infrequently. Hundreds or thousands of years might pass before a nucleus that was hit by one neutron was hit by another. This meant that if the newly created nucleus was radioactive, it decayed into another nucleus before an additional neutron struck it. As a consequence, the s-process could form only certain elements.

"We had an analogy of a river flowing along," said Margaret Burbidge. "If there are a number of holes in the riverbed, then as the river starts to flow, it fills up those holes first." Those metaphoric holes correspond to the s-process elements, elements that do not readily capture additional neutrons and so undergo no further changes once formed. S-process elements are common at or near one of the so-called magic numbers, where a nucleus contains 50, 82, or 126 neutrons. Nuclei with these numbers of neutrons like to be left alone. Examples of such rock-solid isotopes include strontium-88, yttrium-89, zirconium-90 (all containing 50 neutrons), and barium-138 (with 82 neutrons).

As this work on the s-process began, Fowler introduced the Burbidges to Hoyle, and the four began working together on the origin of the elements. Meanwhile, a Canadian, Alastair Cameron, was investigating these issues on his own. "Al deserves a lot more credit than he's ever been given for this work," said Geoffrey Burbidge. "A lot of the processes—particularly the neutron-capture processes—Al understood quite independently of us." Among other feats, Cameron identified the specific nuclear reactions that produced the neutrons driving the s-process.

Cameron was a nuclear physicist, not an astronomer. "My educational knowledge of astronomy was nil," said Cameron. In 1952, while in Iowa, Cameron first stumbled across nucleosynthesis. "I was browsing in the reading room there," he said, "and I saw a copy of what was then called *Science News Letter* and today is called *Science News.* This had a little story in it about how Paul Merrill had discovered lines of technetium in red giant stars. And my immediate reaction was, where on Earth did they get the neutrons to produce that?"

No technetium (atomic number 43) exists naturally on Earth, because the element is radioactive and its longest-lived isotope has a half-life of only 4.2 million years. Consequently, any technetium that the Earth had at birth, 4.6 *billion* years ago, has decayed. If technetium had formed solely in the big bang, the element would not exist in stars either, again because it would have vanished long ago. Yet in 1952, Merrill, an astronomer at Mount Wilson and Palomar Observatories, discovered technetium in twelve red giants, a finding which suggested that the stars themselves had made the element. In the same stars Merrill also found barium and zirconium, two elements that astronomers now recognize arise primarily from the s-process.

Convinced that neutrons had produced these elements, Cameron set out to identify the nuclear reaction that generated those neutrons. His task was not as easy as it may seem, because free neutrons are difficult to make and keep. Although they appear in every atomic nucleus but hydrogen-1, neutrons are the underdog: there are five protons for every one neutron. This is because most of the universe is composed of hydrogen-1, which has no neutrons. Furthermore, free neutrons must dodge various "neutron poisons," species that crave neutrons and swallow them. One neutron poison is hydrogen-1. If a free neutron wanders through a star's hydrogen, the hydrogen-1 grabs the neutron and forms hydrogen-2. Moreover, a free neutron itself is fragile, for if it does not react with something, it soon decays into other particles. In fact, the half-life of a free neutron is only 10.3 minutes.

Faced with these constraints, Cameron sought a reaction that would liberate enough neutrons in red giants to trigger the s-process. "In the course of doing that," said Cameron, "I started going through the possible reactions and soon came to

$$^{13}C + {}^4He \rightarrow {}^{16}O + \text{neutron}$$

and decided that looked like a fairly good bet." Although the reaction began with carbon-13, a rare isotope of carbon, Cameron suggested it could release plenty of neutrons. Furthermore, the reaction occurred in the helium-rich part of the red giant, where there was no hydrogen to capture the neutrons. The free neutrons could therefore hit iron and other heavy nuclei and build up still heavier elements.

"That carbon-13 idea is still a good idea," said Hoyle, "and everybody else missed it." Today, astronomers who compute the conditions inside stars believe that the carbon-13 source powers the s-process in red giants having less than eight solar masses. But Hoyle revealed that Cameron's paper almost did not make it into print, because Subrahmanyan Chandrasekhar, editor of *The Astrophysical Journal*, had received two negative referee reports. As a check, Chandrasekhar sent Cameron's paper to Hoyle. Said Hoyle, "Chandrasekhar had a very good instinct about referees, and he had reservations about their motives. Although I didn't know the names of the referees, I suspected that they had felt that this was a good idea, and they were sorry they hadn't seen it themselves. So I wrote back to Chandrasekhar and said he should publish it." Cameron's paper appeared in 1955.

Five years later, Cameron identified another reaction that supplies neutrons. This one starts when neon-22, a rare isotope of neon, meets helium to create magnesium-25:

$$^{22}Ne + {}^4He \rightarrow {}^{25}Mg + \text{neutron}.$$

Today, the neon-22 neutron source is believed to produce s-process elements in red supergiants having more than eight solar masses.

NUCLEOSYNTHESIS WITH A BANG: THE R-PROCESS

As powerful as the s-process was, it failed to create all the elements heavier than iron, so the work of Burbidge, Burbidge, Fowler, and Hoyle had only just begun. In 1955, the first three members of the group left England for Pasadena, with Hoyle following the next spring. Fowler returned to Caltech, and he persuaded the Burbidges to join him. Logically, since she was an astronomer, Margaret Burbidge should have obtained a fellowship that would have given her access to the telescopes atop Mount Wilson.

But she had tried that once before, in 1947. "The letter I got back implied that I'd been rather stupid in sending in this application," she said, "because I should have known that these fellowships were only available for men." Fowler therefore secured a position for her at the Kellogg Radiation Laboratory. She and Geoffrey used Mount Wilson's telescopes, but the couple had to stay in a cottage separate from the other astronomers, and the observing time was in Geoffrey's name only. While at Mount Wilson, the Burbidges investigated several peculiar stars, including HD 46407, a yellow giant in the constellation Canis Major. HD 46407 had strong lines of barium in its spectrum and was therefore classified as a barium star. The Burbidges found that the star also had overabundances of other s-process elements, such as strontium, yttrium, and zirconium. Indeed, the overabundances matched those predicted by the s-process theory.

Buoyed by these successes, Burbidge, Burbidge, Fowler, and Hoyle began a frenzied quest to explain the origin of every element. "What was marvelous about it," said Geoffrey Burbidge, "was that we all made what we thought were major contributions. Fred was contributing many of the ideas; I was; Willy was contributing the experimental physics; and Margaret was contributing the observational abundance work. We all worked in an office at Kellogg that didn't have windows, and we used to go and work there all day on the blackboard. This went on furiously for about eighteen months. We'd go back and forth, we'd pick up something new and work on it, then decide we didn't know what we were doing and drop it, and then somebody would come in with another idea, and we'd start again. It was a very hectic period."

Of great help was the publication, in 1956, of a table by American scientists Hans Suess and Harold Urey that gave accurate abundances for most of the elements and isotopes. Burbidge, Burbidge, Fowler, and Hoyle showed that they could explain the abundances of the most common elements up to iron through various fusion reactions in stars. In fact, these reactions are the reason that the most common elements are so prevalent—because stars create these elements in large quantities. For example, in the hydrogen-burning stage, a star fuses hydrogen into helium, and during helium burning, it fuses helium into carbon and oxygen. All told, Burbidge, Burbidge, Fowler, and Hoyle found that the various fusion reactions generated huge amounts of oxygen, carbon, neon, nitrogen, magnesium, silicon, and iron, the seven most common metals in the universe.

Suess and Urey's 1956 table also gave abundances for elements heavier than iron. If one plotted these abundances as a function of mass number, peaks and valleys appeared—peaks where certain nuclei were abundant and valleys where other nuclei were rare. Three peaks in the abundance distribution were now understood, for each corresponded to the stable elements, such as zirconium and barium, that the s-process produced. However, near each s-process peak was a second peak, containing nuclei with less mass. These nuclei, the four scientists realized, came from another process, one that involved a rapid rather than a slow flux of neutrons. Burbidge, Burbidge, Fowler, and Hoyle named it the r-process, with r standing for "rapid." Whereas in the s-process hundreds or thousands of years might elapse between successive neutron strikes, in the r-process the neutrons bombarded the nuclei every second or so. Nuclei had no chance to catch their breath before getting hit again, so if a newly formed nucleus was radioactive, it did not have time to decay into another element. Only after the barrage ceased could these neutron-heavy nuclei decay, and when they did, they created the r-process elements, such as gold, silver, and platinum. The r-process turned out to be responsible for most of the elements heavier than iron.

"We were delighted with the r-process," said Margaret Burbidge, "because there were these displaced peaks in the abundance distribution that you couldn't explain with ordinary slow-neutron capture.

What you needed was a rapid flux of neutrons, and we knew that's what happened in hydrogen bombs, where you got as far as californium." Californium is a radioactive element with atomic number 98, and a hydrogen bomb test that was declassified in the 1950s revealed that the rapid neutron bombardments in nuclear explosions could create elements that heavy. But pinning down how the r-process occurred in the Galaxy was not as easy as it had been for the s-process. Because the s-process is slow and gentle, occurring in stars, astronomers could see its products in the atmospheres of those stars, whereas the r-process, which is rapid, was more difficult to locate.

The bomb tests, though, provided a clue, for nuclear explosions resemble miniature supernovae. Since bombs created heavy elements, so should supernova explosions. In addition, Geoffrey Burbidge had noted the similarity between the fading of a 1937 supernova and the decay of a radioactive element. The supernova had exploded in a nearby galaxy named IC 4182, and Walter Baade had tracked the supernova's fading light for nearly two years. Burbidge noticed that this supernova's light declined exponentially, with a half-life of just under two months. Radioactive elements also decay exponentially, which suggested that the supernova's light came from the decay of a radioactive element. Burbidge believed the element to be californium-254, because its half-life is 60 days. Today astronomers know that the idea was right but the isotope wrong, for the supernova's decaying light actually involves two other radioactive isotopes, nickel-56 and cobalt-56, which have half-lives of 6 and 77 days, respectively. Together the decay of these two nuclei mimics that of a single radioactive isotope whose half-life is 60 days.

B^2FH: TO PUBLICATION AND BEYOND

By the summer of 1957, Burbidge, Burbidge, Fowler, and Hoyle had completed their work, producing a lengthy paper which claimed that nearly all the elements had arisen in stars. But there was still a problem. Said Fowler, "We worried, where in God's name are we going to publish this? We knew that when you send in a long paper to some editor, the first thing he did was write back and say, 'I'll publish it if you shorten it.' Well, we didn't want to do that." At a meeting in Pittsburgh, Fowler bumped into Ed Condon, editor of the *Reviews of Modern Physics*. Said Fowler, "I happened to mention to Ed that we had this long paper, and he said, 'Well, that's just what we're looking for.' He needed something for his next issue, and he knew if he sent it out to a referee, that there was so much detail in this paper, the referee would be sure to find something wrong. So he said, 'Send me the manuscript, and I'll publish it without even having it refereed.' " As a result, the paper leapt over the referees and appeared in the October 1957 issue. Astronomers soon began calling the paper B^2FH. "That really caught on," said Fowler. "You'd be surprised how many people, when I meet them, no matter what field they're in, say, 'Oh, you're the F in B^2FH.' "

B^2FH was a personal triumph for Hoyle, who had begun to champion the idea eleven years earlier. But some scientists questioned Hoyle's motives, for he was also the leading proponent of the steady state cosmology. During the 1950s, a major battle raged between proponents of the steady state and those of the big bang. While the big bang cosmology postulated that a gigantic explosion had started the universe expanding, the steady state cosmology maintained that the universe had existed forever. If so, the universe was infinitely old, which implied that its expansion would have thinned it out until the density of its matter was nearly zero. Since the universe's density is not zero, steady state proponents argued that particles spontaneously popped out of nowhere and kept the universe's density constant despite the universal expansion. Hoyle had developed the steady state theory with Hermann Bondi and Thomas Gold in 1948.

Ironically, it was also Hoyle who christened the rival cosmology. "That was in 1950," Hoyle recalled. "I was giving a course of talks on astronomy on the British Broadcasting Corporation, and I coined the phrase 'big bang.' It was one of these things that just stuck. If I had had any sense and patented it, I'd have done very well and made a lot of money."

Popular books claim Hoyle intended "big bang" to be pejorative, but Hoyle disputed that. "The BBC was all radio in those days," he said, "and on radio, you have no visual aids, so it's essential to arrest the attention of the listener and to hold his comprehension by choosing striking words. There was no way in which I coined the phrase to be derogatory; I coined it to be striking, so that people would know the difference between the steady state model and the big bang model."

But a more serious error has seeped into the literature and currently appears as a "fact" in popular books and even technical articles. It is alleged that the great work of B^2FH came about because of the steady state cosmology. Proponents of a big bang could say that the big bang had created the heavy elements, but supporters of the steady state could not, because they did not believe that a big bang had ever occurred. Therefore, it is claimed, the steady state cosmology motivated Burbidge, Burbidge, Fowler, and Hoyle to seek an alternative site for nucleosynthesis.

"That's absolute rubbish," Hoyle said, and Burbidge, Burbidge, and Fowler all confirmed this. "The one thing you've got to get right," Hoyle said, "is this had nothing to do with cosmology at all. It was [J. Robert] Oppenheimer, with whom I never got on very well, who spread the canard that the reason I got onto this theory was to support the steady state cosmology. Anybody who looked at the dates in the literature could see it was rubbish: my first paper on the synthesis of the elements came in 1946, whereas the steady state cosmology didn't come till 1948." Furthermore, only a few sentences in B^2FH's 104-page-long paper even mention cosmology.

"Throughout my life," said Hoyle, "I've done work in many different lines, and it's a matter of principle with me that I keep them in watertight compartments. I don't allow one to influence another." Otherwise, he explained, a bad idea could corrupt a good one. "That's why I never have observed," he said. "I deplore the modern American trend of an observer interpreting his results, because once you start to do that, the observations are compromised. I always much prefer it when somebody who doesn't give a damn about the theory does the observation."

Nevertheless, a few astronomers questioned B^2FH's motives, thinking their work a veiled attempt to bolster the steady state. Said Geoffrey Burbidge, "Willy used to give talks and come back and say, 'Why do they always blame me and say I'm doing this for the steady state?' And I used to say to Willy that some of the antipathy towards you is that they think you are using this as a way of pushing the steady state."

According to Geoffrey Burbidge, some astronomers failed to recognize the big bang's inability to produce heavy elements. "Astronomers never understood the difficulty that the big bang people had with nucleosynthesis, in that you can't get over mass five," said Burbidge. "There is no stable mass five. But the astronomers couldn't understand that, so some of them were quite convinced that the only reason for our developing this theory was that you required it in the framework of the steady state. But it was clear to any good physicist that the mass five was a real problem to get over."

By the time Burbidge, Burbidge, Fowler, and Hoyle published their work, however, most astronomers were already convinced that stars did create the elements. The evidence was overwhelming: old stars were more metal-poor than young ones; the conditions of the big bang did not allow the creation of heavy elements; and peculiar s-process elements had been seen on the surfaces of red giants. The lengthy paper by Burbidge, Burbidge, Fowler, and Hoyle nailed down the idea by showing that stellar nucleosynthesis could account not only for nearly every element in the cosmos but also, and crucially, for its abundance relative to the rest. Today, most of B^2FH still stands, and astronomers still talk about the r-process and the s-process.

Indeed, in relation to its achievements, B^2FH's errors were minor. One concerned helium, the second most abundant element. The universe actually contains far more helium than stars could have produced, a point made by Bondi, Gold, and Hoyle in 1955 and more forcefully by Geoffrey Burbidge in 1958. Although some helium did form in stars, about 80 percent of it was made in the big bang. This scheme for helium production was first worked out in 1964 by Hoyle and Roger Tayler.

B^2FH also missed out on another light nucleus, deuterium (hydrogen-2), as well as the light elements lithium, beryllium, and boron, which are rare and have atomic numbers of three, four, and five. B^2FH could not explain these nuclei, because they are fragile and get destroyed by the high temperatures in stars. The four scientists admitted failure by ascribing these nuclei to the x-process, where x stood for unknown. Today most astronomers believe that all deuterium and some lithium arose during the big bang, a possibility first published in 1967 by Robert Wagoner, Fowler, and Hoyle, while most beryllium and boron formed in interstellar space, where high-energy particles called cosmic rays smashed into heavier elements and tore them apart, creating lighter ones.

THE NOBEL PRIZE

The work of Burbidge, Burbidge, Fowler, and Hoyle was a major accomplishment, and three decades later, in 1983, the Nobel Academy recognized their work—albeit in a way that provoked great controversy. That year, the Nobel prize in physics was awarded to two scientists. Half the prize went to Subrahmanyan Chandrasekhar, for his work on white dwarfs. The other half went to Fowler, and Fowler alone. None of the others who had labored on B^{2}FH shared the prize, not even Hoyle, who had started the whole business.

"Fred was the one who had the original idea of synthesizing elements in stars," said Fowler. "The person who had the original idea certainly should get the prize. Now it's perfectly true I did a lot of work on the equations, and Geoff and Margaret did a lot of work on the astronomy." Fowler said that the omission of Hoyle made receiving the prize a bittersweet experience. "I felt so bad about Fred not sharing in it," he said.

The news came in October 1983, when Fowler was visiting Yerkes Observatory in Wisconsin. Reporters from Sweden had called his home in Pasadena and reached his wife. It was three in the morning California time—"my poor wife," said Fowler—and she told them he was at Yerkes. So the reporters called him there, where it was five in the morning.

Fowler then phoned a friend in Sweden. "He said, 'Yes, it's true; it's all over Stockholm.' I asked him who had shared it with me, and he said Chandrasekhar. And I said, 'Well, did Fred Hoyle win anything?' And he said, 'Of course not.' Just like that."

What happened next is in dispute. Fowler said he thought of declining the prize because Hoyle had not received it. "I actually called Fred," said Fowler, "and he said, 'Don't be a damn fool.' So I accepted the prize."

"That is absolutely not true," said Hoyle. "I was visiting my daughter in Oxford. We had just switched our six o'clock news on, and I was actually riveted, because there was Willy's picture on the tube, and he got it, and that was the first I knew about it. I got it straight from the television screen. It was very brutal, obviously, and I felt bad about it for two or three days, but by the end of the week I had sort of simmered down."

At the time, Geoffrey Burbidge was director of Arizona's Kitt Peak National Observatory, which operates a sister observatory in Chile. "Luckily, I was in Chile," said Burbidge. "I was bombarded by newspaper people trying to reach me in Chile—in fact, they did reach me in Chile—and I simply refused to deal with them. I think it was very unfortunate that at least Hoyle was not recognized in this way. But I have no idea how these things come about."

Burbidge emphasized that B²FH had been a close, four-way collaboration. "Before this happened," he said, "I always said the following about similar situations. If you came down here and I did a piece of work with you that dealt in part with nucleosynthesis, people would immediately deduce that I'd done that part of the paper and you'd done something else. But it could be quite wrong. Only the people who do the work ever know who did what. The whole business of prizes—of singling people out and trying to decide who did what in a paper—is an extremely difficult thing to do."

Said Hoyle, "Fowler came back from Stockholm and he told me a story. I don't know; it sounds very peculiar. He was told that they had an absolute rule that anybody who criticized them never got the prize. And it's certainly true that I had made comments that were not complimentary to them over that pulsar award." Pulsars had been at the center of an earlier Nobel fiasco. In 1967 Jocelyn Bell, a graduate student, discovered the first pulsar, and in 1974 the Nobel prize was awarded— to her advisor.

"The Jocelyn Bell thing was very bad," said Geoffrey Burbidge. "Fred thought it was outrageous, and so did many of my other colleagues." Hoyle wrote a letter of protest that appeared in *The Times* of London. Yet Hoyle's objection to what he and other astronomers saw as a travesty probably cost him the prize.

Some have speculated that Hoyle's advocacy of unpopular scientific theories hurt him. In recent years, he has argued that diseases originate in outer space. And he still does not accept the big bang. In fact, in 1993, Hoyle, Geoffrey Burbidge, and Jayant Narlikar in India revived the steady state cosmology. In the original steady state theory, individual particles were created in intergalactic space to compensate for the decrease in density caused by the universe's expansion. In the new steady state cosmology, matter enters the universe through titanic explosions.

According to Burbidge, the steady state theory still provokes great antipathy from the scientific establishment. "Let me give you an example in *New Scientist*," he said, citing a recent issue in which Narlikar described the new steady state cosmology. "Narlikar writes an article," said Burbidge. "For some reason, they immediately have to ask [big bang proponent David] Schramm to write something in response. Why? They wouldn't do it the other way around—they wouldn't have Schramm write an article and ask me to write something."

In the end, the 1983 Nobel prize hurt the decades-long friendship between Fowler and Hoyle. "It had to, it had to," Hoyle said; "there's no question about that." Fowler and Hoyle no longer see much of each other. "Damn the Swedes," said Fowler. Nevertheless, Hoyle has accepted their decision. "This is all water that's ten years under the bridge," he said. "I just have to leave it to their conscience—if they have any."

1 0

ELS

ORIGINS ELUDE. FOR centuries astronomers have struggled mightily just to discern the Milky Way's size and structure, so deducing how the Galaxy originated billions of years ago might seem impossible. In 1962, however, three astronomers in Pasadena drew on the many discoveries of the 1950s and probed the motions and metallicities of 221 nearby stars to paint a vivid picture of the Galaxy's birth, a picture so compelling—and controversial—that astronomers still discuss and debate it to this day.

THE ORIGIN OF THE GALAXY

The Milky Way began, the Pasadena astronomers said, as an enormous ball of gas that was uniform in density but rapidly collapsing. During this rapid collapse, individual gas clouds condensed and contracted, creating the Galaxy's first stars—the halo stars. Because the gas clouds were then metal-poor, the halo stars themselves were metal-poor; and because the gas clouds were plunging toward the Galactic center, they had extremely noncircular orbits, so the halo stars acquired very elliptical orbits around the Galaxy. During the collapse, some of the halo stars exploded and enriched the gas with metals. As the gas clouds plummeted farther toward the Galaxy's center, they smashed together, their orbits became more circular, and the clouds gathered into a rapidly spinning disk. Now the halo had exhausted its gas, so stars stopped forming there, but they began to arise in the disk, where the gas clouds had settled. Like the gas clouds themselves, the newly formed disk stars had high metallicities and circular orbits. Billions of years later this disk would give birth to the Sun and Earth. But the total time from the formation of the first halo star to the origin of the disk had been brief, only 200 million years—less than 2 percent of the Milky Way's present age, and a mere instant in the total life of the Galaxy.

Today, this model of the Milky Way's formation is as famous as it is contentious. Astronomers refer to it as ELS, after its three originators, Olin Eggen, Donald Lynden-Bell, and Allan Sandage.

Of the three, Sandage is ELS's staunchest supporter. He relishes a challenge, and deciphering the Galaxy's origin from little more than two hundred stars certainly qualified. Likewise, his favorite telescope is not the mighty 200-inch atop Palomar Mountain but Mount Wilson's less powerful 100-inch, with which Edwin Hubble had discovered the expansion of the universe and Walter Baade the two stellar populations. Sandage prefers the 100-inch because it is difficult to operate.

Sandage first encountered astronomy in third grade, when peering through another boy's telescope. "I've known since that time," said Sandage, "that there was no other career." Soon after, Sandage began charting sunspots, watching meteor showers, and observing the flickerings of variable stars. During his junior and senior years at the University of Illinois, he participated in a nationwide star-counting program organized by Bart Bok, who had observers study the Milky Way in different constellations. As part of the project, Sandage examined the constellation Perseus and counted a million stars.

The work was tedious. "I loved it," said Sandage. "I love long-range programs, where you have to have a massive amount of data. And I've always been fascinated with the whole subject of the Milky Way." The star-counting project prepared him well for later work, because he had to determine the stars' apparent magnitudes. He thereby acquired skills in practical astronomy, such as transforming the specks on photographic plates into magnitudes, so when he entered graduate school at

Caltech in 1948, he became the observing assistant to Edwin Hubble. Sandage's advisor was Walter Baade. Thus, Sandage began his career under two of the greatest astronomers of the twentieth century.

But Sandage—who today is embroiled in controversy over both the birth of the Galaxy and the age of the universe—found it difficult to make the transition from amateur to professional. "Astronomy as a profession is different from astronomy as a hobby," he said. "The two aspects of astronomy are orthogonal, it seems to me. Just observing variable stars, for example, or meteors or sunspots is very different from trying to solve equations in stellar dynamics or understanding the Friedmann equation in cosmology. If you're going to be successful as a professional, you have to put away childish things. You have to delve much deeper into the reasons why. You have to understand so much, and the pressure of whether you're capable or not is always there. You question yourself—whether you can do it, whether you can do the work. Do you measure up? We all have heroes, and our heroes, it seems to me, are those people who attain goals that appear to be unattainable. And so, do you measure up?"

The years Sandage was in graduate school witnessed revolutionary advances in the understanding of stars, for only then did astronomers learn that stars evolve from the main sequence to the red giant stage. It all seems so simple now, said Sandage, but that discovery explained the H-R diagram, the cornerstone of stellar astronomy, and enabled astronomers to measure the ages of different star clusters. Furthermore, under Baade's direction, Sandage and Halton Arp used the 200-inch telescope to detect the main sequence in globular clusters. It was faint, because the clusters were old and the bright main-sequence stars had burned out.

"I'm really a stellar evolutionist," said Sandage, who today is actually better known for his work in cosmology, the study of the entire universe. "I'm much more interested in stellar evolution than in cosmology. Stellar evolution was *the* exciting thing for me. It was like the origin of the species, except this was the origin of the H-R diagram. What would we know about the Milky Way, and the formation of it, if the theory of stellar evolution or the explanation of the H-R diagram or the age dating of star clusters were not possible?"

In May 1957 the new developments in stellar evolution, nucleosynthesis, and Galactic structure converged at a conference at the

Vatican. The organizers had invited a small but select cadre of astronomers, including Sandage, to fathom the Milky Way's stellar populations. Baade, father of populations I and II, was there, as were Bertil Lindblad, Jan Oort, William Morgan, Lyman Spitzer, and Martin Schwarzschild. Fred Hoyle and William Fowler also described their work on nucleosynthesis, just five months before the B^{2}FH paper appeared.

"It was perhaps the most electric conference that I've ever attended," said Sandage, who joined the other participants in reviewing the different ages, spatial distributions, kinematics, and metallicities of the stars. "It was the conference where Baade's two absolutely discrete populations with nothing in between broke down into five subgroups." From young to old, these five populations were named extreme population I, older population I, disk population, intermediate population II, and halo population II. Each has more or less survived to the present, albeit with different terminology. What the Vatican participants called extreme population I is simply the newborn stars associated with the spiral arms, such as the bright blue O and B stars. Older population I is really the young thin disk, made of stars younger than about a billion years, and the disk population is the old thin disk, containing most stars near the Sun. Intermediate population II resembles the thick disk, and halo population II is the Galactic halo.

Several years before the Vatican conference, while he was still a graduate student, Sandage first met his future collaborator Olin Eggen —the E of ELS—whom Sandage calls the best stellar astronomer in existence. Said Eggen, "I was at Lick Observatory when Allan was getting his degree. One time he came up with Hubble and Baade to Lick, and we just hit it off."

Since childhood, Eggen had been interested in science, especially chemistry. "I had a chemistry set, bigger than the high school had," Eggen recalled. "One day I released some chlorine gas into the house, and so the set got given to the high school. I decided I had to do something besides chemistry." He chose astronomy—specifically, stellar astronomy. "We don't know anything in astronomy, we really don't —everything's so far away. This is why I stick to stars," said Eggen, who has avoided what many perceive as more glamorous fields, such as the study of other galaxies and the overall universe. "I want to go to my grave at least believing I know something, and not have, the day after the funeral, somebody discover some cosmological principle that changes the whole thing, as is going to happen in the galaxy business."

During the late 1950s, Sandage and Eggen observed metal-poor stars together and published two papers in 1959. During the early 1960s Eggen compiled two star catalogues by using the stars' distances, proper motions, and radial velocities to calculate U, V, W velocities. One catalogue primarily contained stars with low velocities, and the other, stars with high velocities. Eggen began noticing correlations similar to those Nancy Roman had discovered in the mid 1950s—that the stars with high velocities had low metallicities, as measured by the ultraviolet excess she had discovered.

Eggen wanted to use the U, V, W velocities to determine Galactic orbits for the stars, to see which parts of the Galaxy the stars ventured into, and to calculate the stars' orbital eccentricities and thus see how round or elliptical the orbits were. To do so, he needed more than just U, V, W velocities; he needed what physicists call a potential—a mathematical model that describes how mass is distributed throughout the Milky Way—because the gravitational force of the Galaxy's mass pulls on a star and alters its path.

Olin Eggen turned to Donald Lynden-Bell, a British theorist who was visiting Caltech. "Olin asked me for a potential in which one could calculate these orbits," said Lynden-Bell. "I thought a bit and realized that you couldn't get the whole orbit very nicely for any of the potentials that I knew, so I made a little study of what potentials there were that could be solved with things no worse than inverse trigonometric functions. And I gave that to Olin without expecting to be in any way involved in the paper."

Armed with Lynden-Bell's potential, Eggen computed the stellar orbits and discovered two remarkable correlations. First, all the metal-poor stars had eccentric orbits, whereas all the metal-rich stars had circular orbits. Second, many of the metal-poor stars had large W velocities and so were able to climb high above the Galactic plane, whereas all the metal-rich stars had small W velocities and remained near the plane.

Both Lynden-Bell and Sandage were struck by Eggen's findings.

Said Lynden-Bell, "Sandage showed me Olin's correlations and said, 'Look, these must mean something. What do they mean?' " Especially intriguing were the metal-poor stars, whose orbits were quite elliptical. "They had very higgledy-piggledy orbits," said Lynden-Bell. "So Sandage and I had a long series of discussions and came to the conclusion that the Galaxy was formed in infall—that is, these stars were born on infalling orbits." A star that was born while falling toward the Galactic center would shoot past the center and thus acquire an extremely elliptical orbit around the Galaxy.

Eggen, Lynden-Bell, and Sandage ultimately based their conclusions on 221 main-sequence stars, 108 drawn from Eggen's low-velocity catalogue and 113 from his high-velocity catalogue. Following Burbidge, Burbidge, Fowler, and Hoyle, ELS assumed that metallicity marked age, so the most metal-poor stars were the oldest. Because all the metal-poor stars had large orbital eccentricities, orbital eccentricity itself also indicated age: all old stars had large orbital eccentricities, whereas all young stars had low orbital eccentricities.

Eccentricity, in fact, was the key to ELS's model, for Lynden-Bell realized that it was a remnant of a star's birth. During a star's life, he said, the orbital eccentricity changed little. Although a halo star might be 10 or 15 billion years old, its orbital eccentricity still preserved that of the gas cloud that had created it.

There was one catch, however. In order for a star's orbital eccentricity to remain constant, the Galaxy's mass distribution must never have changed rapidly. To clarify this concept, ELS gave an analogy involving the Sun and the Earth. The Earth's orbit around the Sun is nearly circular and so has a small eccentricity. But suppose the mass of the Sun doubled rapidly, meaning faster than the Earth takes to orbit the Sun—one year. Then the Earth would suddenly feel a strong gravitational pull and be yanked toward the Sun, and the Earth's orbit would become very elliptical and highly eccentric. If, on the other hand, the Sun's mass doubled slowly—over a period of many years—it turns out that the Earth's orbit would remain circular.

But this was just an illustration. ELS did not propose that those stars which currently have eccentric orbits were born with circular ones. "The paper's often been misinterpreted," said Lynden-Bell. "We gave an example, that if the Galaxy fell together very quickly, then you could increase a star's orbital eccentricity. A lot of people later thought that this was our reason for generating the high eccentricities. It was not. We actually believed that these stars were *born* onto eccentric orbits, because the gas clouds had not yet settled down into circular orbits."

The argument was subtle. If the halo stars once had circular orbits around a larger Galaxy of gas that later collapsed and increased the stars' orbital eccentricities, then that original stable Galaxy must have been held up by the pressure of its gas. But if the gas pressure was great enough to prevent the initial collapse of the Galaxy, the pressure was also great enough to prevent the collapse of any part of the Galaxy that might form stars. Thus, ELS argued, the Galaxy was never a large, stable ball of gas. Instead, it was born in infall, and the first stars—the metal-poor stars of the Galactic halo—were *born* on highly elliptical orbits that they have preserved to this day.

Because of the high orbital eccentricities of the halo stars, ELS also argued that the collapse of these gas clouds was rapid. The orbits of the stars reflect the paths of the gas clouds that created those stars, so the halo stars' elongated orbits demonstrated that the gas clouds had been falling freely toward the Galaxy's center. In modern parlance, such a rapid collapse is called a "free-fall," because the gas clouds did not collide with or impede one another.

A free-fall collapse, such as that which ELS envisioned for the halo, proceeds in less time than a star takes to complete one orbit around the Galaxy. Since the Sun orbits the Galaxy about once every 200 million years, the collapse was thus complete in no more than that length of time. As a result, said ELS, all the stars and globular clusters in the halo formed in under 200 million years. To support this, ELS noted that the five globular clusters that then had well-determined H-R diagrams—M3, M5, M13, M15, and M92—seemed to be of the same age. ELS admitted that the cluster ages were not accurate enough to test their prediction fully, but at least the ages did not contradict it.

The Galaxy's collapse came to an abrupt end for a good reason: the halo stars, which had already formed, continued on their eccentric orbits, but the gas clouds did not. Stars are small, and nothing could stop them. The stars did not collide with other stars, and any star that encountered a gas cloud plowed through it. The halo stars therefore continued to revolve around the Galaxy on their original orbits.

But it was a different story for the gas clouds, which are large and can extend over hundreds of light-years. As these gas clouds came together, they were so big that they inevitably smashed into one another. The collisions smoothed out the clouds' eccentric orbits and made them more circular, and the clouds assembled into a spinning disk. By now, the halo had stopped giving birth to stars, since all the gas was in the disk, where new stars began to arise. The disk stars were metal-rich—since high-mass halo stars had enriched the gas through nucleosynthesis—and on circular orbits, like the gas clouds from which the stars formed. Finally, because the collapse was so rapid, said ELS, the whole sequence—from the first halo star to the first disk star—took only 200 million years.

ELS thus synthesized a vast array of information into a grand picture of the Galaxy's origin. According to ELS, the halo and disk had formed in fundamentally different ways. The halo was created in a free-fall collapse, whereas the disk formed through what is today called "dissipation," because the gas clouds hit one another, lost energy, and settled into a disk. Since then, said ELS, stellar orbits have not changed much. The halo stars still have the eccentric orbits on which they were born, and the disk stars still have the circular orbits on which they were born. (Although ELS did not mention it, the disk stars' orbits have changed slightly, because the stars are perturbed by interstellar clouds, as Lyman Spitzer and Martin Schwarzschild had suggested during the early 1950s.)

"Many of these individual bits and pieces had been known before," said Sandage. "Preceding ELS had been the fantastic work, much underrated, of Nancy Roman on the high-velocity stars. As you know, Nancy is the one that discovered the ultraviolet excess in the high-velocity stars, and she also calculated U, V, and W space motions. But that's as far as she took it. Nevertheless, the whole seeds of ELS belong to the planting that Nancy Roman put in the ground. She certainly influenced us."

Initially, however, ELS influenced almost no one. "It's not like an opening night on Broadway," Eggen said. "You don't read the reviews in the next day's newspaper." Said Sandage, "ELS was a sleeper.

Hardly anybody took any notice of it for some ten years, and that surprised us terribly. But what surprised us even more was how it took off after that."

Yet when ELS awoke, it was in the midst of a raging controversy—for a new theory had emerged that was challenging the orderly picture that ELS had painted of the Galaxy's origin.

1 1

GALAXY IN CHAOS

ON FEBRUARY 5, 1963, three months after publication of the epic
ELS paper on the Galaxy's origin, astronomers found that what they
thought was a bunch of kittens was really a band of fierce mountain
lions. Since 1960, a new breed of "star" had puzzled astronomers, for
these stars differed from others in two ways. First, they emitted radio
waves, so astronomers had dubbed them "quasars," short for quasi-
stellar radio sources. Second, and even more baffling, their spectra were
indecipherable. Allan Sandage himself had investigated the quasars,
but neither he nor anyone else recognized the lines in their spectra.
Then, in early 1963, Maarten Schmidt at Caltech realized that the
mysterious spectra were actually ordinary spectra showing enormous
redshifts, which meant that the quasars were not in the Galaxy but
billions of light-years beyond. Moreover, to be seen from such a huge
distance, quasars must be incredibly luminous, outshining everything
else in the universe.

With Schmidt's discovery, astronomers around the world swung
their telescopes toward the quasars and away from the Milky Way.
Then, in 1967, Jocelyn Bell discovered the first pulsar, and in 1971
other astronomers found the first black hole candidate. The study of
the overall Milky Way was neglected, ELS slept, and sixteen years
elapsed before a new model of the Galaxy's origin appeared. Some

considered the new model a modification of ELS, others a challenge to it. Rather than the smooth and orderly picture that ELS had envisioned, in which the Galaxy was born from a single collapsing entity, the new model proposed that the Milky Way had formed chaotically, as dozens of small galaxies smashed together. If this theory was right, the Galaxy's origin had not the grandeur of a classical symphony but the reckless fury of a rock concert.

"Prior to our 1978 paper," said Leonard Searle, one of the new model's creators, "the field was not very lively. There was a prevailing view of how things had happened, and no one had looked very critically at it. Our paper was critical and hypothetical in nature—it didn't prove anything, but it suggested some possibilities that other people hadn't yet formulated. It stimulated a lot of work. Half the people wanted to confirm it, half the people wanted to shoot it down."

Born in London, Searle now directs the Carnegie Observatories in Pasadena, a few hours' drive from the 200-inch telescope atop Palomar Mountain. There he and Robert Zinn, then a young postdoc fresh out of Yale, conducted the observations that led them to propose the new theory of the Galaxy's origin. Zinn had become interested in astronomy after a neighbor showed him Saturn through a telescope, but when Zinn explored the sky on his own, his favorite objects were the globular clusters. It was an appropriate choice, for Searle and Zinn would use the globulars to support their theory, just as sixty years earlier Harlow Shapley had used them to overthrow the small, Sun-centered Kapteyn universe.

PEERING AT DISTANT GLOBULARS

With the 200-inch telescope, Searle and Zinn measured the metallicities of globulars at different distances from the Galactic center. A few of the clusters they examined were bright and nearby, but most were faint, distant, and unexplored. Some were on the outskirts of the Galaxy, and two even resided beyond the edge of the Milky Way's disk, for they were over 100,000 light-years from the Galactic center. According to prevailing wisdom, these remote clusters should have had extremely low metallicities, less than 1 percent of the Sun's. This was because the Galactic halo was thought to have a metallicity gradient—

that is, objects in the outer halo were more metal-poor than those closer in.

A metallicity gradient reflects how the halo formed. In a rapid, free-fall collapse, as ELS had envisioned for the halo, the halo should *not* have a metallicity gradient. During a free-fall collapse, gas clouds do not hit one another and so do not alter their orbits around the Galaxy. Stars that formed from these clouds therefore follow the same orbits as the clouds, no matter when the stars arose. Stars that formed later will be more metal-rich, because nucleosynthesis has enriched the gas with metals; but these stars' orbits will resemble those of stars born earlier, since the clouds' orbits will be unchanged. Therefore, the metal-poor and metal-rich halo stars will have similar orbits and distributions, and no metallicity gradient arises.

In contrast, during a slow, dissipative collapse, a metallicity gradient does occur, because the gas clouds hit one another, alter their orbits, and settle down to a plane, all the while becoming more metal-rich. Imagine a metal-poor gas cloud that starts with zero velocity two thousand light-years above the Galactic plane, forms some stars, and then falls toward the plane. The stars will dive through the plane and again journey two thousand light-years from it. But the cloud will not. Instead, it hits other clouds and slows down. As a result, suppose the cloud can travel no farther than a thousand light-years from the Galactic plane. Any stars it forms will travel no farther than this and will be more metal-rich than the first group of stars, since the cloud itself is. Thus, there is a metallicity gradient, because stars that stay near the plane are more metal-rich.

The science, then, is straightforward: a slow collapse produces a metallicity gradient, whereas a rapid collapse does not. But the history of this idea is more complicated. Since ELS stated that the halo had formed in a rapid collapse, they implied that no metallicity gradient exists in the halo. But the ELS paper itself actually said nothing about a halo metallicity gradient. Moreover, in 1969 Allan Sandage, after observing additional stars, claimed the halo did have a metallicity gradient. He split his most metal-poor stars into two groups, one ranging in metallicity from 4 to 13 percent of the Sun's, the other ranging from 0.2 to 4 percent. He found that the more metal-poor of these halo groups had, on average, W velocities nearly twice as high as those of the more metal-rich group. This meant that the most metal-poor stars ventured farther from the plane and traveled farther out into the halo than the more metal-rich halo stars. This implied that the outermost halo was more metal-poor than the halo near the Sun; in short, the halo had a metallicity gradient. As a result of Sandage's 1969 work, many astronomers mistakenly thought that ELS themselves had claimed that the halo had a metallicity gradient.

"I grew up with that picture of how the Galaxy formed," said Zinn. "That whole idea of the formation of the Galaxy had in it a metallicity gradient, and that was based on misreading ELS, or Sandage's 1969 paper, or a combination. That's the thing I was taught in graduate school." Consequently, when Searle and Zinn began measuring the globulars' metallicities, Zinn expected to find a correlation between distance and metallicity, with the most distant clusters being extremely metal-deficient. He even thought they would determine just how steep the halo's metallicity gradient was.

But the actual data would prove Zinn's expectation wrong. Searle and Zinn measured the metallicities of yellow, orange, and red giants in nineteen globular clusters. One of the first surprises came after Searle and Zinn examined their farthest globular, NGC 7006, which lies some 120,000 light-years from the Galactic center in the constellation Delphinus. NGC 7006 turned out to be metal-poor, but no more so than most other globulars. Its metallicity was around 3 percent of the Sun's, similar to that of nearby globulars like M13, a great cluster in the constellation Hercules. Other globulars told the same story. Metallicity did not decrease as one proceeded farther out into the halo.

"When we didn't find a metallicity gradient in the outer halo," said Zinn, "we said, 'Uh-oh. We got a result here that's inconsistent with what we think we already know, namely ELS and Sandage's 1969 extension of ELS.' " They pulled out Sandage's 1969 paper and scrutinized it. Sandage had claimed that more metal-poor halo stars had higher W velocities than more metal-rich halo stars. But Searle and Zinn found that Sandage's more metal-rich halo group actually included some disk stars, which had low W velocities and therefore depressed the mean W velocity of the metal-rich halo group. Using Sandage's data, Searle and Zinn concluded that the W velocities of those stars with metallicities less than 10 percent of the Sun's did not

increase with decreasing metallicity. Instead, the W velocities remained the same, consistent with no metallicity gradient in the halo.
This was also consistent with ELS.

"So sometime in 1977," said Zinn, "we had in hand that we didn't
have a gradient in metal abundance, and then we sat around thinking
about that. And I think it was Leonard who said, 'Is there a gradient in
anything else?'"

THE SECOND PARAMETER

It turned out that there was another gradient, and it involved the globular clusters' Hertzsprung-Russell diagrams, which track the evolution

Figure 14. The horizontal branch consists of stars that are burning helium,
instead of hydrogen, at their cores. The H-R diagram of a metal-poor globular
cluster has a so-called blue horizontal branch, because the horizontal-branch
stars are blue, white, or yellow.

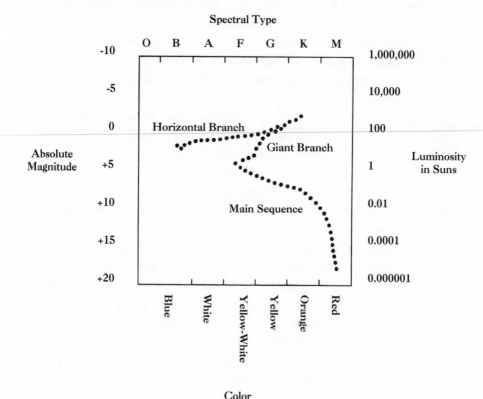

of the clusters' stars. Because the globulars are old, their blue and white main-sequence stars died long ago, and the brightest main-sequence stars are yellow, like the Sun. As these yellow stars run out of the hydrogen they are burning at their cores, they expand into orange and red giants. The giants have masses similar to the Sun's, since they evolved from stars like it. But in contrast with the Sun, the giants now burn hydrogen in a layer surrounding the core.

As a giant burns hydrogen outside its core, the core gets hotter and hotter until the helium there ignites. When this happens, the star shrinks, fades, and heats up, but by how much depends on the star's metallicity. Although metals make up only 2 percent of a metal-rich star's mass, they absorb light of certain wavelengths and dramatically affect the star's color—just as a little salt can completely change a

Figure 15. In contrast, a more metal-rich globular cluster has a red horizontal branch: the horizontal-branch stars are bunched together in the yellow part of the diagram, and none of the stars is blue or white.

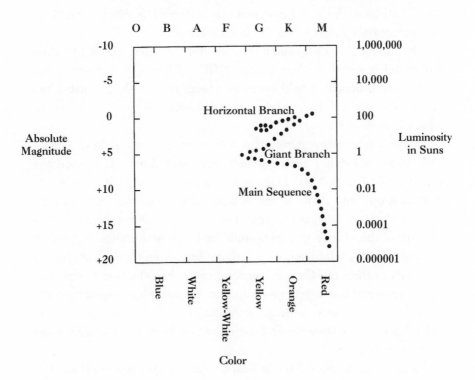

meal's taste. When a metal-rich star like the Sun ignites helium, it becomes orange. But when a metal-poor star ignites helium, it gets hotter than a metal-rich star and turns either white or blue. Over time, as it burns its helium, this star cools at a nearly constant luminosity and eventually becomes yellow. Thus, on the H-R diagram, these metal-poor core-helium-burning stars form a horizontal line from blue to yellow. Astronomers call this line the horizontal branch. The most famous horizontal-branch stars are the RR Lyrae stars, which are white or yellow-white and pulsate because horizontal-branch stars of these colors are unstable. Whether stable or not, though, every horizontal-branch star is burning helium at its core.

The important point is that the overall color of a globular's horizontal branch depends strongly on the cluster's metallicity. In a fairly metal-rich globular, the horizontal branch is yellow; in a metal-poor globular, the horizontal branch is considered blue, because it contains blue and white stars as well as yellow ones. For example, the fairly metal-rich globular cluster 47 Tucanae, whose metallicity is 20 percent that of the Sun, has a yellow horizontal branch (which astronomers call "red"), whereas the metal-poor globular cluster M13, whose metallicity is 2 percent solar, has a blue horizontal branch.

But some globulars violate this rule. For example, Searle and Zinn had found that their farthest cluster, NGC 7006, was so metal-poor that its horizontal branch should have been blue. But in 1967, Sandage and Robert Wildey had observed the remote cluster and discovered its horizontal branch to be red.

"I can remember going off and looking at every globular cluster H-R diagram I could get my hands on," said Zinn. "I counted the number of blue and red horizontal-branch stars in each cluster and plotted it up—and there, lo and behold, was a gradient." All globulars that lay closer to the Galactic center than the Sun obeyed the metallicity rule: if the cluster was metal-rich, its horizontal branch was red; if the cluster was metal-poor, its horizontal branch was blue. But clusters at distances from the Galactic center greater than the Sun's often violated the metallicity rule. Two clusters in the outer halo might have the same metallicity but completely different horizontal branches, one red, the other blue. So some second parameter, other than metallicity, must be responsible.

Zinn brought the result to Searle's office. Said Searle, "Our data presented particularly sharp evidence for the existence of this so-called second parameter, and we got to bothering about what it might be. And many bottles of Jack Daniel's later, we came up with the ideas that we presented in the paper."

There were plenty of possible explanations for the second parameter. A horizontal-branch star's helium abundance could affect its color, as could its carbon abundance, nitrogen abundance, oxygen abundance, and even its rotation. But Searle could see no reason that any of these should vary from the inner halo to the outer halo when the metallicity did not.

He chose something else. "We speculated that the second parameter was age," he said, "and we wondered whether the prevailing view that the Galaxy's halo had formed in a short period of time might not be in error." Age affects a globular cluster's horizontal branch because the younger the cluster, the more massive are its stars that are evolving into giants and horizontal-branch stars. And it turns out that the more massive a horizontal-branch star, the redder it is. Searle and Zinn calculated that two clusters having the same metallicity but different horizontal-branch colors differed in age by over a billion years. The cluster with the red horizontal branch was at least a billion years younger.

That contradicted ELS, who had said that the halo formed during a rapid collapse and that all the globulars had arisen within 200 million years of one another. It also seemed to contradict Searle and Zinn's other finding, that the outer halo had no metallicity gradient. If globular clusters differed in age, then the collapse must have been slow; but if the collapse had been slow, the halo should show a metallicity gradient, which it did not.

"Leonard deserves the lion's share of the credit here," said Zinn. "He synthesized a picture and put together the things that seemed to be contradictory: there is no metallicity gradient and yet things happened slowly."

"The picture that we thought could account for both," said Searle, "is one in which, before the Galaxy formed, there were a number of clumps—subgalaxies—which were its precursors. In each of these clumps, star formation and cluster formation occurred, and these clumps of material fell together to form the Galaxy. It is just a more messy version of the simplest picture that you could imagine."

In the Searle and Zinn model, only the inner halo collapsed in the rapid way that ELS had said, for there the globular clusters' horizontal-branch colors matched their metallicities, indicating that the clusters had similar ages. But the outer halo—that beyond the Sun—was formed not from a single gas cloud, as ELS had said, but from dozens of small galaxies that crashed chaotically into the Milky Way. These galaxies may have resembled the ten that today orbit the Milky Way: the two Magellanic Clouds and the eight dwarf galaxies like Sculptor and Fornax. Such galaxies may have been the building blocks of our Galaxy. Indeed, the galaxies that now orbit the Milky Way may be the survivors of this process, the lucky few that escaped the ravenous Galaxy as it was swallowing other galaxies and growing into a giant. In this, the satellite galaxies metaphorically resemble comets and asteroids, debris left over from the formation of the solar system.

Searle and Zinn's model could explain both of the features they had found in the globular clusters. Because different galaxies had formed at different times, the globular clusters that one galaxy brought to the Milky Way differed in age from globular clusters that other galaxies brought. And because the process was chaotic and messy, no metallicity gradient arose in the halo.

"The beauty of astronomy," said Zinn, "is that when you get some new data, you have license to extrapolate to grand pictures. Other sciences, which are much more highly developed, don't have that fun. And Leonard and I had a lot of fun coming up with a new picture."

To an extent, the Searle and Zinn model was a product of its times. Astronomers had once believed that galaxies lived their lives in isolation from one another. But in 1972, American scientists Alar and Juri Toomre championed the idea that galaxies often collide. The Toomre brothers ran computer simulations showing that galactic collisions could explain the features of several peculiar galaxies. Their most famous example was a pair of interacting galaxies in the constellation Corvus that had drawn out long strings of stars from each other, earning the pair the nickname the Antennae. The 1978 Searle and Zinn model, then, extended galactic collisions and mergers to the Milky Way itself.

But initial response to the Searle and Zinn scenario was mixed. "I got a lot of heat about Searle and Zinn and some other things I'd written as being contrary to ELS," said Zinn. "The other criticism that one heard occasionally, not spoken directly to one's face, was, 'That

Searle and Zinn thing is crazy; it's a terrible paper; how could they claim all these things from their observations?' " Zinn recalled an encounter with one famous astronomer who called the Searle and Zinn paper "awful." "You guys claimed all kinds of things," this astronomer said, "and you had no right to do that."

Zinn added, "I have a tenured job now, and I can look back at it, but in the beginning you worried about being labeled as some kind of clown because you'd written a foolish paper. But then again, there were a few people who right away were saying that this was something new and important. The fact that people were talking about it made one happy, even if it was negative things, because a lot of papers are ignored when they're published."

SEARLE AND ZINN MEET ELS

Searle and Zinn had made one claim that, in principle, was testable. The second parameter was age, and two globular clusters with the same metallicity and different horizontal-branch colors should differ in age by over a billion years. For example, NGC 7006, the remote globular with the red horizontal branch, should be younger than other globulars.

In theory, a star cluster's age can be determined from its H-R diagram. Bright, massive main-sequence stars die before less massive ones do, so the older a cluster is, the fainter its main sequence. The cluster's brightest main-sequence stars, which are about to turn into giants, constitute what astronomers call the main-sequence turnoff, and as a cluster gets older, its main-sequence turnoff gets fainter, moving down on the H-R diagram. So in principle, one can measure the relative ages of clusters simply by comparing their H-R diagrams.

In practice, however, several difficulties arise. First, more than just age affects a cluster's H-R diagram. For example, metallicity alters the brightness of the main-sequence turnoff. A metal-poor globular has a brighter main-sequence turnoff than a metal-rich globular of the same age. Also, interstellar dust lying between Earth and a cluster dims the cluster's light. And the main-sequence stars in globulars are faint, so observational errors inevitably cause trouble. Consequently, during the late 1970s and early 1980s, the age difference between two globulars could not be pinned down to better than a few billion years. This meant that one could not distinguish between the ELS model, which predicted that all globulars were born within 200 million years of one another, and the Searle and Zinn model, which predicted that the globulars had an age spread of over a billion years.

"For *years,*" said Zinn, "people would measure one age after another for a globular cluster, and even though the error bars were 3 or 4 billion years, they would always come to the point: 'Well, yet again, we've measured the age of another cluster; it's the same as these other ones, and therefore there's no age spread in the halo. Hence, ELS is right and Searle and Zinn are wrong.' Sometimes they'd say that without even noticing that they were comparing clusters of the same horizontal-branch [color], so Searle and Zinn would say you wouldn't get a detectable age spread there anyway. That was a bit annoying."

Finally, in the late 1980s, several astronomers took aim at the globulars with improved instruments and demonstrated that at least some globulars were several billion years younger than others. The key celestial protagonists were a pair of clusters named NGC 288 and NGC 362. In three ways, these globulars are identical, which made comparing their H-R diagrams easier and provided an ideal test for measuring their age difference. First, both clusters have nearly the same metallicity, 10 percent of the Sun's. Second, they both lie in the same part of the sky, so both can be observed by the same person with the same telescope on the same night. Third, little interstellar dust lies between them and us, so their light suffers only a small amount of absorption en route to Earth. However, the two clusters differ in one important respect: NGC 288 has a blue horizontal branch and NGC 362 a red one. If Searle and Zinn were right, NGC 288 should be older. In 1989, Michael Bolte, now at Lick Observatory, compared the clusters' main-sequence turnoffs and reported that NGC 288's was fainter, which meant that the cluster was older than NGC 362, just as Searle and Zinn had predicted. In fact, Bolte estimated that NGC 288 was 3 billion years older than its near-twin.

In 1991, a leading practitioner of the art of globular cluster dating gave a talk at Berkeley. A few minutes before his presentation, he was asked what the age spreads among globulars said about ELS. He looked as if the questioner were crazy. "ELS? ELS is dead."

Such statements rile Allan Sandage, who is ELS's strongest advocate. "I get angry," he said. "Every time there is a globular cluster that's somewhat older or younger, they say, 'Aha, ELS is dead; Searle and Zinn is there.' That's simply not true." The reason, Sandage explained, is that the Galaxy's collapse time depended on the density of gas. The greater the density, the faster the collapse, so denser parts of the primordial Galaxy collapsed before other parts. But ELS had simplified the problem by assuming uniform density throughout the early Galaxy, which gave a single collapse time of 200 million years.

Lynden-Bell made the same point. "ELS was very much a first cut at this problem," he said. "We thought of the collapse of a uniform sphere, but I don't think any of us really believed that it was all uniform. It would have been what you would now call a computer job if we hadn't made it homogeneous. Had we been thinking about the collapse of a nonuniform sphere, then of course the central part takes a shorter time to collapse, because the mean density there is greater, and the outer parts would collapse much later. But at that time, nobody had thought very much about galaxies forming by dribs and drabs. I would see Searle and Zinn as a modification of what we wrote, and probably correct."

"Searle and Zinn is ELS with noise," said Sandage. "They have a hierarchy of collapse times, so the noise is the density fluctuations in the initial proto-Galaxy." As a result of such fluctuations, he explained, a few globulars that are much younger or older than the mean do not negate ELS. Nevertheless, Sandage believes that the globulars have a small age spread and that the metal-rich globulars are, on average, no younger than the metal-poor ones. "If I were to describe the way the Galaxy formed now," said Sandage, "I would not use precisely the same model that we had in ELS. Certainly the proto-Galaxy was not smooth; it was lumpy. That's why I believe that Searle and Zinn is ELS with noise."

But Zinn sees a much greater contrast between the two models. "Fundamentally, they're different pictures," he said. "ELS's model is for a single object undergoing collapse. Our picture is there is a Galaxy that has collapsed, which we call the inner halo and which underwent rapid collapse, and then on top of that there is accretion of stuff that produces an outer halo that, in the mean, is younger and has a larger dispersion in age than the rapidly collapsed core."

Since the work of ELS and Searle and Zinn, Sandage has continued to study high-velocity stars near the Sun. Assisting him were astronomers Charles Kowal and Gary Fouts. In 1987, Sandage and Fouts analyzed 1125 stars and reaffirmed ELS. Three years later, in 1990, Sandage published a paper defending ELS against the attacks that had been made on it. Sandage's 1990 paper included this tongue-twister: "I propose to state what I perceive ELS to have said, what they did say but should not have, what they did not say but should have, and how the critics can be answered concerning what they say ELS said, which, indeed, they did not." Sandage now pointed out, for example, that a rapid, ELS-style collapse would not produce a metallicity gradient in the halo.

Such talk doesn't sit well with Zinn, though. "He's having his cake and eating it too," said Zinn. "In his 1990 paper, where he takes shots at us critics, he's saying ELS never predicted a metallicity gradient in the halo, so the fact that people don't find it now doesn't contradict ELS. That's true. But it does contradict Sandage (1969), and it does contradict Sandage and Fouts (1987). So you can see," said Zinn, with understatement, "that there's a little bit of contention."

OTHERS JOIN THE FRAY

Although Searle and Zinn reached their conclusions after studying globular clusters, these objects have not been the only weapons in the battle over the Galaxy's origin. Individual stars in the halo—"field stars," in the vernacular of astronomers—have played an even larger role. While determining the ages of field stars is difficult and usually impossible, they are crucial to understanding the halo because they outnumber globular cluster stars a hundred to one. Furthermore, many halo stars lie close enough to Earth for astronomers to determine distances and proper motions. These data, plus the stars' radial velocities, then give the U, V, W velocities that allow astronomers to calculate the stars' orbits around the Galaxy.

One of the leaders in the effort to study the halo's field stars has been Australian astronomer John Norris, who during the 1980s struck several blows against ELS. Norris has spent most of the last thirty years at the Australian National University's Mount Stromlo and Siding Spring Observatories, where Olin Eggen was director from 1966 to

1977 and where Allan Sandage spent the year 1968 to 1969. Of Sandage, Norris said, "Even though we have since come to quarrel scientifically, he was and remains to some extent one of my idols."

Norris wandered into the controversy over the Galaxy's origin by accident, while he was searching for metal-poor stars. In 1981, Howard Bond, then at Louisiana State University, claimed that the Galaxy had a shortage of extremely metal-poor stars, those with less than 0.1 percent of the Sun's metallicity. Bond called these supposedly missing stars "population III," said that they preceded the halo stars, and claimed that the missing stars had produced the small amounts of metals found in the halo.

Norris set out to look for these metal-deficient stars, in part to see what clues they might hold to the origin of the elements. "What was driving us was to find stars more metal-poor than 0.1 percent the solar metallicity and see what we could learn about early nucleosynthesis," he said. "And Bond had been asking questions: where is population III? He was saying there were no stars with metallicities less than 0.1 percent the Sun's."

With Michael Bessel and Andrew Pickles, Norris investigated 309 candidate metal-poor stars and determined the distances, metallicities, and velocities for many of them. None of the stars had a metallicity under 0.1 percent of the solar value, but many of them did have low metallicities. According to ELS, all these metal-poor stars should have had eccentric orbits around the Galaxy; in fact, it was precisely the high orbital eccentricities of the metal-poor stars that had led ELS to their rapid collapse model for the Galaxy's origin. But as Norris, Bessel, and Pickles computed the orbits of the metal-poor stars, they found a startling result. Of 95 stars whose metallicities were less than 10 percent of the Sun's, 19 were on fairly circular orbits. ELS had found no such stars at all. "This strikes at the basis of the ELS model of the very rapid collapse of the Galaxy," Norris and his colleagues wrote in a 1985 paper.

The reason for the difference boiled down to how ELS had found their metal-poor stars. ELS had used a catalogue of high-velocity stars to discover those with few metals, so all of ELS's metal-poor stars had high velocities and hence high orbital eccentricities. "ELS had a kinematically selected sample," said Norris, "so they were blind to metal-poor stars of low eccentricity, because these wouldn't turn up in high-velocity samples." In contrast, the metal-poor stars that Norris, Bessel, and Pickles studied had been selected with no advance knowledge of the stars' kinematics, so they were a fairer sample of the Galactic halo.

In a 1986 paper, Norris presented further evidence against ELS. According to ELS, when the halo formed, it was spinning slowly, but it spun faster and faster as it collapsed. Thus, as a group, the oldest and most metal-poor halo stars should revolve more slowly around the Galaxy than the slightly younger and more metal-rich halo stars; so the average V velocity—which measures how fast stars revolve around the Galaxy—should decrease as the stars' metallicities decrease. But Norris found that stars with all metallicities less than 4 percent solar had the same average V velocity: −190 kilometers per second. On average, therefore, all the metal-poor stars revolved around the Galaxy at an equally slow pace. This suggested that ELS were wrong in claiming that the halo stars had led to the formation of the disk. Instead, Norris's research supported Searle and Zinn, for his work suggested that the halo stars had come from other galaxies that struck the Milky Way when it was young.

Meanwhile, another astronomer was also turning up evidence against ELS. In 1981, Bruce Carney, of the University of North Carolina at Chapel Hill, started a project to observe some nine hundred high-velocity stars within a thousand light-years of the Sun. His main colleague in this work was David Latham, an astronomer at the Harvard-Smithsonian Center for Astrophysics. The results of the Carney-Latham survey, as published in 1990, contradicted ELS and supported Searle and Zinn.

"Sandage isn't even speaking to me at the moment," said Carney. "The last paper I sent to him came back stamped RETURN TO SENDER on the envelope." Their dispute goes back many years. "Ever since my thesis, we have sort of been on opposite sides of scientific issues," said Carney, who completed his dissertation in 1978. "The main result was that the metal-poor globular clusters looked to be older than the metal-rich ones, and Sandage didn't find that palatable, since from ELS he would have believed that they were all the same age."

Nevertheless, when Carney began the Carney-Latham survey, he thought ELS was right. "It was a beautiful concept," he said. "It was not only quantitatively consistent but philosophically pretty. It seemed to be what you expect an ensemble of gas and stars to do. It starts sort of like a proto–solar system—a big, slowly rotating ball that collapses, flattens out, forms a disk—and *voilà*, you've got a disk system out of what had been a halo." Although his own and others' work had suggested that the time scale was longer than ELS had said, Carney otherwise considered ELS valid. And he maintains a high regard for Sandage. "He is a very worthy adversary," Carney said, "and a man of very broad knowledge. When you write a paper that's challenging what he's done, you know you've got to have it backed up by good data and a thorough analysis."

The Carney-Latham survey turned up the same problems for ELS that Norris had: metal-poor stars on circular orbits, stars that ELS had said did not exist; and no correlation of the metal-poor stars' V velocities with metallicity, which argued against ELS's view that the halo formed as a single gas cloud that had spun faster as it collapsed. In fact, according to Carney's team, ELS's whole idea that the halo evolved into and gave birth to the disk was wrong. As Carney and his colleagues wrote in 1990, "[A]ll the data are consistent with the high-velocity, low metallicity halo population having had a chemical and dynamical history almost independent of the Galactic disk. We propose that the halo was assembled from mergers of small satellites with the Galaxy along the lines discussed by Searle and Zinn."

A QUESTION OF BALANCE

Despite the work of John Norris and Bruce Carney, Allan Sandage continues to believe that ELS is basically correct. "It's certainly true that we cannot maintain every detail of the ELS picture as we put it out in 1962," said Sandage. "All of these new results and interpretations enrich the picture. The following four main points, however, are what ELS said, and I think those still hold: the Galaxy formed by collapse; the collapse was rapid; correlations exist between the stellar orbits and the metallicity; and the disk is dissipative. Now if you want to ask if every detail of the ELS picture is right, the answer is absolutely no, but that's progress."

Yet in just the last few years even ELS's greatest critics have found evidence that supports the ELS model, convincing them that they can no longer explain their data solely in terms of Searle and

Zinn. "My working hypothesis is that both ELS and Searle and Zinn hold some truth," said Norris. "ELS were strong on the collapse, in which there should be some dependence of kinematics on abundance. On the other hand, Searle and Zinn were strong on the idea that a lot of material was accreted, and in that material there should be no dependence of kinematics on abundance. Some of the early work I did, from 1985 to 1990, demonstrated that Searle and Zinn had a fair amount of truth, because for low abundances there seemed to be very little dependence of kinematics on abundance. But more recently I've come to believe that Searle and Zinn can't be the full picture." Norris's new work, published in 1994, suggests that metal-poor stars may show a weak dependence on kinematics. He therefore believes that ELS may explain the kinematics of some metal-poor stars and Searle and Zinn the rest. "I don't know whether it's right or not," said Norris, "but that's where I'm at now."

Carney's latest observations also show signs that ELS may be at least partially correct. "When the first set of papers came out, from 1987 to 1990," said Carney, "all we found were faults with ELS, in that clearly there was not a homogeneous evolution from the halo into the disk. Now it's becoming a little more interesting." He and Latham have expanded their survey, adding five hundred stars with less extreme kinematics. The expanded Carney-Latham survey, said Carney, shows hints that the halo consists of two components. "The first is the halo we saw before," he said, "something that doesn't seem to have any dynamical relation to the disk population, presumably because it originated from protogalactic fragments *à la* Searle and Zinn. Those were captured by the Galaxy and so really didn't play any role in the formation of the disk. But there also seems to be another halo component that does have disklike kinematics." This second component, which ranges in metallicity from 0.5 percent to 10 percent of the Sun's, may be the structure that collapsed dissipatively to form the present disk, as ELS said.

"Philosophically," said Carney, "I still think that, at some level, ELS must be right. The disk didn't spring full-grown from nowhere. It presumably was larger and more turbulent in its early days and probably got stirred up by a lot of star formation and supernovae before settling down into its more or less quiescent state now." As for Searle and Zinn, Carney thinks they are probably right, too. "I don't think they're the whole story," he said, "but the signs for accretion of pieces that formed around the Galaxy are pretty compelling. If I had to bet, I would say that half or more of the globular clusters came from things that originally had nothing to do with the Galaxy."

To Zinn, the issue is not who is right but that both ELS and Searle and Zinn stimulated so much work. "ELS set the ball rolling," said Zinn. "It's a classic paper. When you figure that ELS were looking at only a couple hundred nearby stars, it's remarkable that their model has anything right at all. ELS clearly had to be only a pioneering study, and what they discovered in their data was there's a disk and there's a halo. They showed that through observations of stars near the Sun you could deduce something about the structure of the disk and halo.

"If ELS is mostly wrong, I think it's immaterial. Maybe it matters a lot to Allan, but it doesn't matter to the field. It's just like it doesn't really matter to me if somebody tomorrow shows that Searle and Zinn is wrong. Our motivation was to present something new, hopefully right, but at least interesting, and in that way push the next guy on to the next step."

Eggen takes a similar view. "We had opened a can of worms, and when different people look into a can of worms, they see different worms," he said. "I'd be sorry if everything one did was a finished product, because then we'd soon be out of business. Just the fact that ELS is still discussed, thirty years later, is something. Good God," he said, "in this business, thirty years is a lifetime."

As for Sandage, he remains ELS's strongest supporter. At the end of his 1990 paper, he listed six conclusions concerning the Galaxy's origin and evolution, most of which reiterated ELS. In his sixth and final conclusion, though, Sandage admitted, "The only conclusion not in dispute is that 'the Galaxy has not always been as we see it today.' "

12

THE THICK DISK

ASTRONOMY DID NOT always thrive in New Zealand. The first time anyone there earned a doctorate in the subject was 1979, and it went to someone who had never met an astronomer. "I had never even been to a senior-level astronomy lecture," said Gerard Gilmore. "I had no formal training in astronomy at all." Nevertheless, during the 1980s, Gilmore would proclaim that a new stellar population existed in the Milky Way, a disk population that extended above and below the conventional disk of the Galaxy. This so-called thick disk would provide a new window on the Milky Way's past—and add fuel to the fiery controversy over the Galaxy's birth.

A NEW STELLAR POPULATION

As a graduate student in physics at New Zealand's University of Canterbury, Gerard Gilmore had planned to study quantum mechanics, but he tried it and hated it. In search of a new field of study, Gilmore learned of a neglected observatory nearby and began to explore it. Astronomy, he soon discovered, was a lot more exciting than physics, and he devised a thesis project to measure the variability of quasars, conducting observations on clear nights and reading about astronomy on cloudy ones. No astronomers were there, and he had no thesis advisor. He met his first astronomer six months after receiving his doctorate.

"Nobody told me all these things that I should have known, like what the Galaxy is like," said Gilmore. "Everything I learned I learned by reading journals, so I learned from the bottom up what a great deal of dissension and disagreement there is about most things in astronomy, rather than learning from textbooks, which give you a completely wrong impression that answers are precise and well understood. So I went into astronomy taking for granted that most things are either wrong or not half as well understood as you think they are."

After moving from New Zealand to Scotland, Gilmore continued his quasar work. He had no interest in either stars or the Milky Way. "I was a typical extragalactic astronomer," he said. "I never had any intention of looking at a star in my life."

Stars got in the way. Gilmore was tracking the changing brightnesses of quasars by examining photographic plates taken on different nights. The plates came from the 48-inch U.K. Schmidt Telescope in Australia. The telescope was owned by the Royal Observatory in Edinburgh, where Gilmore held a job and which maintained an archive of all the plates that had ever been taken with the telescope. Each plate had a dozen of the quasars Gilmore was interested in but also abounded with some 10 million stars in our Galaxy that lay between us and the quasars.

"While I started working away on these dozen quasars," said Gilmore, "it immediately became apparent to me that it was pretty stupid to throw away 10 million minus twelve images and just look at the twelve. I should try to do something with the 10 million as well."

Gilmore and a graduate student named Neill Reid began looking at the stars. Many of the plates were of the south Galactic pole, the column of space perpendicular to and below the plane of the Galaxy. Here one peers through the Galactic disk much as a geologist drills through strata of sediment, so a view of this region probes the vertical structure of the Milky Way's disk. But this was not the reason Gilmore had so many plates of the south Galactic pole; rather, the pole had served as a test field that astronomers used to calibrate the telescope.

Reid was interested in the red stars on Gilmore's plates. Most of these stars were red dwarfs, and together Gilmore and Reid began a

North Galactic Pole

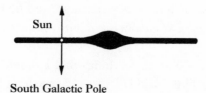

Sun

South Galactic Pole

Figure 16. The north and south Galactic poles are the directions where we look perpendicular to the Galactic disk. Any point along the column of space represented by either arrow on this diagram is considered to be "at" that particular Galactic pole.

series of papers titled "New Light on Faint Stars." Reid counted the number of red stars having different apparent magnitudes, and Gilmore modeled the distribution of the red stars by assuming they belonged to what was then called the disk and today is called the thin disk. The thin disk population has a scale height of about 1000 light-years, which means that the average thin disk star lies this far from the Galactic plane. Gilmore's model for the Galaxy fit the red-star counts well.

But the plates also recorded yellow stars, and the yellow-star counts told another story, for these stars lay much farther from the Galactic plane than they should have if they all belonged to the thin disk. "I spent about a year trying to make that go away," said Gilmore, who had based his star-count model on what he found in astronomy books. "But I was unable to make it go away, so I became convinced that the Galaxy was right and the books were wrong." In a 1983 paper, the third in their series, Gilmore and Reid proposed that many of the yellow stars represented a new Galactic population that was more spread out than the thin disk and had a larger scale height. Gilmore and Reid called the new stellar population the thick disk, and they put its scale height at 4700 light-years, over four times that of the thin disk. Extrapolating their data to the Galactic plane, Gilmore and Reid estimated that the thick disk contributed 2 percent of the stars near the Sun. Together, the two disk components—thin disk and thick disk— explained the star counts at the south Galactic pole. The thin disk contributed all the red stars, which were intrinsically faint and therefore must be nearby in order to be seen, while the thin and thick disks took care of the yellow stars, which were intrinsically brighter and could be seen at greater distances. "I didn't really think it was such a big deal," said Gilmore. "It was paper three in a series that was mostly

National Optical Astronomy Observatories (NOAO)

NOAO

As viewed from Earth, the Milky Way (above) looks like a river of stars running across the sky. If we viewed the disk of the Milky Way edge-on from a distance of millions of light-years, our Galaxy would resemble the spiral galaxy NGC 4565 (below).

M51, the Whirlpool Galaxy, is the most spectacular spiral galaxy visible from Earth and was the first in which astronomers detected spiral structure. In 1994, a supernova exploded in M51, ejecting oxygen and other heavy elements into the galaxy. This is the same process that has allowed intelligent life to develop in the Milky Way.

The Andromeda Galaxy (above) is the largest member of the
Local Group; the Milky Way ranks second. M33 (below), another
spiral galaxy, is the third largest member of the Local Group.

NOAO

Courtesy of Paul Hodge

The Large Magellanic Cloud (above) and Ursa Minor (below) are both satellite galaxies of the Milky Way, orbiting our Galaxy the way moons circle a planet. The Large Magellanic Cloud is bright, shining with a tenth of the Milky Way's luminosity. In contrast, Ursa Minor is so faint that it is almost impossible to see, even on this negative photograph.

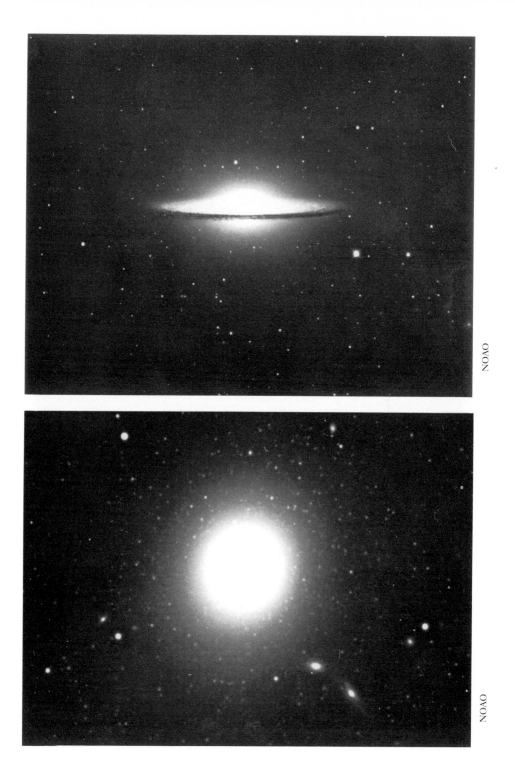

Astronomers first detected galactic rotation in the Sombrero Galaxy (above). The elliptical galaxy M87 (below) shows the fuzzy appearance that typifies its class.

NOAO

NOAO

NOAO

NOAO

Star clusters come in two types. The first, exemplified by the Pleiades (above), are called "open clusters" because their stars are spread out, whereas the second, exemplified by M5 (below), are the tightly packed "globular clusters." In the Milky Way, open clusters are near the Galactic plane while globular clusters usually lie far above or below it.

NOAO

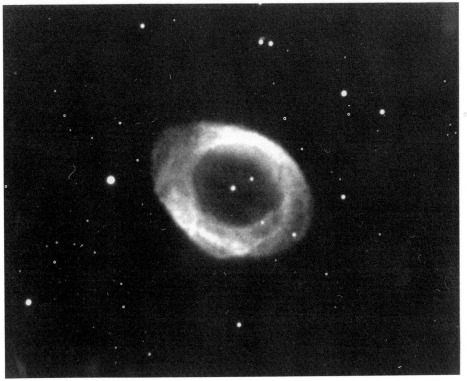

NOAO

A star dies either violently or gently. In the first case, the star explodes as a supernova, shooting its remains into space and forming a supernova remnant like the Crab Nebula (above). In the second case, a star ejects its outer atmosphere, creating a planetary nebula such as the Ring Nebula (below).

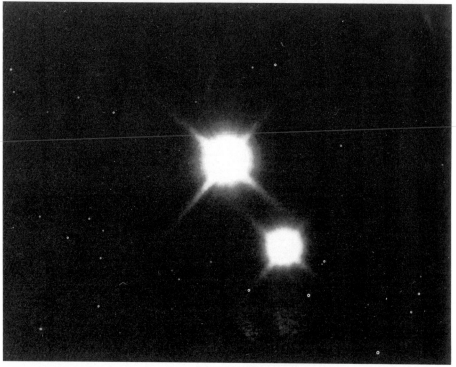

Debris from supernovae and planetary nebulae gathers in space and can give birth to new stars, a process that occurs today in the Orion Nebula (above). The Sun formed in this way, as did each of its neighbors. The Sun's nearest neighbor is the triple star Alpha Centauri, whose two brightest stars peer at Earth in the lower photograph.

about low-mass stars. If it had been ignored, I probably would have given it up and gone off and done something else."

The paper was anything but ignored. "The reaction was quite hostile," said Gilmore. "And because people were hostile to it, I became determined to prove to myself that it was right. As I became convinced it was right, I became obstreperous and determined to prove to other people that it was right as well."

INTO THE THICK OF THINGS

By championing the existence of the thick disk, Gerard Gilmore entered battle with one of the titans of American astronomy, John Bah-

Figure 17. This close-up of the Galactic disk shows the thin disk, the thick disk, and the lower part of the stellar halo. According to current data, the thick disk is about 3.5 times thicker than the thin disk. At any time, though, a few thick disk stars are passing through the thin disk, and a few stars from the Galactic halo are darting through the thick and thin disks.

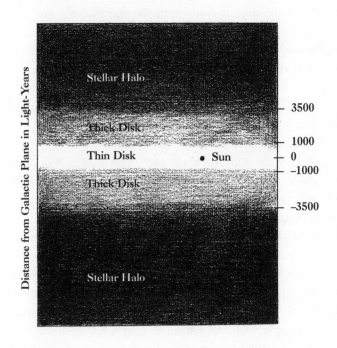

call, an astronomer at the Institute for Advanced Study in Princeton. An ardent supporter of NASA's Hubble Space Telescope, Bahcall had just worked out a detailed Galaxy model with Raymond Soneira. This model predicted the numbers, colors, and brightnesses of stars the Hubble Space Telescope would see in different directions, and the Bahcall-Soneira Galaxy model had no thick disk. Like Gilmore, Bahcall originally had no interest in the Milky Way; he and Soneira had constructed their Galaxy model only because they wanted to know how many foreground stars might masquerade as distant galaxies and thereby complicate Hubble's observations.

As a result of his work on the Milky Way, John Bahcall saved the Hubble Space Telescope from disaster. To observe an object, the telescope had to point toward, or "guide on," stars that lay in nearly the same direction as the desired object. Such stars are called guide stars, and without them the telescope could not point steadily toward anything. To lock on to guide stars, the guide-star system had to be powerful enough to see them. The more powerful the system, the better, because the more stars it could see, the more likely one lay near an object to be studied. In building the guide-star system, the developers of the Hubble Space Telescope had relied on a reference book by Clabon Allen that gave the number of stars with different apparent magnitudes.

Yet during a meeting at Caltech in 1980, Bahcall realized he could use his Galaxy model to check Allen's work. "On the plane, I did a crude calculation," said Bahcall, "and I got an answer that was very different from the one in Allen's book." The answer was disturbing: it showed that Allen had overestimated the number of stars, which meant the Hubble Space Telescope's guide-star system was too weak to work. "I communicated this to the project," said Bahcall, "which rejected it on the grounds that, one, Ray and I didn't know what we were talking about; two, it would be impossibly expensive to do anything about; and three, we didn't have independent data." But later work confirmed Bahcall's result, so the guide-star system was inadequate and had to be redesigned. "That was a real crisis for the project," said Bahcall. "It left a lot of people with red faces and cost a fair amount of money."

But with the 1983 Gilmore-Reid paper proposing that the Galaxy had a thick disk, Bahcall's own work was under attack, for the Bahcall-Soneira Galaxy model represented star counts with only a disk—a thin one—and a halo. A gigantic debate ensued, with Gilmore and Bahcall slugging it out, each arguing that his Galaxy model better fit the observed star counts. Gilmore was especially combative. In 1984 he wrote: If the parameters derived by Bahcall & Soneira are a valid representation of our Galaxy, three conclusions follow:

(1) All those authors of recent star count data who have analysed their data have done so incorrectly . . .

(2) Our Galaxy is unique . . .

(3) The direct evidence for a population . . . with intermediate kinematic properties implies a weak or non-existent correlation between stellar [W] velocity and [the stellar distribution perpendicular to the Galactic plane].

This last point would have contradicted the well-known fact that the greater a star's velocity perpendicular to the Galactic plane, the farther above and below the plane it can travel.

The following year saw another sharp exchange, after Gilmore, Reid, and a colleague, Paul Hewett, had obtained more star-count data. They wrote, "No published model of the stellar distribution in the Galaxy is in acceptable agreement with these data." Bahcall and his colleague Kavan Ratnatunga immediately responded: "The Bahcall-Soneira Galaxy model is in good agreement with the observations of Gilmore, Reid and Hewett." Meanwhile, other observers of star counts weighed in, some supporting Gilmore, others Bahcall, and the controversy raged on.

"This thing became like a tabloid newspaper headline," said Gilmore, "and it got blown out of all proportion. John's an outstanding scientist, he'd done an amazing number of things, and he was an extremely high public profile person. And then suddenly there was this little random postdoc with a funny accent coming from somewhere far away standing up and saying he was wrong." Gilmore said that Bahcall's opposition to the thick disk was intense. "It was vigorous and it was consistent," said Gilmore. "He wrote me great long letters saying I was wrong, and he wrote letters to my boss saying I was wrong, too."

In his papers, Gilmore abbreviated the Bahcall-Soneira Galaxy as the BS model. "That Gilmore," one astronomer remarked. "Real subtle, isn't he?" Today Gilmore calls the abbreviation an accident, for it does not mean in Britain what it does in America. Nevertheless, to American astronomers, it perpetuated the notion of a war between the two.

One battle was set to occur in May 1989, when Gilmore and Bahcall were to speak at a conference in Connecticut on the gravitational force perpendicular to the Galactic plane, another area where the two had reached opposite conclusions. But the promised showdown never took place, because on his way to Connecticut Bahcall was mugged in New York's Grand Central Station and could not attend. "What I want to know," one astronomer said, "is where Gilmore was at the time."

The true situation between the two rivals, said Gilmore, was quite different. "In fact, John and I got on extremely well," he said. "It was a scientific disagreement in the ideal way in which science is supposed to work. You can like a person as an individual but profoundly disagree with their logical deduction."

Bahcall, however, saw the controversy differently. "Gerry is not like most of the scientists that I have known," said Bahcall. "He's more of a debater than a person who was seeking to find out together with his colleagues what the best answer was." Bahcall added, "In his view, I was sort of the establishment. And he was at that time a relatively less-known individual, and he felt that, in order to scramble up the ladder, he had to knock off the guy that was ahead of him on the ladder. So I had a feeling that a lot of his sarcasm in both his talks and his papers had to do with the knock-the-guy-off-the-ladder syndrome. I didn't pay a lot of attention to it."

One thing both Gilmore and Bahcall agree on is the unreliability of star counts, which initially were all either side had to go on. "Don't believe star counts at face value," said Gilmore. "Kapteyn got a spectacularly wrong answer by taking star counts in isolation. The distribution of stars one sees on the sky is the product of a large number of functions and parameters, and there are too many of them to get all the information back out just from star-count data."

It was, of course, an old lesson. Jacobus Kapteyn's incorrect model for the Galaxy was based on star counts, and Bart Bok had used them in his unsuccessful attempt to spot the Milky Way's spiral arms. Star counts only prolonged the thick disk controversy, for they determined Galactic structure so poorly that Gilmore and Bahcall could fiddle with other parameters and fit the same star-count data with two different models, one with a thick disk and the other without.

"There's a limited amount that star counts can tell us," said Bahcall. "When it comes to subtle differences, like whether there's an additional thick disk that maybe contains only 10 percent of the stars and overlaps in space with the other two components, it's very hard to separate that out with pure star counts."

THROUGH THICK AND THIN

What was needed to resolve the controversy between Gilmore and Bahcall was stronger data, data that transcended star counts and probed the stars' distances, metallicities, and kinematics. In 1987, Eileen Friel, then a graduate student at Lick Observatory, reported on the distances and metallicities of stars in the constellation Serpens. In this direction, one's line of sight makes an angle of 51 degrees with the Galactic plane. Therefore, most nearby stars in Serpens are members of the thin disk, distant stars are members of the halo, and stars in between would be members of the thick disk—if it exists.

Instead of just counting stars, Friel obtained spectra of over 200 stars with apparent magnitudes between 11.5 and 14.5. She compared her results with the predictions of the Bahcall-Soneira model and found that the model came up short. If Bahcall and Soneira's work were correct, she would have seen 59 halo stars, but she found only 8 stars with the low metallicities characteristic of the halo. At the same time, she saw 22 stars with metallicities between those of the halo and the thin disk—stars that were likely members of the thick disk. The stars the Bahcall-Soneira model had predicted would be luminous metal-poor halo giants lying far away turned out to be less luminous intermediate-metallicity giants that lay closer to the Galactic plane, in the thick disk. From her observations, Friel concluded that the Galaxy had a thick disk, but it was thinner than Gilmore and Reid had proposed. She put its scale height at around 3000 light-years.

As had Gilmore and Reid, Friel found the thick disk by looking away from the Galactic plane, to regions where thick disk stars dominated. But if the thick disk existed, at any time a few members of the population should also be passing through the solar neighborhood. Such stars would have velocities between those of the low-velocity thin disk stars and those of the high-velocity halo stars, since thick disk stars travel higher than the former but stay closer to the plane than do the latter.

In 1987, Allan Sandage and Gary Fouts reported that the thick disk showed up in their survey of the nearby stars, and in 1989 Bruce Carney, David Latham, and their colleague John Laird came to a similar conclusion with their survey. Stars in both surveys have high proper motions, so the samples were biased toward stars with high velocities. Because of this, thick disk and halo stars should be overrepresented. Still, one could examine the metallicities of stars having different W velocities, because thin disk stars are metal-rich, halo stars are metal-poor, and thick disk stars are in between. Although all the stars presently lie near the Sun, their W velocities carry them to different distances above and below the plane. Stars with small W velocities remain near the plane and should be primarily thin disk stars; stars with greater W velocities travel a few thousand light-years from the plane and should be primarily thick disk stars; and stars with the largest W velocities venture many thousands of light-years from the plane and are primarily visitors from the Galactic halo.

Three populations appeared in the data. Many of the stars with the lowest absolute W velocities had solar metallicities and so were members of the thin disk. However, many of the other stars had a metallicity around 30 percent solar and belonged to the thick disk. A few stars appeared at a metallicity of 3 percent solar, which corresponded to the few halo stars with low W velocities.

At greater W velocities, the three peaks in the metallicity distribution—corresponding to the thin disk, thick disk, and halo—continued to appear, but the relative numbers of stars in each population changed. The thick disk showed up strongest among stars with absolute W velocities between 20 and 60 kilometers per second, for most of these stars travel over a thousand light-years from the Galactic plane. At still higher W velocities, both teams found, as expected, that stars with halo metallicities dominated, for these stars travel beyond not only the thin disk but the thick disk as well.

From this analysis, Sandage and Fouts, and Carney, Latham, and Laird concluded that the thick disk actually exists, and in 1990 Bahcall himself appeared to concede defeat. Bahcall was the third author on a paper (with Stefano Casertano and Kavan Ratnatunga) that analyzed the Sandage-Fouts and Carney-Latham data and found that a population between the thin disk and the halo was indeed necessary to account for the data. But Casertano, Ratnatunga, and Bahcall refused to call this intermediate component the thick disk. To them, it was the "old disk."

"I was the referee on that paper," said Gilmore. "I suggested that they point out that their data were in excellent agreement with everybody else's, and I didn't like what they called the things. What John calls the old disk is what everybody else in the community calls the thick disk."

Nevertheless, through his terminology, Bahcall made a point: though the thick disk exists, it differs from the one Gilmore and Reid had proposed in 1983. "It's thinner, and it weighs less," said Bahcall. Gilmore and Reid had derived a scale height for the thick disk of 4700 light-years, but the best data today suggest the scale height is smaller, around 3500 light-years. Bahcall considers the difference significant; Gilmore does not.

However one views it, the thick disk is over three times thicker than the thin disk. The latter, though, is more complicated, because thin disk stars of different ages have different scale heights. Astronomers traditionally divide the thin disk into two subpopulations, the young thin disk and the old thin disk. The young thin disk consists of stars that are younger than a billion years and includes all the bright blue O and B stars, such as Regulus and Spica, most red supergiants, such as Betelgeuse and Antares, and many of the bright white A stars, such as Sirius and Vega. But the young thin disk accounts for only about 10 percent of the entire thin disk. The vast majority of thin disk stars are in the old thin disk and have ages between 1 and about 10 billion years. The Sun is an old thin disk star, as are most of its neighbors, such as Alpha Centauri.

Because interstellar clouds have kicked them around, old thin disk stars have a greater scale height than young thin disk stars. The scale height of the young thin disk is about 350 light-years; the scale height of the old thin disk is 1000 light-years. Even this latter number is tiny, however, compared with the thin disk's diameter of 130,000 light-years. The thin disk therefore lives up to its name. The thickness of

TABLE 12-1: STELLAR POPULATIONS IN THE SOLAR NEIGHBORHOOD

Population	Percentage of Stars Near Sun	Scale Height (Light-Years)	Velocity Dispersions (Km/Sec.)			Mean V Vel. (Km/ Sec.)	Mean Metal Abund. (Sun = 1)	Examples —Stars and Clusters
			U	V	W			
Young Thin Disk	9%	350	15	10	10	0	1.0	Sirius, Vega; Pleiades, Hyades
Old Thin Disk	87%	1000	35	25	20	−10	0.8	Sun, Alpha Centauri; M67, NGC 188
Thick Disk	4%	3500	65	50	40	−30	0.3	Arcturus; 47 Tucanae
Halo	0.1%	—	130	100	90	−200	0.02	Kapteyn's Star, Groombridge 1830; M13, M92

any disk population is twice its scale height, since the scale height measures the distance from the Galactic plane, whereas the thickness is the distance from one side of the Galactic plane to the other. For the thin disk, then, the thickness is 2000 light-years, which is only 1/65 its diameter. Stack two compact disks atop each other and you have a miniature model of the thin disk.

By contrast, the thick disk is about 3.5 times thicker than the old thin disk. Because it is thicker, the thick disk has a greater W velocity dispersion. The young and old thin disks have W velocity dispersions of 10 and 20 kilometers per second, respectively, but the thick disk has a W velocity dispersion around 40 kilometers per second. This is still much less than that of the halo, whose W velocity dispersion is 90 kilometers per second. Both spatially and kinematically, then, the thick disk falls between the thin disk and the halo. The same applies to the thick disk's metallicity. Best estimates for this population's metallicity put it at about 25 percent of the Sun's, less metal-rich than the thin disk, whose metallicity is solar, but more so than the halo, whose metallicity peaks at 2 percent solar.

Because the Sun lies in the thin disk, most of its neighbors are thin disk stars. But the thick disk contributes 3 to 5 percent of the stars near the Sun, and one of them is a spectacle: the orange giant Arcturus, 34 light-years away, probably belongs to the thick disk. Arcturus's metallicity is 30 percent of the Sun's, similar to that of the thick disk. The star's kinematics also mark it as a thick disk star. Although Arcturus has an unremarkable W velocity—+4 kilometers per second—the star's V velocity of −103 kilometers per second gives Arcturus a fairly eccentric orbit around the Galaxy. This is just what one would expect of a star whose properties lie between those of the thin disk, where stars have nearly circular orbits, and those of the halo, where stars have extremely elongated orbits. The negative V velocity indicates that Arcturus is currently near apogalacticon, its farthest point from the Galactic center, whereas the Sun is near perigalacticon, its closest point to the Galactic center. The two stars are thus passing like ships in the night. Arcturus normally lies much closer to the Galactic center than it is now, and the Sun normally lies farther.

FIRST SIGNS OF THE THICK DISK

Although the thick disk sparked great controversy during the 1980s, astronomers had actually seen it before. In 1957, the participants at the Vatican conference on stellar populations had divided the Milky Way into five populations, one of which matched what astronomers now call the thick disk. But the Vatican participants gave it the awkward name "intermediate population II," and they failed to appreciate its importance.

"Many things are before their time," said Sandage, who attended that conference. "Also, the courage of the young is much greater than the courage of the old, and Gilmore and Reid were very young." It further helped, said Sandage, that they had given the thick disk a name. "Naming something," he said, "is a long way down the road to accepting it. Gilmore and Reid made a central discovery, and they deserve complete credit for it and for understanding its significance."

Even before the Vatican conference, the thick disk had staged an appearance, though it perplexed Sandage and other astronomers. In 1951, Philip Keenan and Geoffrey Keller of Perkins Observatory in Ohio first reported on the spectral types, luminosities, and distances of what they considered high-velocity stars. In their final sample, Keenan and Keller included 83 stars having total velocities exceeding 85 kilometers per second; that is, the combination of the stars' U, V, and W velocities was greater than this speed. One of their stars was Arcturus. By Walter Baade's definition, such high-velocity stars belonged to population II and should have resembled the stars of globular clusters. In particular, Keenan and Keller's stars should have had an H-R diagram that matched those of the globulars, with faint main-sequence stars and luminous orange giants. Using the stars' spectral types and luminosities, Keenan and Keller plotted an H-R diagram, published in 1953.

"When that came out," said Sandage, "Baade was astounded, Arp and I were astounded, and it looked as if all of these simple ideas were not right." Keenan and Keller's H-R diagram indeed showed a faint main sequence, but the giants were faint as well, so it did not resemble the H-R diagram of a globular cluster.

With hindsight, though, that puzzling result now makes sense. "All those stars," said Sandage, "are thick disk stars. Keenan and Keller's high-velocity stars are not really high-velocity stars but intermediate-velocity stars." Because they belong to the thick disk, they had higher metallicities than halo stars, and because metal-rich giants are dimmer than metal-poor ones, Keenan and Keller's H-R diagram differed from that of a globular cluster.

In 1962, however, ELS—Eggen, Lynden-Bell, and Sandage—had missed the thick disk in their study of nearby stars, and Sandage now explained why. ELS had selected their stars from two star catalogues, one containing mostly low-velocity stars that were members of the thin disk, the other containing high-velocity stars that were members of the halo. Since the thick disk had an intermediate velocity, ELS missed it.

Gilmore suspects that extragalactic astronomers were to blame for "losing" the Milky Way's thick disk. When these astronomers studied faint galaxies, they could see only a (thin) disk and a halo. Those observations influenced Bahcall and Soneira, who had based their model of the Milky Way on the assumption that our Galaxy resembled other galaxies.

"It was an interesting example of information being lost rather than built up," Gilmore said. "Extragalactic astronomers expected things *a priori* to be simple population I and population II, whereas people who had grown up knowing the detailed information about our Galaxy had always known it was much more complicated than that." Gilmore cites as an example Jan Oort, who believed in the thick disk as soon as Gilmore told him about it. Ironically, Oort had been the referee on the first Bahcall-Soneira paper.

In the years before Gilmore and Reid's paper, the thick disk had appeared in other studies. For example, in 1979, David Burstein, then at Lick Observatory, coined the term when he detected a "thick disk" in several other galaxies, although it is now unclear whether this component matches the Milky Way's own thick disk. In 1982, William Hartkopf of Georgia State University and Kenneth Yoss of the University of Illinois observed stars at the north and south Galactic poles and found two groups, one metal-rich, with a W velocity dispersion of 22 kilometers per second, the other somewhat metal-poor, with a W velocity dispersion of 44 kilometers per second. Most stars in the first group are members of the thin disk, and most in the second are members of the thick disk.

"The main effect of our 1983 paper," said Gilmore, "wasn't so much a new discovery as it was to appreciate that it was unexpected. The thick disk was independently discovered by loads of people. The discovery itself was not dramatic. It was the popularization of it that had the big effect."

A Window into the Galaxy's Past

The thick disk not only represents a major structure in our Galaxy but also holds clues to the Milky Way's origin and evolution. To understand where the thick disk fits into Galactic history, though, astronomers must first know the thick disk's age. Several arguments indicate that it is old, having existed for roughly 10 billion years. The thick disk star Arcturus is old, for its mass is similar to the Sun's, yet the star has already evolved into a giant. Other signs point to an old thick disk as well, for astronomers have found no young or even moderately old main-sequence stars in this population.

A further way to ascertain the age of the thick disk is through its globular clusters. Most globular clusters—about 75 percent of them—are members of the Galactic halo. These clusters form a spherical distribution around the Galaxy and have metallicities of less than 10 percent of the solar value. But a second population of globulars, the remaining 25 percent of clusters, also exists. They lie in a disk and are more metal-rich. This dichotomy was first recognized in the 1950s.

Unfortunately, investigations during the 1970s disputed this result and claimed that the globulars did not constitute two discrete systems. In 1985, however, Robert Zinn analyzed 121 globular clusters and proved once and for all that they do indeed form two separate systems. One is a spherical halo population in which metallicities range from 0.3 to 10 percent of the Sun's, and the other is a disk system where the mean metallicity is 30 percent of the solar value, similar to the thick disk's metallicity. As specific examples, the great cluster in the constellation Hercules, M13, belongs to the halo population, and 47 Tucanae belongs to the thick disk population.

Zinn originally put the scale height of the disk globulars at only 1600 light-years, but later work reported in 1989 by Zinn's student Taft Armandroff upped this to 3600 light-years, which matches present estimates for the scale height of the thick disk. Metal-rich globulars like 47 Tucanae are therefore members of the thick disk, so astronomers can use these clusters' H-R diagrams to determine the thick disk's age. The H-R diagrams indicate that the clusters are roughly 10 billion years old, because they have no bright main-sequence stars. This in turn means that the thick disk itself is old. Whatever phenomenon created the thick disk therefore happened long ago in the Galaxy's history, so the thick disk opens another window on the Milky Way's ancient past.

Theories explaining the origin of the thick disk abound. Said Sandage, "I would like to believe that the thick disk is the part of the Galaxy that still conforms to classical ELS-type collapse." In his view, the thick disk represents the short-lived phase in the Milky Way's past when the Galaxy made the transition from the halo to the thin disk. If so, the thick disk formed after the halo, as gas clouds from the halo, which had been in a free-fall collapse, hit one another and settled into a plane. These gas clouds gathered first into a thick disk, where they formed the stars we see there today, and then the gas clouds fell farther into a thin disk. If the thick disk originated this way, it should have a metallicity gradient, with thick disk stars near the plane and the Galactic center more metal-rich than those farther from either.

Bruce Carney draws a more dramatic history of the thick disk, one that fits Searle and Zinn's chaotic model for the Galaxy's origin rather than ELS's stately one. Carney believes the thick disk is the remains of another galaxy that strayed too close to the Milky Way and was swallowed by it. "If there was a galaxy like the Large Magellanic Cloud that was cannibalized by the Milky Way in the past," said Carney, "that galaxy's stars would get spread out into a disk—just as the Large Magellanic Cloud will when it falls into the Galaxy in a few billion years." In Carney's view, the luckless galaxy was one of the many that Searle and Zinn believe crashed into and merged with the Milky Way. If Carney is right, the thick disk should not have a metallicity gradient. This agrees with the Carney-Latham survey, which indicates that thick disk stars with high W velocities have the same metallicity as thick disk stars with low W velocities. Furthermore, most thick disk globular clusters lie much closer to the Galactic center than the Sun, yet they have metallicities equal to thick disk stars near the Sun. This, says Carney, also argues that the thick disk has no metallicity gradient.

John Norris—who, like Carney, has challenged ELS—promotes another, more skeptical view. "I think you've got to be very careful when you say there are two components here of different origin," he said. "The point that I've emphasized is that you can explain a lot of what you see in terms of a continuous configuration." In Norris's view, the thick disk may not be a discrete population of stars separate from the thin disk. Instead, it may simply be an extension, or "tail," of the thin disk, consisting of thin disk stars that have somewhat greater velocities and scale heights. A helpful analogy is a rainbow, in which no sharp line demarcates one color from the next, so the colors form a continuum from red to violet. The thin disk itself is a continuum of stars with different ages, scale heights, and velocities. Although astronomers speak of the young thin disk and the old thin disk, they recognize that the one blends into the other. These are therefore subpopulations, not separate populations.

If Norris is right about the thick disk, its stars may simply be thin disk stars that have been scattered to larger velocities and scale heights by encounters with interstellar clouds, an idea he put forth in 1987.

However, this idea does not explain the thick disk globulars, since they are too massive to have been kicked around by interstellar clouds. Furthermore, present-day interstellar clouds cannot scatter stars up to the velocities of thick disk stars. This latter problem might be solved, though, if the Galaxy's interstellar clouds were at one time more massive than they are today. In 1991, Norris and Sean Ryan suggested that the thick disk may also reflect the initial stages of the entity that collapsed to form the thin disk, as in the ELS model, and in agreement with Sandage's view.

The question of whether or not the thick disk forms a separate population is crucial, said Gilmore. "If there really is a discontinuity between the thin disk and the thick disk," he said, "then it takes a specific physical process to create it. The thick disk is a group of stars that have a common origin and so should be thought about as a coherent, discrete thing, which needs to be explained. But if the thick disk is just the tail of the thin disk, that doesn't mean very much."

In the last few years, Gilmore and Rosemary Wyse of Johns Hopkins University have probed the thick disk directly by obtaining spectra of stars that lie several thousand light-years above the Galactic plane, where thick disk stars outnumber both thin disk and halo stars. Their study, by far the largest ever of the thick disk, determined the distances, velocities, and metallicities of some 2500 stars. These findings, as well as other recent data, have convinced Gilmore that the thick disk is indeed a separate population.

"It's looking surprisingly so," he said. "There are only tails that overlap rather than bodies that overlap—but only on very, very recent data. Until maybe a year ago, you really couldn't tell. So it looks as if the simplest possible model, the discrete model, is turning out to be more correct than it had any right to be. Generally, as data get better and better, things get more and more complex, but in this case they're getting more and more clear."

In the past, Gilmore suggested that the thick disk was the transition stage from the halo to the thin disk, as in an ELS-type model. But the new data have changed his mind. "All the evidence is supporting a merger," he said, "and the shock of that merger is a fossil for us to puzzle over now." He added, "It should be said that this disproves almost every other speculation I've had on the subject."

Gilmore therefore agrees with Carney that a galaxy like the Large

Magellanic Cloud smashed into the Milky Way when it was young and formed the thick disk. But the two astronomers disagree on precisely how the collision created the thick disk. Gilmore believes that the collision kicked what were once stars in the thin disk to greater scale heights—thereby puffing up what was then the thin disk—and that these former thin disk stars now make up today's thick disk.

In contrast, Carney thinks most of the stars in the present thick disk already existed in the victim galaxy before it hit us. "The thin disk had to have been puffed up at some level," said Carney, "but my own interpretation is that the thick disk is mostly a bunch of stars that fell into the pre-existing Galactic stellar disk. The collision may have helped star formation go on in the Milky Way, so some of the stars formed as a consequence, but most of them were originally part of the victim." Thus, in Carney's view, a star like Arcturus probably came from another galaxy.

Whatever the case, Gilmore noted that collisions and mergers were once common. "In the early universe, this stuff was happening all the time. It was raining galaxies in those days, and raining dwarf galaxies onto some other galaxy was a normal day's weather." When the Milky Way was young, the universe was small, since it had only just begun to expand from the big bang. Therefore, the universe was more crowded and galaxies collided more often. Many astronomers now believe that quasars, which were a prominent feature of the early universe, occurred when two large galaxies collided and flared up.

Quasars, of course, had led Gilmore and Reid to the discovery, or rediscovery, of the thick disk, for quasars had motivated Gilmore to inspect the plates that had captured stars from the thick disk. And what happened to those quasars? "The quasars sort of fell through the gap," Gilmore confessed, "and I've never actually got back to doing quasars again." The Milky Way, it seems, was just too interesting.

13

SHIFTS IN THE GALACTIC WIND

EVERY TIME A supergiant star exhausts its fuel and explodes inside a spiral arm; every time a red giant casts a planetary nebula into the Galactic sea; every time a white dwarf receives too much mass from a companion and annihilates itself: every time one of its stars dies, the Milky Way Galaxy gains some of the heavy elements the star forged during its life. That life may have been brief, with the star dying near its birthplace, or as long as the Milky Way is old. But each star, upon its death, contributes in its own way to the Galaxy that gave it birth. Some offer oxygen, others carbon and nitrogen, still others iron. These elements float into space, where they can enrich interstellar clouds in which the Galaxy will create new stars and planets. One observatory has a poster that says, in part, "Keep up the good work, especially all you OB stars and planetary nebulae."

THE POWER OF ABUNDANCE RATIOS

Because different types of stars provide different elements to the Galaxy, astronomers can investigate past stellar generations through the abundance patterns preserved in stars that are still alive. During the early 1950s, astronomers discovered that old stars had lower metallicities than young ones and concluded that the Galaxy's metallicity had increased with time. But today astronomers can go much further, probing the details of the Galaxy's evolution in an attempt to detect shifts and changes in the Galactic wind.

"The grand and glorious goal that we're after," said Christopher Sneden, an astronomer at the University of Texas at Austin who has spent over twenty years divining stellar abundances, "is that we'd like to be able to characterize the chemical evolution of the Galaxy from day one to the present. To do that, we have to tell how the ratios of the elements in stars have changed with position, with time, with velocity —with all the different evolutionary stages of our Galaxy." By seeing how the abundance of each metal varies from old stars to young, astronomers can deduce which types of stars arose at various epochs in the Galaxy's past.

For years, however, astronomers did not exploit this opportunity. Instead, they lumped all metals together and assumed that each heavy element tracked every other—for example, if a star had half the Sun's abundance of iron, the star was presumed to have half the Sun's abundance of oxygen, carbon, nitrogen, and every other metal as well.

"If you look at metal-rich stars, that's not a bad approximation," said Sneden. "You'll get pretty bored, and you won't get much funding to keep observing solar-metallicity stars, because you almost always get the same answer." These metal-rich stars belong to the thin-disk population, whereas clues to the Galaxy's ancient past lie primarily in the stars of the thick disk and the halo.

Astronomers traditionally determine a star's metallicity by measuring its iron abundance, because iron is a complex atom that absorbs light at many different wavelengths and so produces dark lines all over most stars' spectra. The plethora of lines makes iron easy for spectroscopists to measure, so when astronomers mention a star's "metallicity," they are almost always referring to the star's abundance of iron.

Yet iron hardly accounts for all, or even most, of a star's metallicity. "When you look at the Sun," said Sneden, "to a first approximation, it's hydrogen, helium, and oxygen, and everything else, including iron, is a trace. The number of oxygen atoms is twenty-five times the number of iron atoms." A glance at the top ten elements in the universe confirms the point. The most common elements are the two lightest, hydrogen and helium, which were created in the big bang. Of the

TABLE 13-1: TOP TEN ELEMENTS

Atomic Number	Element	Abundance by Number (Hydrogen = 1.00000)	Produced Mostly By	Ejected into Galaxy By
1	Hydrogen	1.00000	Big bang	Big bang
2	Helium	0.097	Big bang	Big bang
8	Oxygen	0.00085	Helium burning in high-mass stars	Type Ib, Ic, and II supernovae
6	Carbon	0.00036	Helium burning in red giants	Planetary nebulae
			Helium burning in high-mass stars	Type Ib, Ic and II supernovae
10	Neon	0.00012	Carbon burning in high-mass stars	Type Ib, Ic, and II supernovae
7	Nitrogen	0.00011	Hydrogen burning in main-sequence stars and red giants	Planetary nebulae
12	Magnesium	0.00004	Neon burning and carbon burning in high-mass stars	Type Ib, Ic, and II supernovae
14	Silicon	0.00004	Oxygen burning in high-mass stars	Type Ib, Ic, and II supernovae
26	Iron	0.00003	Type Ia supernovae	Type Ia supernovae
			Type Ib, Ic, and II supernovae	Type Ib, Ic, and II supernovae
16	Sulfur	0.00002	Oxygen burning in high-mass stars	Type Ib, Ic, and II supernovae

metals, the champion is oxygen, with iron ranking a distant seventh, behind not only oxygen but also carbon, neon, nitrogen, magnesium, and silicon. By looking at only a star's iron abundance, astronomers therefore miss most of the star's metallicity.

Because it is so abundant, oxygen would seem a better choice for measuring metallicity. Unfortunately, it is especially difficult to observe. Even though it is more abundant than iron, oxygen is a relatively simple atom, with only eight electrons to iron's twenty-six, so it produces few spectral lines. In fact, most studies of oxygen have relied on just two sets of spectral lines. One is a pair in the red located at wavelengths of 6300 and 6363 angstroms, the other a triplet in the infrared at wavelengths of 7772, 7774, and 7775 angstroms.

But even these few spectral lines held startling clues to the Galaxy's past, for the oldest stars turned out to harbor much higher oxygen abundances than astronomers had expected. The first hints came in 1966, when two astronomers in England, Elizabeth Gasson and Bernard Pagel, measured the oxygen abundance of Arcturus, the bright orange giant that is now recognized to be part of the thick-disk population. Arcturus's iron abundance is only about 30 percent of the Sun's, but Gasson and Pagel found that its oxygen abundance was higher, more than one would have predicted from its iron abundance. That meant Arcturus had a higher oxygen-to-iron ratio than did the Sun.

The following year, a team of astronomers led by Peter Conti at Palomar Observatory in California reported oxygen abundances in over a dozen giant stars. Conti's team confirmed the high oxygen-to-iron ratio in Arcturus and found similar ratios in other mildly metal-poor giants, such as Algeiba, a beautiful double star in the constellation Leo. But the astronomers failed to find oxygen in their most metal-deficient star, a yellow giant in the constellation Boötes named HD 122563.

HD 122563 finally yielded its secrets several years later, when Christopher Sneden and his colleagues took a closer look at it. "That's my favorite star," said Sneden. "I've analyzed it to death, because it's the brightest very metal-poor star in the sky. Every other very metal-poor star is at least a magnitude fainter." HD 122563 is a sixth-magnitude halo star that lies about a thousand light-years from the Earth and has an iron abundance only 0.2 percent of the solar value, which makes the star more metal-deficient than even the most metal-poor globular cluster. Sneden began studying the star while he was a graduate student in Texas. Attempts prior to Sneden's had relied on photographic plates to record the star's spectrum, but Sneden and his colleagues armed themselves with modern electronic detectors that captured more light and picked up the star's weak spectral lines. In 1974, he and Texas astronomers David Lambert and Lynne Ries reported the detection of oxygen in HD 122563. According to Sneden's team, the star's oxygen-to-iron ratio was a startling four times higher than the Sun's.

Sneden and his colleagues wondered whether their result applied throughout the Galactic halo or whether it was specific to just this one star. A further problem was HD 122563's giant nature, which meant that the star no longer generated energy by fusing hydrogen into helium at its core. Instead, the star was fusing hydrogen outside its core, or fusing helium into carbon and oxygen, or doing both. Perhaps the high oxygen abundance on HD 122563's surface stemmed from the star's own nuclear fusion. If so, the surface composition no longer reflected that of the material from which the star formed, and the high oxygen-to-iron ratio was a false clue to the conditions that prevailed long ago.

Unlike giants, main-sequence stars don't pose such problems. They convert hydrogen into helium only at their centers, so their surfaces remain unaltered and accurately reflect the Galaxy's past composition. But main-sequence stars are also fainter, which makes them more difficult to observe. Nevertheless, during the late 1970s, Sneden, Lambert, and Rodney Whitaker detected oxygen in thirteen main-sequence stars. Some were metal-rich, like the Sun, but most were metal-poor. Two of the metal-poor stars had already achieved notoriety, for they were HD 19445 and HD 140283, the two stars that Lawrence Aller and Joseph Chamberlain had used nearly thirty years earlier to overthrow the conventional belief that all stars shared the Sun's metallicity.

When Sneden and his colleagues measured the oxygen abundances in the metal-poor stars, they found high oxygen-to-iron ratios in all but one. This result, published in 1979, has been repeatedly confirmed by other astronomers, who find that halo stars have oxygen-to-iron ratios three to four times greater than the Sun's. Because they are metal-poor, the stars still have less oxygen than the Sun, but they have far more than one would have predicted from their iron abundance.

SUPERNOVAE: MARKERS OF THE GALAXY'S EVOLUTION

The high oxygen-to-iron ratio in the Galaxy's oldest stars is an important clue to its past. These stars are three times older than the Sun; during the ancient era when they arose, the Milky Way must have been producing significantly more oxygen relative to iron than it has since. Astronomers have found the reason for this shift by considering the two major varieties of supernovae. One arises from short-lived stars that ejected large amounts of oxygen into the Milky Way almost as soon as the Galaxy was born; the other variety comes from long-lived stars that blew up later and gave the Galaxy iron.

Regrettably, supernova nomenclature obscures the point, because astronomers recognize four types of supernovae, which are designated type Ia, type Ib, type Ic, and type II. At first glance, type Ia, Ib, and Ic supernovae seem similar, but they are only superficially so. All these supernovae lack hydrogen in their spectra, whereas the spectra of type II supernovae display strong lines of hydrogen.

As it turns out, the real difference between various supernovae involves the stars they arise from, and different exploding stars eject different elements into the Galaxy. Type Ia supernovae represent one major variety, which produces iron; type Ib, Ic, and II supernovae represent the other, which produces oxygen. All types of supernovae explode in the Galaxy today, but only the type Ib, Ic, and II supernovae occurred in the short-lived stars that enriched the Milky Way first. A type Ib, Ic, or II supernova arises from a high-mass star, one born with more than eight solar masses. It begins life as a blue O- or B-type main-sequence star, fusing hydrogen into helium at its core. When its core runs out of hydrogen, it burns hydrogen outside the core and expands into a supergiant. Soon the helium core ignites and helium burning occurs, fusing the helium into carbon and oxygen. When the core runs out of helium, helium burning continues in a shell surrounding the core. Carbon then starts to burn at the star's core and creates neon and magnesium. Meanwhile, hydrogen and helium still burn in layers surrounding the core.

Soon the star begins burning neon into oxygen and magnesium, then oxygen into silicon and sulfur, and finally silicon and sulfur into iron and other elements with similar mass. At this point the star resembles an onion. Hydrogen burns in the outermost layer, helium in a deeper layer, carbon in a still deeper layer, neon in a deeper layer, oxygen in a deeper layer, and finally silicon and sulfur at the star's core.

But silicon and sulfur do not burn for long. A star with twenty solar masses burns silicon and sulfur into iron for only two days, and then the party is over: iron does not burn, so the star runs out of energy, collapses, and explodes. If the star had previously blown off its outer hydrogen envelope or lost it to a companion star, then no hydrogen will appear in the supernova's spectrum and astronomers will classify it as either type Ib or Ic. If the star has retained its hydrogen envelope, strong hydrogen lines will mark the supernova's spectrum and astronomers will classify it as type II. The famous supernova that exploded in the Large Magellanic Cloud in 1987 was type II, and the 1994 supernova that exploded in the spiral galaxy M51 was type Ic.

During such a supernova, the inner part of the star, which is made of iron, becomes a neutron star or black hole, so this material never enriches the Galaxy. However, the material in the star's envelope, which is rich in oxygen, is set free into the Milky Way. Thus, when a massive star explodes, the Galaxy receives a huge shot of oxygen but only the small amount of iron that is forged in the explosion itself. The 1987 supernova in the Large Magellanic Cloud gave that galaxy roughly 1.6 solar masses of oxygen but only 0.075 solar masses of iron. This works out to 75 oxygen atoms for every iron atom, or an oxygen-to-iron ratio that is 3 times that of the Sun.

High-mass stars live only a few million years, so almost as soon as the Galaxy began forming stars, these oxygen-ejecting supernovae started exploding. Other stars that formed in the early Galaxy therefore inherited high oxygen-to-iron ratios, and astronomers still see these stars today, in the Galactic halo.

The rapidity with which oxygen can enter the Galaxy was shown dramatically in 1992, when Katia Cunha and David Lambert, at the University of Texas at Austin, observed newborn stars in the constellation Orion. The stars differed in age by only 8 million years, but Cunha and Lambert discovered that some of the youngest stars have as much as 40 percent more oxygen than the oldest, indicating that supernovae have enriched the Orion star-forming clouds with large amounts of oxygen during just the last 8 million years.

In contrast, hundreds of millions or billions of years must elapse before the other variety of supernova, which creates iron, explodes. A type Ia supernova arises when a white dwarf exceeds the Chandrasekhar limit of 1.4 solar masses. The average white dwarf has only 0.55 to 0.60 solar masses and so is far below the Chandrasekhar limit.

But if a white dwarf circles another star, that star can transfer material to the white dwarf, gradually increasing its mass. Most white dwarfs consist of carbon and oxygen. When the white dwarf exceeds 1.4 solar masses, its carbon begins to fuse in a runaway nuclear reaction that will annihilate the white dwarf itself. The end product of this deadly nuclear runaway is nickel-56, which is radioactive with a half-life of 6 days, and decays into cobalt-56. The cobalt is also radioactive, with a half-life of 77 days, and decays into iron-56, a stable isotope that does not decay into any other. Iron-56 is the most common iron isotope in nature, and most of it comes from type Ia supernovae.

Once the type Ia supernova creates iron, that iron floats freely into the Galaxy, since no remnant of the exploding star survives to trap the metal with its gravity. A single type Ia supernova produces about 0.6 solar masses of iron, an impressive accomplishment for a star that is only 1.4 times more massive than the Sun. The explosion also sets free some of the oxygen that was in the white dwarf, but the oxygen adds up to only about 0.14 solar masses. This works out to an oxygen-to-iron ratio that is only a thirtieth of the Sun's. Once type Ia supernovae began to explode, they therefore lowered the Galaxy's oxygen-to-iron ratio—even as they were increasing the Milky Way's overall metallicity.

However, it takes a great deal of time for a type Ia supernova to explode, because a star becomes a white dwarf long after its birth, and then only slowly accumulates enough mass to exceed the Chandrasekhar limit. Hundreds of millions of years probably elapsed before the Galaxy's first type Ia supernova exploded, and by then the other supernovae had given oxygen a commanding head start over iron. The present value of the oxygen-to-iron ratio, which lies between that produced by the two main varieties of supernovae, is a mixture of both processes. This mixture manifests itself in the Sun, where nearly all of the oxygen and some of the iron came from type Ib, Ic, and II supernovae, and most of the iron came from type Ia supernovae.

If this model is right, then other elements spewed out by type Ib, Ic, and II supernovae should show the same behavior in halo stars that oxygen does. For example, high-mass stars also create large amounts of neon, magnesium, and silicon, which are respectively the third, fifth, and sixth most common metals in the universe. Neon is difficult to study, because it produces few spectral lines, but the magnesium-to-iron and silicon-to-iron ratios in metal-poor stars are indeed high, for the same reason the oxygen-to-iron ratio is: magnesium and silicon were ejected into the young Galaxy by short-lived high-mass stars that gave the Milky Way little iron.

In contrast, astronomers have discovered that two other important metals, carbon and nitrogen, track iron almost perfectly. Carbon is the basis of all terrestrial life, and nitrogen makes up most of Earth's atmosphere. They are, respectively, the second and fourth most abundant metals in the universe and lie just below oxygen on the periodic table. But whereas the oxygen-to-iron ratio in metal-poor stars is high, the carbon-to-iron and nitrogen-to-iron ratios are the same as in the Sun. This immediately reveals that most carbon and nitrogen entered the Galaxy slowly, as iron did, and so did not arise from the short-lived high-mass stars that produced oxygen.

Because the abundances of carbon and nitrogen track iron, it might seem that all three elements originated from the same place— type Ia supernovae—but this is wrong. Calculations indicate that type Ia supernovae eject little carbon and almost no nitrogen. According to the calculations, a type Ia supernova gives off only 0.03 solar masses of carbon and less than a millionth of a solar mass of nitrogen.

Instead, astronomers believe that most carbon and all nitrogen came not from supernovae but from more ordinary stars that did not explode—those born with less than eight solar masses. After these stars evolve into red giants, they fuse helium into carbon (and oxygen), and this carbon can float to the top of the star's atmosphere. If large amounts of carbon surface, so many lines from carbon compounds appear in the star's spectrum that astronomers call the giant a carbon star. When the star ejects its outer atmosphere and forms a planetary nebula, the carbon enters the Galaxy, and astronomers indeed observe that many planetary nebulae have high abundances of carbon. Most of the carbon atoms now in our bodies were once part of a planetary nebula and, before that, a red giant. Some carbon, however, did come from the high-mass stars that exploded as supernovae, since high-mass stars do produce carbon during their lives.

The same is not true for nitrogen: high-mass stars produce hardly any of it, so nitrogen arises almost entirely from stars that do not explode. Nitrogen is a byproduct of hydrogen burning. In a main-sequence star with more than 1.5 times the mass of the Sun, and in giant stars of all masses, hydrogen burning proceeds through what astronomers call the CNO cycle, in which carbon, nitrogen, and oxygen catalyze hydrogen into helium by participating in the nuclear reactions that convert the first element into the second. During the CNO cycle, the total number of carbon, nitrogen, and oxygen atoms remains the same, but the nuclear reactions gradually transform the carbon and oxygen into nitrogen. Thus, during both the main-sequence stage and the red giant stage, a star can create nitrogen. Like carbon, the nitrogen can float into a red giant's atmosphere and be cast into the Galaxy when the star creates a planetary nebula.

In recent years, Christopher Sneden and other astronomers have also begun the difficult task of investigating the rare trace elements that are much heavier than iron. "It's important to remember just what 'trace element' really means," said Sneden. "When we talk about a metal-poor star, we're already talking about a star that is very clean of anything but hydrogen and helium. I often get the feeling, when I'm looking at exotic elements like europium, which has almost no abundance compared with something like carbon, that I'm counting one or two atoms in the entire star. And we're using these trace elements to say magnificent things about what went on in the early Galaxy."

Most elements heavier than iron arise in the r- and s-processes. The r-process occurs when a rapid flux of neutrons bombard nuclei in a supernova. The more gentle s-process occurs when a slow flux of neutrons strike nuclei inside a red giant. Therefore, r-process elements should have entered the Galaxy before s-process elements, because supernovae exploded almost as soon as the Galaxy formed, whereas red giants evolved from longer-lived stars and so took time to eject material.

Difficult though they are, recent observations seem to confirm this expectation. To spectroscopists, the most prominent r-process element is europium (atomic number 63) and the most prominent s-process element is barium (atomic number 56), because these are easier to detect in stellar spectra than other r- and s-process elements. Most very metal-poor stars—those with metallicities less than 1 percent of the Sun's, such as HD 122563—have a barium-to-europium ratio that is only between a fifth and a tenth of the solar value. This suggests that in its earliest years, when the Galaxy was most metal-deficient, the r-process elements had already begun to enter the Milky Way, whereas few if any s-process elements had.

Yet the observations also hint that the europium-to-barium ratio is not exactly the same from one metal-poor star to another equally metal-poor star. This may reflect observational error, or it may indicate that just one or two supernovae, each producing a slightly different mix of heavy elements, altered the chemistry of the clouds from which metal-poor stars in different regions of the Galaxy formed.

SEARCHING FOR
METAL-DEFICIENT STARS

To investigate the earliest years of the Galaxy, astronomers must first discover the most metal-poor stars of all—a herculean task, for such stars are scarce. Because the Sun lies in the thin disk, most nearby stars are metal-rich, and in the solar vicinity the halo contributes only one star in a thousand. Moreover, most halo stars have metallicities around 2 percent of the Sun's, and astronomers probing the dawn of the Milky Way's life need stars that are much more metal-poor—with metallicities that are less than 1 percent of the Sun's, and preferably less than 0.1 percent.

In recent years, Timothy Beers of Michigan State University has waded through millions of stellar spectra and fished many such metal-poor stars out of the celestial sea. By casting a large net, Beers and his colleagues have caught hundreds of stars with metallicities that are less than 1 percent of the Sun's and dozens of stars with metallicities less than 0.1 percent. Their work, published in 1992, has yielded half of all the most metal-deficient stars now known and has added greatly to the understanding of the ancient Galaxy. Yet when Beers was a graduate student, he had little but contempt for the study of stars.

"I just assumed that most stellar work was not very interesting," said Beers, who earned his doctorate in 1983 for observing galaxy clusters. "My perception was that most of what we were going to learn about stars we already knew. It didn't seem like something you'd want to spend a career doing."

While a postdoc at Caltech, though, Beers looked into a survey started by George Preston and Stephen Shectman to ferret out the rare metal-poor stars. Preston and Shectman's photographs of the sky recorded crude spectra of some 30,000 stars. The crude spectra captured the ionized calcium lines at wavelengths of 3934 and 3968 angstroms, the two most prominent metal lines in the spectra of stars like the Sun. Preston and Shectman found that about one star in a thousand had weak calcium lines, which meant the stars probably had metallicities of less than 10 percent of the solar value. For these stars, Preston and Shectman obtained better spectra to estimate the actual metallicity.

At the time, other astronomers were claiming that extremely metal-deficient stars, those with less than 0.1 percent of the Sun's metallicity, did not exist, a "fact" that gave rise to fanciful theories concerning the Galaxy's early evolution. The most popular postulated that the extremely metal-poor gas that once existed could not have formed small stars that would survive to the present. Instead, according to this logic, all the first stars were massive and short-lived, which is why none exists today. When these massive stars died, they enriched the halo's gas, from which the halo stars that survive today supposedly formed.

As Beers examined the spectra that Preston and Shectman had taken, he soon discovered that these ideas were wrong. "We started finding all these very interesting stars," said Beers. "Not only was the technique working, but we were discovering far more metal-deficient stars than anyone thought were present." And those stars captivated him. "As I started to play with the spectra, I realized that there was an enormous amount yet to be learned about the halo of the Galaxy," he said, noting that many astronomers had studied the halo by observing the globular clusters, which number less than two hundred. "The thing that was unsatisfying to me about the globulars was that the sample was small and will always be small, so there are many questions that will never be answered. When I was sitting here with this computer disk full of a thousand or so spectra of stars, I started to realize that these stars were just as good tracers of the chemical history and the kinematics of the halo as the globular clusters. And we can find thousands more."

Nevertheless, the project required perseverance. "You really are playing a numbers game," said Beers, who estimates that he has scanned 3 or 4 million crude spectra to find the rare metal-poor stars. "The pristine gas that was once available in the Galaxy didn't stay pristine very long. It doesn't take more than one or two supernovae to pollute the gas. So if you want to find these metal-deficient stars that are telling you about the Galaxy's very early history, you have to look through an enormous number of stars that are not chemically interesting."

The work of Beers, Preston, and Shectman indicates that extremely metal-deficient stars are rare, but no rarer than simple models of the Galaxy's evolution would predict. According to these models, the distribution of metal-poor stars with different metallicities should be uniform. Suppose you started with stars whose metallicities are less than 1 percent of the Sun's, which are already rare. Only a tenth of those stars will have metallicities below 0.1 percent solar, and only a tenth of those stars will have metallicities of less than 0.01 percent solar.

As an analogy, suppose you start with a sample of Americans who were born during the 1980s. Such people constitute roughly a tenth of the population of the United States. Now suppose you want an American who was born during a specific year in the decade, for example, 1988. Only one in ten people in your starting sample would meet this criterion. Suppose further that you want only people born during January of 1988. Things get even tougher, and they get tougher still if you want only people born on January 1, 1988. To find even one such person, you'd need to sample tens of thousands of Americans.

Difficult though their search was, Beers, Preston, and Shectman succeeded in unearthing a horde of metal-deficient stars. To quantify the stars' metallicities, Beers's team used a technique that Australian astronomer John Norris had developed to estimate metallicities from the ionized calcium lines. The 1992 paper listed over 400 stars with metallicities less than 1 percent of the Sun's. Of those stars, 70 had metallicities less than or equal to 0.1 percent solar, and 3 had metallicities of less than 0.01 percent. To date, the most metal-poor champions have metallicities around 0.006 percent of the Sun's.

"Some of these stars are so metal-deficient," said Beers, "that there are as many atoms of ionized calcium in interstellar space between us and the stars as there are present on the star itself." So far he sees no lower limit to stellar metallicities, for the number of extremely metal-deficient stars he and his colleagues have found agrees with what he expected, given the total number of stars they have observed. But he said, "If we do hit a wall in metallicity below which, no matter how hard we try, we absolutely, positively cannot find any stars, then we

will know that something turned on at a certain metal abundance, either in the star formation or in the process of how metals are created."
Plenty of uncharted territory lies ahead, for Beers estimates that only 3
percent of all the metal-deficient stars brighter than fifteenth magnitude have been found.

Such ancient stars not only preserve clues about the early Galaxy
but also shed light on the origin of the universe itself. "Things have
really come full circle," said Beers. "Some of the best clues we have
about cosmology are coming from high-resolution spectra of the most
metal-deficient stars." Using stars for cosmology is a powerful method
for digging into the heart of the big bang. The oldest stars formed
before supernovae had enriched the Galaxy and given it much metallicity, so such stars consist of nearly pure material from the big bang
and harbor the three elements—hydrogen, helium, and a little lithium
—that it created. The abundances of these light elements probe the
exotic conditions that prevailed during the first few minutes of the
universe's life. In fact, when Beers, Preston, and Shectman began their
project, they hoped to uncover a star with almost no metallicity at all—
a pure chunk of the big bang. They have not yet succeeded, but that is
hardly surprising, since such stars must be extremely rare. But they
might well exist, somewhere in our Galaxy.

"Sometimes I lie awake at night and wonder if somebody already
has the spectrum of the star that we've been looking for," said Beers.
"They don't recognize it, because it's not what they're looking for, and
the star gets erased from the computer disk."

SOMETHING FROM (ALMOST) NOTHING:
THE NEUTRINO-PROCESS

Meanwhile, as the search continues, other astronomers are exploring
new ways to create the elements that the Galaxy has produced. In
recent years, Stan Woosley, a supernova expert at the University of
California at Santa Cruz, has suggested that some of the rarer elements
in the Milky Way, such as fluorine and boron, may arise from a supernova's neutrinos—neutral particles that have little or no mass and
travel at or near the speed of light.

Woosley has long been interested in chemistry, the origin of the elements, and explosions. "When I was young," he said, "I liked to make explosives out of things I could get at the drugstore that I can't get anymore. One day the shed blew up. My parents didn't like that very much."

Today, Woosley still makes things blow up, but in a computer rather than a shed. He and his colleagues model the evolution of high-mass stars and study the stars as they explode, to see which elements supernovae eject into the Galaxy. These calculations indicate, for example, that high-mass stars produce most of the Galaxy's supply of oxygen, neon, magnesium, and many other elements.

Woosley believes supernovae also drive what he calls the neutrino-process, a new nucleosynthetic sequence in which neutrinos convert one element into another. Neutrinos might seem unlikely protagonists for nucleosynthesis, since they normally pass through everything in their path. Each second, trillions of neutrinos from the Sun pass through your body, but they hardly ever interact with it.

However, when a high-mass star collapses and explodes, it shoots out 10^{58} neutrinos in just a few seconds. Such neutrinos were detected on Earth in 1987, when the supernova exploded in the Large Magellanic Cloud. If only a tiny fraction of these neutrinos disturb the debris the star ejects, they can transform some of the ejected elements into new ones. Because neutrinos interact so weakly, the neutrino-process can produce only small amounts of elements. Therefore, it cannot contribute much to elements that are already common, such as oxygen, but it may create a significant fraction of elements not readily produced elsewhere.

"In the past, people have focused most of their efforts on trying to make the most abundant isotopes, like oxygen and iron," said Woosley. "What we've come to find in the late 1980s and early 1990s is that the rarest isotopes can actually be diagnostics of some very interesting physics, because they take special circumstances to make. It's very easy to make iron, for example; but it turns out to be very hard to make fluorine or boron."

The first investigation of the neutrino-process came during the 1970s, when Russian scientists looked into it. But supernova models then were not realistic, and later studies questioned the entire concept. In 1988, however, Woosley and Wick Haxton, then at the University of

Washington in Seattle, used improved models and proposed that the neutrino-process could produce one of the more mysterious elements in nature, fluorine.

Fluorine at first seems familiar, since it is present in everyday products such as toothpaste and fluoridated water. But compared with elements of similar mass, fluorine is rare. If one represented each row of the periodic table by building houses whose sizes reflected the elements' abundances in the universe, fluorine would be a shack squeezed between two mansions. Fluorine has atomic number nine and lies between oxygen (atomic number eight) and neon (atomic number ten), the first and third most abundant metals in the cosmos. Yet fluorine doesn't even make the list of the top twenty metals.

Because it is rare and lies just below an element—neon—that is prevalent in the same high-mass stars that explode and create neutrinos, fluorine is a perfect candidate for the neutrino-process. According to Woosley, fluorine originates when a supernova's neutrinos storm through the exploding star's neon shell. Most neon is neon-20, and a neutrino can strip off a proton to produce fluorine-19, the only stable isotope of fluorine. Alternatively, the neutrino can rip a neutron off neon-20 to create neon-19, which is radioactive and decays into fluorine-19.

Another rare element, boron (atomic number five), may arise in a similar way, because it lies just below the common element carbon (atomic number six). Most carbon is carbon-12, and a neutrino can remove a proton or neutron to make either boron-11 or carbon-11. However, just seconds after the neutrinos penetrate the carbon shell, the supernova shock wave strikes and destroys much of the newly formed boron, because the element is fragile. But carbon-11 is tougher and survives the shock. Since carbon-11 is radioactive with a half-life of twenty minutes, it quickly decays into boron-11 after the shock passes.

This method of boron production breaks with conventional theory. Astronomers have long believed that boron originates in space, where high-energy particles called cosmic rays smash into heavier atoms— especially carbon, nitrogen, and oxygen—and split them into lighter atoms, such as boron. This theory, though, has a problem that the neutrino-process may solve. Boron has two isotopes, boron-11 and boron-10, and the former is four times more common than the latter. But the cosmic-ray theory for boron production predicts that the ratio of boron-11 to boron-10 should be only 2.5. Since the neutrino-process makes boron-11 but little boron-10, it could augment this ratio to the observed value.

In 1992, however, the detection of fluorine in several stars raised questions about whether neutrinos had actually created this element. In that year, Alain Jorissen, Verne Smith, and David Lambert, of the University of Texas at Austin, reported that they had discovered fluorine in orange and red giant stars, a finding that meant the giants themselves were manufacturing the fluorine through nuclear reactions that did not involve neutrinos. Nevertheless, Stan Woosley believes that the neutrino-process produces most of the fluorine in nature. He notes that the total amount of fluorine produced by giant stars depends on the number of stars that reach the right stage of evolution, a number he says is uncertain.

Whatever the case, a simple test might ultimately confirm or refute the neutrino-process. If fluorine and boron are indeed created when high-mass stars explode and expel huge numbers of neutrinos, then the abundance of fluorine and boron should track that of other elements that high-mass stars produce, such as oxygen. Just as the oxygen-to-iron ratio in metal-poor stars exceeds that in the Sun, the fluorine-to-iron and boron-to-iron ratios should, too. If instead fluorine and boron track nitrogen, which red giants produce, then fluorine and boron probably originate in those stars rather than from the supernova-induced neutrino-process. Performing this test, however, will be difficult, for it requires detecting fluorine and boron in stars of different metallicities. Both elements are rare, and neither produces prominent spectral lines; so for now, the issue remains unresolved.

Having glimpsed the clues that different elements offer in the stars near the Sun, astronomers can apply their knowledge to other regions of the Milky Way and even to galaxies beyond it. The next two chapters leave the solar neighborhood and journey to other parts of the Galaxy, first exploring the complex environment surrounding the Galaxy's center and then venturing to the most remote outposts of the Milky Way.

14

THE GALACTIC
METROPOLIS

THE MILKY WAY Galaxy is mighty and immense, and both its center and edge lie far, far from Earth. From end to end the Milky Way's disk, which radiates most of the Galaxy's light, stretches 130,000 light-years. A photon that set out during the peak of the last ice age and raced at light speed across the disk would by now have covered only a seventh of its extent. The Sun and Earth lie within this disk, 27,000 light-years from the Galaxy's center, or about 40 percent of the way from the center to the edge. If the Milky Way were a huge metropolis, the Sun and Earth would reside in an inner suburb.

Yet, from what we see in the night sky, we glimpse only a fraction of the Milky Way's magnificence. The most distant first-magnitude star in the sky is the white supergiant Deneb, 1500 light-years away. Carve a large sphere in the Galaxy by scooping out all stars closer than Deneb and gather them like colorful jewels extracted from a celestial mine. In your arms would be 50 million stars, some shining red, others yellow, and still others white, orange, or blue. Nearly every star you can see in the night sky lies within this huge sphere: Antares, Betelgeuse, Sirius, Alpha Centauri, Polaris, Vega, Rigel, Arcturus, and all the other stars that set the sky afire. Sprinkled among these gems are millions of red, orange, and white dwarfs that glow so feebly they do not appear in even the largest star catalogue.

But this enormous sphere, which paints almost the entire night sky, is just a tiny brush stroke on a vast mural, for it represents less than a thousandth of the volume of the Galactic disk. To fully understand the Milky Way, astronomers must venture away from the solar neighborhood and explore other parts of the Galaxy's realm. Especially important are the Galactic bulge, the dense swarm of old stars that surrounds the Galactic center; and the Galactic center itself, the point about which all else in the Milky Way revolves, a point that may harbor a ravenous black hole with a million times more mass than the Sun. Together, the Galactic bulge and center constitute the heart of the Galactic metropolis.

PROBING THE INTERSTELLAR MEDIUM

To investigate the Galactic bulge and center, however, astronomers must gaze through the Galactic disk, because both the Sun and the Galactic center lie in the plane of the disk. But peering through the disk is difficult—not only because the disk is large, but also because it is choked with interstellar gas and dust that obscure the stars. Interstellar gas does produce radio waves, though, and these penetrate the gas and dust so that astronomers on Earth can detect and study the radiation. As a consequence, most of what astronomers know about the disk's large-scale structure actually comes from their observing material in the interstellar medium—the space *between* the stars—which thwarts the passage of ordinary starlight.

These observations reveal that interstellar gas and dust account for 5 to 10 percent of the disk's mass. This material comes from the Milky Way's supernovae, novae, planetary nebulae, and stellar winds, as well as from debris that falls into the Galaxy from outside. On average, the interstellar medium in the disk contains only one atom per cubic centimeter. By contrast, terrestrial air has 25 quintillion—that is, 25 billion billion—molecules per cubic centimeter, so the density of the interstellar medium is so low that it would pass for a perfect vacuum on Earth. In fact, by terrestrial standards, even the densest parts of the interstellar medium are incredibly tenuous. But the Milky Way is so large that all these interstellar atoms add up to several billion solar masses.

Every element appears in the interstellar medium, but the most common is hydrogen. Most interstellar hydrogen is made of individual hydrogen atoms, which broadcast radio waves that are 21 centimeters long. Astronomers first detected this radiation in 1951. Most of the remaining interstellar hydrogen consists of hydrogen molecules (H_2), in which two hydrogen atoms join. Because it is fragile, molecular hydrogen exists only where gas and dust are dense enough to shield the molecules from ultraviolet radiation, which tears them apart.

The two main types of interstellar hydrogen—atomic and molecular—have different distributions. Atomic hydrogen is more common in the outer Galaxy, while molecular hydrogen is more common in the inner. About 70 percent of the Milky Way's atomic hydrogen lies farther from the Galactic center than the Sun does, and atomic hydrogen even extends beyond the stars of the Galactic disk. In contrast, 90 percent of the molecular hydrogen lies nearer the Galactic center than the Sun, with little in the outer Galaxy.

Unfortunately, molecular hydrogen is hard to study, because unlike atomic hydrogen it does not emit radio waves. But carbon monoxide (CO), another molecule that co-exists with molecular hydrogen, broadcasts readily at a wavelength of 2.6 millimeters. This radiation was first picked up in 1970. Although carbon monoxide is much less abundant than molecular hydrogen, the former traces the latter, so astronomers can study molecular hydrogen by studying the distribution of carbon monoxide.

Through studies of carbon monoxide, astronomers have found that most molecular gas resides in huge complexes called giant molecular clouds, with lesser amounts of gas in small molecular clouds. A typical giant molecular cloud has 200,000 solar masses and is 150 light-years across. Although still tenuous by terrestrial standards, molecular gas is sufficiently dense that clumps of it can collapse and form new stars. Indeed, a single giant molecular cloud can spawn hundreds or thousands of stars. The best-known stellar nursery, the Orion Nebula, is a small part of the nearest giant molecular cloud, 1500 light-years away. In the Orion Nebula, a few hot O and B stars have torn the electrons off hydrogen atoms, creating a region of ionized hydrogen—an H II region—that is visible to the naked eye. Other molecular clouds, smaller than the giant variety, lie closer to Earth and can also create stars. All these molecular clouds make the Milky Way a vigorous, star-forming galaxy, one that each year gives birth to about ten new stars.

Interstellar gas and dust gather in the spiral arms, thereby allowing astronomers to use both atomic hydrogen and giant molecular clouds to map the Galaxy's spiral structure. When William Morgan's team made optical observations of H II regions in the early 1950s, the astronomers found the Sagittarius spiral arm, which lies closer to the Galactic center than the Sun; the Orion arm, which includes the Sun; and the Perseus arm, which lies beyond the Sun. Since then, studies of interstellar gas and dust have confirmed the existence of the Sagittarius and Perseus arms, but the Orion arm is more controversial. Some astronomers suspect it is only an outgrowth of another arm and call it the Orion spur instead. The dispute illustrates the irony that ours is the one spiral galaxy in the universe that astronomers will never really be able to see. Nevertheless, from present observations, most astronomers believe that the Milky Way is not a breathtaking spiral like the galaxy M51 but a more ragged and less beautiful spiral.

In 1991, however, French astronomer Françoise Combes emphasized how uncertain the distances of giant molecular clouds are and showed how this uncertainty blurs the Galaxy's spiral pattern. To prove her point, Combes imagined that she lived in M51 and was using its giant molecular clouds to map the galaxy's spiral arms. She assumed distances as uncertain as those in the Milky Way and plotted the clouds' positions. She did not get the beautiful spiral galaxy astronomers know as M51; instead, she got a mess. The result suggests that the Milky Way could be a galaxy as stunning as M51 and astronomers on Earth would not know it—just as M51's residents might not know the beauty of their home galaxy.

Although the Milky Way is a spiral galaxy, the cause of the spiral remains a mystery. Spiral arms stretch from the inner Galactic disk into the outer, but the outer disk takes longer to revolve than the inner. Therefore, the spiral arms should be destroyed by the differing rates of revolution, in the same way that stirring a cup of coffee disperses the cream. The most popular resolution to this problem was proposed by Bertil Lindblad in Sweden and developed during the 1960s by American astronomers Chia Lin and Frank Shu. According to this theory, the spiral arms represent regions of somewhat enhanced density called density waves, analogous to traffic jams on a highway. Within a density wave, the force of gravity is a bit stronger than elsewhere. The density waves have a spiral pattern, and they revolve around the Galaxy. But they revolve more slowly than the stars and gas do, so as the stars orbit the Galaxy, they pass in and out of the density waves. Gas and dust do likewise, but the density waves affect them more, squeezing them so that they become denser and form giant molecular clouds, which then give birth to stars. These newborn stars include a few blue stars that have high masses and do not live long enough to leave the spiral arm that created them. This explains why blue supergiants line the spiral arms but do not exist elsewhere.

Although they are part of a great pattern, these massive stars also create irregularities. When a high-mass star explodes, it clears out the interstellar material near it. The Sun lies in such a clearing, a "local bubble" that extends for a hundred light-years in most directions and up to a thousand in some. Supernovae can also blast gas into the Galactic halo. In fact, a few young stars exist in the halo and may have arisen when interstellar clouds collided there to form stars.

APPROACHING THE GALACTIC BULGE

Interesting though it may be, interstellar material is little more than muck that astronomers must look through when they explore regions near the Galactic center. Every July, in the evening sky, the central region of the Milky Way comes into view, even though it is shrouded behind the gas and dust in the constellation Sagittarius, where the band of light called "the Milky Way" is widest and brightest. Sagittarius features eight stars arranged in the shape of a beautiful celestial teapot. West of them, in the neighboring constellation Scorpius, beckons the bright red supergiant Antares.

Yet all these stars are nearby, lying on the inner edge of the Orion spiral arm. Antares is just five hundred light-years away, and it marks the fringe of the Orion arm the way a lighthouse warns navigators of a dangerous shore. Beyond Antares lies darkness, the darkness of the interarm region between the Orion and Sagittarius arms. Plenty of stars shine there, but none is as luminous as the blue and red supergiants that light the spiral arms. Three thousand light-years onward is the edge of the mighty Sagittarius arm. There lie young stars, star clusters, and nebulae, such as the beautiful Lagoon and Trifid nebulae.

If one were to journey still closer to the Galactic center, one would encounter two changes that have nothing to do with the spiral arms. First, the stellar density—the number of stars per cubic light-year—would increase, doubling every 10,000 light-years, because stars are more common in the inner disk than in the outer. Second, the mean metallicity would also increase, because the thin disk, unlike the halo, has a metallicity gradient. Its inner parts formed more stars, some of which have enriched the interstellar medium with metals.

Well past the Sagittarius arm lies the ultimate goal of a journey to the Galactic hub—the Galactic center itself, 27,000 light-years from Earth. But the Galactic center's exact distance is unknown, and astronomers use various methods to measure it. One of the most accurate relies on RR Lyrae stars in the Galactic bulge. Their apparent brightness, coupled with their known intrinsic brightness, reveals the Galactic center's distance. Thirty years ago, astronomers believed that the distance was somewhat greater than 27,000 light-years; ten years ago, they favored a somewhat smaller distance. Today, the "official" distance is 28,000 light-years, although many astronomers currently favor a value of 26,000 light-years. Splitting the difference and assuming the Galactic center to be 27,000 light-years away, one can calculate that a star like the Sun at the Galactic center would have an apparent magnitude of +19, if no gas or dust lay in front of it. This is so faint that a large telescope would be required to see it. In reality, though, gas and dust block the view so much that a Sunlike star at the Galaxy's center would appear a trillion times fainter, with an apparent magnitude of +49—far beyond the reach of the largest telescopes on Earth. Even a star like Deneb, one of the most luminous in the Galaxy, would have an apparent magnitude of +37, still too faint for the largest telescopes to see.

Surrounding the Milky Way's center is the Galactic bulge, a swarm of 10 billion old stars that extends some 3500 light-years in all directions around the center. In a picture of an edge-on spiral galaxy, the bulge is the blob at the galaxy's center that extends above and below the narrow disk.

In the Milky Way, though, the bulge has so far yielded few of its secrets. "The bulge is one of the least studied stellar populations in the Galaxy," said Columbia University's Michael Rich. "Observing it is hard work." More than gas and dust hinder would-be explorers of the bulge; its location does, too. The bulge lies in the southern sky, whereas most astronomers live in the northern hemisphere. And because of the bulge's great distance, even its brightest stars are faint. Moreover, its high stellar density and its distance make individual stars appear jammed together, impeding efforts to measure colors, brightnesses, and spectra of separate stars. Indeed, as viewed from Earth, most bulge stars nestle inside a circle whose radius is only eight degrees on the sky, a circle little larger than the Sagittarius teapot. Confronted with these obstacles, most astronomers, said Rich, have directed their interests—and telescopes—elsewhere.

One of the first astronomers to surmount the challenge and peer into the Galactic bulge was the man who had discovered the stellar populations—Walter Baade. In 1945, two years after he spotted stars in Andromeda's bulge, Baade searched for stars in the bulge of the Milky Way. He probed a break in the interstellar gas and dust near the spout of the Sagittarius teapot. This region, now called Baade's window, lies four degrees south of the Galactic center, so Baade's line of sight passed within 1800 light-years of the Galaxy's heart. Baade chose the field in part because it contained a globular cluster, NGC 6522, whose stars helped him quantify how much light the intervening dust absorbed.

From 1945 to 1949, Baade scrutinized this clearing and discovered dozens of RR Lyrae stars. These stars yielded an accurate distance to the Galactic center and were also part of the stellar population in the bulge. Since nearby RR Lyrae stars were members of population II— stars with high velocities that took them far above and below the Galactic plane—Baade concluded that the Milky Way's bulge was also a population II system, like the Galactic halo, and that the bulge might be the innermost part of the halo, because the eccentric orbits of halo stars would ferry them into the bulge at perigalacticon.

Although Baade had great insight, his conclusions were only partially correct, and complications soon developed. As astronomers discovered during the early 1950s, nearby population II stars are old and metal-poor; therefore, if the bulge were also population II, it should have been old and metal-poor. But in 1956 William Morgan obtained spectra of Andromeda's bulge and surprised his colleagues by finding that it was metal-rich. Soon after, he found that the Milky Way's bulge was also metal-rich. And in 1958, Jason Nassau and Victor Blanco discovered over two hundred red M-type giants in the bulge, stars that are rare in globular clusters but common in metal-rich populations. The Galactic bulge thus seemed to be what could not be, according to then prevailing views of Galactic evolution: old but metal-rich.

In the years since, several astronomers have attempted to investigate it further. Michael Rich began his study of the bulge in 1980, the year after he finished college. He worked with Lick Observatory's Albert Whitford, who, during the 1940s, had helped pioneer the study of the bulge by carrying out observations at infrared wavelengths that penetrate obscuring gas and dust. Now Whitford and Rich looked at 21 orange K-type giants in Baade's window. Using an optical telescope in Chile equipped with modern instruments, Whitford and Rich obtained the stars' spectra, and Rich liked what he saw. "They were very interesting stars," said Rich. "I'd never seen anything so strong-lined and so unusual." In 1983, Whitford and Rich reported that the metallicities of the stars spanned a broad range. Some of the bulge stars had metallicities below a tenth of the solar value, but other stars were five to ten times more metal-rich than the Sun, a finding that was unprecedented for stars in the solar neighborhood. So intriguing were these stars that Rich continued to study them after he entered graduate school at Caltech. In 1988, he reported metallicities for 88 bulge stars and confirmed the earlier result. The bulge has a wide range of metallicities, from about a tenth of the solar value to ten times the solar value.

"That's the key thing to understanding the bulge population," said Rich. "It isn't one globular cluster with one abundance. Compared with an individual cluster, it's very complex. And even compared with the Galactic halo, it's complex." The different metallicities give rise to different types of stars. Metal-poor stars can become RR Lyraes, such as those Baade had seen, whereas metal-rich stars evolve into M-type red giants that are rare in globular clusters, which are metal-poor. Yet all the stars in the bulge are old; it has no young blue or white main-sequence stars.

At first sight, an old, metal-rich population seems to contradict ideas of Galactic evolution. After all, in the solar neighborhood, old stars are metal-poor, having formed when the Galaxy, or at least the solar neighborhood, had low metallicity. Yet the bulge is both old and metal-rich, so some other factor must have been operating.

The other factor, explained Rich, was the bulge's high density, which triggered intense star formation and rapid Galactic enrichment. "In the high-density environments at the centers of galaxies," he said, "it's possible to have many generations of supernovae very quickly, and that builds up the metals fast. A hundred generations can occur in 100 million years, and that's a very short time. So you can have a very, very metal-rich population very rapidly." Furthermore, because the bulge lies near the center of the Galaxy, metals ejected by supernovae got trapped by the Galaxy's strong gravity and became incorporated into subsequent generations of bulge stars.

Although astronomers once thought the bulge was simply the inner part of the halo, most astronomers today, including Rich, believe the bulge is a distinct population. The bulge rotates, whereas halo stars, as a group, do not. Also, the bulge is much more compact, and its metallicity does not resemble that of the metal-poor halo.

All these characteristics hold clues to the bulge's formation. Some astronomers believe that the bulge was the first part of the Galaxy to take shape, serving as the kernel whose gravity attracted additional gas clouds, which formed the rest of the Galaxy. If so, the bulge should contain the oldest stars in the Galaxy, as old as, and perhaps older than, the globular clusters. Other astronomers, including Rich, suspect that the bulge formed billions of years after the halo, from gas shed by the halo. Rich bases this conclusion on stars in the bulge that, while old, appear to be several billion years younger than the globulars. These stars are Miras—pulsating red giants—with long pulsation periods, some exceeding 700 days. Long periods imply that the stars are big and bright, which in turn implies that the stars are somewhat more massive than stars in globular clusters. Since massive stars have shorter lifetimes, these stars must be somewhat younger than stars in globulars, where Miras have shorter periods and lower masses.

The bulge may also help determine the exact type of galaxy the Milky Way is. Spiral galaxies come in two main varieties, normal spirals and barred spirals. A normal spiral has a round bulge that resembles an orange; a barred spiral's bulge is oval and resembles a watermelon. Recent observations suggest that the Galactic bulge is not round. If it is sufficiently nonspherical, the Galaxy would be a barred spiral, a possibility first suggested in 1963 by Gérard de Vaucouleurs, an astronomer at the University of Texas at Austin.

THE GALACTIC CENTER

For many astronomers, though, the real action lies not in the Galactic bulge but in what its stars circle, the heart of the metropolis. Like a city's downtown, the Galactic center is a crowded collection of a huge number and variety of objects. Within just one light-year of the Galactic center are millions of stars of all colors, caught in the stranglehold of what may be a massive black hole.

It is just this complexity that attracts Farhad Yusef-Zadeh, an Iranian-born radio astronomer at Northwestern University. "I like the Galactic center because it has a lot of phenomena in such a small area," said Yusef-Zadeh. "For me, it's always leading to something different. I started with a problem there, and then I got into something else and then moved on to the next and the next. It wasn't planned to be that way, but the more you get into it, the more you find that there are so many different things you can do with the Galactic center."

The Galactic center offers unique insight into the nuclei of other galaxies. In the Milky Way, said Yusef-Zadeh, "You can study structures of one light-year, even a tenth of a light-year. You can't do that in other large galaxies." That's because even the nearest such galaxy, Andromeda, is ninety times farther away than the Milky Way's center. "You can get a much more detailed picture of a massive black hole at the Galactic center than at any other place in the universe," he said. Massive black holes probably lie at the centers of most giant galaxies, and they also power the great luminosities of the quasars, heating infalling material so that it radiates profusely.

Nearly everything that astronomers know about the Galaxy's inner sanctum has come from radio observations, such as those by Yusef-Zadeh, and from infrared observations. Both radio and infrared radiation penetrate the gas and dust cloaking the Galactic center. Visible light, by contrast, does not: if a trillion photons of yellow light set out from the Galactic center and raced toward us, only one would make it to the Earth.

Historically, radio astronomers reached the Galactic center long before their infrared colleagues. In 1932, when Karl Jansky discovered extraterrestrial radio waves, his strongest source lay near Sagittarius, home of the Galaxy's center. These radio waves arise from electrons in interstellar material and therefore let astronomers study the gas at the Galactic center. Three decades later, in 1966, two astronomers at Caltech, Eric Becklin and Gerry Neugebauer, first detected the Galactic center's infrared glow, a form of radiation that comes from stars. Infrared wavelengths are longer than those of visible light, and cool stars—such as red M-type giants—actually emit more infrared radiation than visible light, just as they emit more red light than blue. At the Galactic center, hotter stars also appear brightest in the infrared, because their visible light is absorbed by the intervening dust that infrared radiation penetrates. Radio and infrared observations therefore complement each other, the former probing the gas and the latter the stars. However, radio views are sharper and clearer, because radio astronomers can combine separate antennae into huge arrays that mimic a single giant dish and reveal fine details in celestial objects.

Since 1990, radio and infrared astronomers have painted a picture of the Galactic center so detailed that it is easy to lose the forest for the trees. Within one light-year of the Galaxy's center, however, three "trees" stand taller than the rest: a source of radio waves named Sagittarius A* (pronounced "ay star") that marks the very heart of the Galaxy and may be a massive black hole; a cluster of blue stars near Sagittarius A* named IRS 16 that hurls material into the black hole; and a red supergiant named IRS 7 that gets blasted by IRS 16's ejecta and has a tail like that of a comet.

SAGITTARIUS A*: THE GALAXY'S BIGGEST BLACK HOLE?

Of the three objects, Sagittarius A* has provoked the most excitement —and the most controversy. "Sagittarius A* is definitely a unique object in the Galaxy," said Yusef-Zadeh. The point in the Galaxy about which all else orbits, Sagittarius A* emits five times more energy at radio wavelengths than the Sun does at visible ones, making Sagittarius A* the strongest radio source in the vicinity of the Galactic center.

Sagittarius A* is probably a black hole, and its radio waves are the scream of a hot disk of gas that orbits the black hole before plunging in. This accretion disk is hot both because its material has converted gravitational energy into heat and because the material rubs against other orbiting material. The infalling material feeds the black hole and increases its mass. The black hole probably formed long ago, when one of the stars at the Galactic center collapsed, and it then had only a few solar masses, but it grew more and more massive as it swallowed nearby stars and the debris shed by more distant ones. Today, Sagittarius A* may harbor as much as a million solar masses. That a massive black hole lies at the Galactic center was first suggested by British astronomers Donald Lynden-Bell and Martin Rees in 1971.

But it is not yet certain that Sagittarius A* is a black hole. "You want the proof," said Yusef-Zadeh, "and unfortunately we don't have the proof. It's difficult to prove a massive black hole, period, anywhere. But at the Galactic center, there is definitely a lot more hope than elsewhere."

The trouble is that black holes, by their very nature, are tough to study. Even the existence of an ordinary black hole, which forms from the collapse of one star, is difficult to prove. Although the Galaxy probably contains roughly 100 million black holes, astronomers have found only four good candidates in the Milky Way, including Cygnus X-1 and A0620-00. In these systems, a normal star circles an invisible one so quickly that the invisible star's mass must be that of a black hole.

Although black holes have more mass than the Sun, they are not especially large. Sagittarius A*'s radio waves emerge from a region that is less than the diameter of the Earth's orbit, which is just a few light-*minutes* across. Furthermore, the radio source is elongated, like a black hole's accretion disk seen edge-on. In 1991, a team of German astronomers discovered an infrared source at the position of Sagittarius A*. If this infrared radiation comes from Sagittarius A*, it further supports the existence of an accretion disk, since the radiation is blue before it gets reddened by dust. The blue color signals something hot, like a black hole's accretion disk.

If Sagittarius A* is a black hole, however, it far outweighs an ordinary one like Cygnus X-1. Astronomers know this because Sagittarius A*'s proper motion shows the object to be at rest with respect to everything else in the Galaxy. To remain so still, Sagittarius A* must be fairly massive. Most astronomers, including Yusef-Zadeh, believe Sagittarius A* contains about a million solar masses. The best evidence for such a high mass is that stars and gas near Sagittarius A* whirl around it extremely fast. Their high speed probably arises because they feel the strong gravitational pull of a million-solar-mass black hole. However, those astronomers who do not believe in the existence of such a massive black hole claim that most of this mass resides in a dense cluster of stars near Sagittarius A* and that Sagittarius A* itself has only a few hundred solar masses. Some astronomers have even argued that stars alone can explain everything and that no black hole of any mass exists.

To settle the dispute, in 1992 Mark Wardle at Northwestern University and Yusef-Zadeh proposed a test that invokes Einstein's general theory of relativity to measure the mass of Sagittarius A*. Einstein's theory says that an object's gravity can bend light coming toward an observer from a source behind the object. Normally, light diverges, spreading out in all directions from its source, but the gravity of an intervening object can bend diverging light rays together so that, as viewed from Earth, two or more images of the light source appear in the sky, each the result of a separate light ray. This so-called gravitational lensing was first seen in 1979, when astronomers spotted two images of the same quasar that sit side by side in the constellation Ursa Major. There are two images because a large galaxy lies between Earth and the quasar, and it lenses the quasar's light.

In the same way, said Wardle and Yusef-Zadeh, Sagittarius A* should gravitationally lense the light coming from stars behind it. "We have a very interesting situation right at the Galactic center," said Yusef-Zadeh. "The density of stars is very large, and they're all surrounding what may be a million-solar-mass black hole. These stars have very high velocities, so the chance of seeing one of these stars go behind this massive black hole is very large. Gravitational lensing theory then tells us that this radiation passing through the space-time curvature of the black hole will get amplified." If such lensing is de-

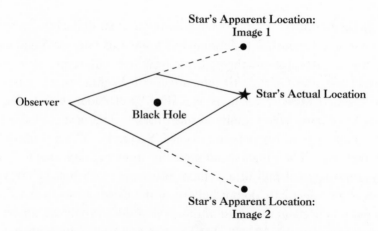

Figure 18. A large mass, such as a black hole, bends the light of a more distant object so that an observer can see more than one image of the light source. This is called gravitational lensing.

tected, it would reveal whether Sagittarius A* has a million solar masses, as most astronomers suspect, or only a few hundred. Yusef-Zadeh believes that the infrared is the best region of the spectrum in which to perform the test, but it requires better resolution than infrared telescopes currently achieve. However, Yusef-Zadeh thinks that infrared astronomers will attain the necessary resolution in the next five to ten years.

In fact, in just the last five years, infrared astronomers have stormed into the Galactic center with sensitive arrays of infrared detectors that provide sharp views of its stars. In 1992, astronomers in Germany led by Andreas Eckart and Reinhard Genzel swept through the Galactic nucleus and detected 340 individual stars within 1.3 light-years of Sagittarius A*. These stars are presumably the brightest members of the dense cluster that inhabits the Galaxy's central light-year.

"That was an excellent piece of work," said Yusef-Zadeh. "For a long time, from radio work, we knew a lot about the gas. But this infrared-array technology has been able to tell us something about the stars very near the center. It's definitely the way to go."

IRS 16 AND IRS 7

Lying just over a tenth of a light-year from Sagittarius A* is a second major actor in the Galactic hub, a bright object named IRS 16 ("IRS" for "infrared source"). During the 1970s, when infrared astronomers first catalogued it, IRS 16 appeared to be just one object. More recently, however, the high-resolution infrared pictures have revealed it to be a cluster of at least 25 stars. Although their light looks red because of the intervening dust, most of IRS 16's stars are actually blue, which suggests that they are blue supergiants similar to Rigel. A blue supergiant loses mass in a stream of particles called a wind, and the stars of IRS 16 blow a wind that ejects several Earth masses a day. The wind from IRS 16 may fuel Sagittarius A*'s accretion disk, replenishing it with material to compensate for that which the putative black hole swallows.

This accretion of material distinguishes the Milky Way from Andromeda, whose center may harbor a black hole over ten times as massive as Sagittarius A*. Still, despite the greater mass, Andromeda's black hole packs less than a tenth the radio punch of Sagittarius A*. "I think it all depends on how gaseous a galaxy is," said Yusef-Zadeh, since more gas means more accretion by the black hole and a more luminous accretion disk. "Andromeda seems to be very pristine—very clear of gas—compared with the nucleus of our Galaxy." Because of this, Sagittarius A* outshines its counterpart in Andromeda, and the wind from IRS 16 supplies Sagittarius A* with much of the material the black hole snares.

But the best indicator of IRS 16's wind comes from the Galactic center's brightest infrared source, IRS 7. One of the most remarkable stars at the Galactic hub, IRS 7 is a red supergiant similar to Antares that lies less than one light-year from the Galactic center. IRS 7 is therefore caught in the gravitational lock of Sagittarius A* and gets blasted by the wind from IRS 16.

Like all red supergiants, IRS 7 loses mass in a wind, but IRS 7's own wind is weak, with a velocity of only 25 kilometers per second. By contrast, IRS 16's wind travels at 700 kilometers per second. This powerful wind smacks into IRS 7's and pushes it away, producing a long tail on IRS 7 that points away from IRS 16. The same phenomenon occurs in the solar system, where the solar wind buffets material ejected by a comet, creating a tail that points away from the Sun. IRS 7's cometary tail was discovered in 1990, from radio observations conducted by Yusef-Zadeh and Mark Morris, of the University of California at Los Angeles, and independently from infrared observations made by a team led by Caltech's Gene Serabyn.

Together, Sagittarius A*, IRS 16, and IRS 7 form a triad that resounds throughout the Galactic center. As a likely black hole, Sagittarius A* is the gravitational anchor about which IRS 16 and IRS 7 pivot, while IRS 16 blows a strong wind that fuels Sagittarius A*'s accretion disk and gives IRS 7 a cometary tail.

IRS 16 and IRS 7 reveal something else, too. Since the former contains blue supergiants and the latter is a red supergiant, they must have large masses, which suggests that the Galactic center has spawned new stars in just the last few million years. But the gas now near the Galactic center is so disturbed that it probably cannot condense into new stars. Perhaps the gas was denser a few million years ago, and the newborn stars blasted it away. Or perhaps the high-mass stars formed in another way. In 1993, Andreas Eckart's team suggested just such a possibility. Because the stellar density near the Galactic center is so high, with stars typically separated by only a hundredth of a light-year, the stars frequently smash into each other, creating massive stars that ultimately evolve into blue and red supergiants.

Outside this region lie three arms of ionized gas that look like spiral arms but stretch for only a few light-years. Encircling the arms is a rotating ring of neutral gas—the so-called circumnuclear ring—that extends out to about 30 light-years. And surrounding the ring is a series of parallel filaments, each about 130 light-years long but only 3 light-years wide, which jut up perpendicular to the Galactic plane and twist around one another. These filaments probably trace a strong magnetic field that engulfs the Galactic center the way radiation belts surround the Earth.

Beyond the bustling hub of the Galaxy is the rest of the Milky Way: the Galactic bulge, the Sagittarius arm, and the Sun, over 27,000 times farther from the Galactic center than the luckless star IRS 7.

Beyond the Sun lies the Perseus arm, and beyond that the edge of the Galactic disk, 65,000 light-years from the Galaxy's center.

But the Galaxy does not end there. Past the edge of the Galactic disk lies a vast domain that surrounds the disk, emits little light, but inexplicably harbors most of the Milky Way's mass. So great is the Galaxy that some of its stars orbit it more than 100,000 light-years from its center, in the outermost parts of the Galactic halo. At still greater distances are at least ten other galaxies, colonies in a mighty empire that are caught in and ruled by the Milky Way's gravitational grip. This remote wilderness, spanning over a million light-years, is the Galactic frontier.

1 5

FRONTIERS

IN THE CONSTELLATION Leo, over 700,000 light-years from Earth, two dwarf galaxies patrol the Milky Way's dark outer boundary. These remote galaxies, Leo I and Leo II, harbor millions of diverse stars, yet both galaxies obey the Milky Way's command. Like two sentries guarding the portals of intergalactic space, Leo I and Leo II constitute the Milky Way's farthest contingent, the most distant outposts in the Galactic empire.

Spanning over a million light-years and extending far beyond the fringes of the Galactic disk, the Milky Way's empire is vast and its holdings many. The most prominent colonies in the Galactic empire are ten satellite galaxies, including Leo I and Leo II, which revolve around the Milky Way just as moons circle a giant planet. A few globular clusters reside at the Galaxy's frontier, as do individual stars—mere flotsam on a deep dark sea.

And the Galactic frontier *is* dark, the few stars shining within it exceptions that prove the rule. The space beyond the Galactic disk cradles something no one can see, mysterious "dark matter" whose gravitational force rules the Milky Way's colonies. The satellite galaxies feel the force of this dark matter, and their motions help prove that this substance outweighs the rest of the Milky Way.

Dark matter forms a halo around the entire Galaxy. To distinguish this halo of dark matter from the familiar halo of old stars closer in, astronomers call one the stellar halo and the other the dark halo. Because it makes up the majority of the Milky Way's mass, the dark halo implies that most other large galaxies also consist primarily of dark matter. Moreover, if dark matter is sufficiently common in the universe, the gravitational attraction of all this dark matter will one day halt the expansion of the universe and cause it to collapse.

SATELLITE GALAXIES

Of the satellite galaxies that plow through this dark halo, the two nearest, biggest, and brightest are the Large Magellanic Cloud and the Small Magellanic Cloud. In fact, these irregular galaxies are so famous that many people mistakenly think the Magellanic Clouds are the only galaxies that orbit our own. Both galaxies were named for Portuguese explorer Ferdinand Magellan, but he was hardly the first to notice them, since the Clouds are bright enough to be seen with the naked eye. Unfortunately, the Magellanic Clouds lie so far south that they are not visible from the United States, Europe, or Japan, which means that many people have never seen them.

The Magellanic Clouds are jewels in the Galactic crown, pumping out more light than all the other satellites combined. The Large Magellanic Cloud lies 160,000 light-years from the Milky Way's center, two and a half times farther than the edge of the Galactic disk. The Small Magellanic Cloud is just past its big brother, 190,000 light-years from the Galactic center. The Large Cloud shines with a tenth of the Milky Way's luminosity and the Small Cloud with a sixtieth. The larger galaxy has 6 billion solar masses and perhaps 10 billion stars; the smaller has about a billion solar masses and maybe 2 billion stars.

Popular books and even a few astronomers denigrate the Magellanic Clouds by calling them small galaxies, but the Clouds are actually larger and brighter than most other galaxies in the universe. Of the thirty galaxies in the Local Group, the Large Magellanic Cloud is the fourth brightest—behind Andromeda, the Milky Way, and M33—and the Small Magellanic Cloud's rank varies from five to eight, depending on who compiles the data. The supposed smallness of the Magellanic Clouds is entirely psychological, stemming from our living in and comparing them with a giant galaxy—our own. Of course the Magellanic Clouds are smaller than the Milky Way; but then, so is most every galaxy.

The Clouds' brilliance and beauty reflect their vibrancy, for both galaxies abound with gas that gives birth to hordes of new stars. In 1987 one of the young stars in the Large Magellanic Cloud exploded, producing the brightest supernova since 1604, when a star exploded in the Milky Way. However, because the Magellanic Clouds are smaller than the Galaxy, they have had fewer supernovae during their lives, so the metallicity of young stars in the Large and Small Magellanic Clouds is only 60 and 30 percent of the solar value, respectively.

Beyond the distance of the Magellanic Clouds lie at least eight other galaxies that orbit ours, ragtag dwarf galaxies that have recently attracted far more attention than their small size would seem to merit. "It's always of interest to study extremes," said Carlton Pryor, an astronomer at Rutgers University who has observed the Milky Way's dwarfs. "We can learn things about galaxies in general by studying the least of galaxies. Because the dwarfs are close, they're among the least

Figure 19. The Milky Way rules an empire that spans over a million light-years and includes at least ten satellite galaxies.

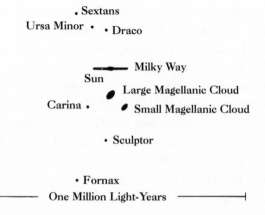

· Leo I   · Leo II

· Sextans
Ursa Minor ·   · Draco

━━●━ Milky Way
Sun
        ● Large Magellanic Cloud
Carina ·   ● Small Magellanic Cloud

· Sculptor

· Fornax
├────── One Million Light-Years ──────┤

luminous galaxies about which we have any knowledge." Unlike the Clouds, the dwarfs contain millions rather than billions of stars, and the faintest dwarfs emit less light than the single brightest star in the Milky Way. These galaxies will remain faint, too, because they have little if any gas from which to form new stars.

Yet dwarfs outnumber all other types of galaxies put together. "The fact that they're quite common around our Galaxy argues they're probably quite common everywhere," said Pryor, noting that most Local Group galaxies are dwarfs. "We're seeing a common but very hard-to-study type of galaxy. Because of their low luminosity, we can find things like Ursa Minor and Draco only around our own Galaxy and a few other nearby galaxies."

TABLE 15-1: SATELLITE GALAXIES

Galaxy	Distance from Galactic Center (Light-Years)	Year Found	Absolute Visual Magnitude	Velocity Toward or Away from Galactic Center[1]
Large Magellanic Cloud	160,000	—	−18.1	+76
Small Magellanic Cloud	190,000	—	−16.2	+22
Ursa Minor	215,000	1954	−8.9	−88
Draco	250,000	1954	−8.6	−95
Sculptor	255,000	1938	−10.7	+74
Sextans	295,000	1990	−9.4	+76
Carina	350,000	1977	−9.3	+7
Fornax	440,000	1938	−13.1	−34
Leo II	720,000	1950	−9.9	+16
Leo I	890,000	1950	−12.0	+177

[1] Velocities are in kilometers per second. Positive velocities indicate galaxies moving away from the Galactic center; negative velocities indicate galaxies moving toward the Galactic center.

Ursa Minor and Draco are the two closest dwarfs, lying 215,000 and 250,000 light-years from the Galactic center. Like all Milky Way dwarfs, Ursa Minor and Draco take their names from the constellations in which they lie. Both galaxies were discovered in 1954, possess a few million old stars, and have luminosities comparable with that of a globular cluster. Draco is the dimmest galaxy known, shining only four times more brightly than the star Rigel. A bit farther than Ursa Minor and Draco, at a distance of 255,000 light-years, is the more luminous Sculptor system, which was the very first dwarf galaxy discovered. Harlow Shapley reported it in 1938, after he took a closer look at what he thought might be only a fingerprint on a photographic plate.

"For a long time," said Pryor, "the dwarfs were a real backwater. Nobody studied them at all. They just seemed to be little galaxies with old stellar populations: a lot like the globular clusters, but farther away and harder to study. There were a lot of other interesting things going on—big galaxies: ellipticals, spirals—which were certainly much prettier and easier to study." Few people have ever seen a picture of a Milky Way dwarf, because these unphotogenic galaxies rarely make it into popular books and magazines. The galaxies are subtle. Not only are they faint, but the few stars they possess are spread out from one another, so seeing a galaxy in a photograph of Ursa Minor or Draco takes effort and imagination. In fact, for many years Paul Hodge, of the University of Washington in Seattle, was one of the few astronomers who investigated these obscure objects. As early as the 1950s he was examining photographs of the dwarfs and studying their properties.

Farther away than Sculptor lie the Sextans dwarf, discovered by computer in 1990, and the Carina dwarf, discovered in 1977. Beyond these galaxies is the greatest dwarf, Fornax, which Shapley had also reported in 1938. Lying 440,000 light-years away, Fornax puts the other seven dwarfs to shame, emitting more light than all of them combined and outshining Rigel 250 times. The galaxy even has five globular clusters of its own. In the past, some astronomers argued that the dwarf galaxies themselves might be diffuse globulars, but Fornax suggests otherwise. Galaxies have globulars, but globulars don't have globulars.

The farthest dwarfs are the aforementioned Leo I and Leo II, 890,000 and 720,000 light-years away. Leo I is the brighter of the two but is harder to observe, since it lies just a third of a degree from the constellation's brightest star, Regulus, whose glare interferes with the galaxy's light. Of all the dwarfs, Leo I has the highest radial velocity. The galaxy is speeding away at 177 kilometers per second and may eventually escape the Milky Way's grasp.

Giant galaxies other than the Milky Way also rule their own empires of satellite galaxies. The nearest large galaxy, Andromeda, lies 2.4 million light-years away and governs seven or eight other galaxies. Andromeda's empire seems to be greater than ours, for M33, the third largest Local Group galaxy, may actually be one of its colonies, and most of Andromeda's other satellites outshine most of ours. Andromeda probably reigns over additional galaxies that are as faint as the Milky Way's dimmest but that have not been discovered because of their greater distance.

All the Milky Way's colonies—from the spectacular Magellanic Clouds to the humble Draco dwarf—yield information about the Galaxy's formation. These objects may resemble those which coalesced to build the Galaxy, as Leonard Searle and Robert Zinn had envisioned. "I like the Searle and Zinn picture," said Pryor. "In that picture, the dwarfs we see today have something to do with the pieces that came together to form our Galaxy. At the time they were forming our Galaxy, the dwarfs were rather different, certainly more gas-rich. The dwarfs we see today are probably the tail end, the remnants of the things that didn't get collected up."

DARK MATTER

The dwarfs also probe the Milky Way's mass and its content of dark matter. If the dwarfs are gravitationally bound to the Galaxy, their velocities should reveal the Galaxy's mass. The faster the dwarfs move, the more massive the Milky Way must be in order to hold on to them. Astronomers who study the motions of the satellites have concluded that the Galaxy has roughly a trillion times the mass of the Sun.

As it turns out, this implies that the Milky Way abounds with dark matter. To quantify the amount of dark matter, divide an object's mass by its luminosity. The greater the resulting mass-to-light ratio, the darker the object. The Sun, by definition, has a mass-to-light ratio of 1. Brighter stars have lower mass-to-light ratios, because most are somewhat more massive but much more luminous. Conversely, fainter stars have high mass-to-light ratios, because most have somewhat less mass but give off much less light. Since most stars are less luminous than the Sun, the mass-to-light ratio in the solar neighborhood is about 3.

But the overall Galaxy has a much greater mass-to-light ratio. As determined from the motions of the satellite galaxies, the Milky Way's mass is roughly a trillion times that of the Sun, whereas the Galaxy's luminosity is only 15 billion times the Sun's. Dividing the first number by the second gives the Milky Way a mass-to-light ratio of 67. This far exceeds the value in the solar neighborhood and means the Galaxy is full of dark matter.

Most of this dark matter lies in the outer part of the Galaxy. Astronomers know this from the speeds with which visible stars orbit the Milky Way, for the star's velocities reveal how much mass lies inside their orbits. The faster the orbital speed and the farther the star, the greater the mass within the orbit. The Sun, for example, lies 27,000 light-years from the Galactic center and revolves around it at 230 kilometers per second. From these numbers astronomers calculate that the Milky Way, in order to hold on to the Sun, must have 100 billion solar masses inside the Sun's orbit. Since the Milky Way's total mass is roughly a trillion solar masses, 90 percent of the mass must lie *outside* the Sun's orbit, most of it in the dark halo.

Dark halos were first detected around other spiral galaxies. In 1969 American astronomers Vera Rubin and W. Kent Ford discovered that stars in the outer part of Andromeda's disk revolve fast, and since then astronomers have found the same behavior in the Milky Way. The first to recognize the ramification of this orbital behavior was Australian astronomer Kenneth Freeman, also in 1969. Freeman reasoned that if the disk contained most of the Galaxy's mass, stars at the edge of the disk should revolve more slowly than the Sun, because they feel a weaker gravitational pull. This pattern holds within the solar system: far-off Pluto moves more slowly around the Sun than the Earth does, because nearly all mass in the solar system lies in the Sun, inside the orbits of both Earth and Pluto. But in the Milky Way, stars in the outer disk revolve about as fast as the Sun. This implies that most of the Milky Way's mass does not lie within the edge of the Galactic disk but beyond it, in the dark halo. Clouds of neutral hydrogen gas confirm this finding, because they extend beyond the disk's edge and also revolve fast.

Additional evidence for the existence of the Milky Way's dark halo comes from the high-speed halo stars darting through the solar neighborhood. During the 1980s, the Carney-Latham survey of the Galactic halo turned up a number of such stars. One of them—a subdwarf named Giclas 233-27, in the constellation Lacerta—had a record-breaking radial velocity of 583 kilometers per second toward the Sun and displaced the previous radial-velocity champion, found in 1943. Some of Giclas 233-27's high radial velocity results from the Sun's motion around the Galaxy toward the star, but most reflects the star's high velocity with respect to the Galaxy. To prevent this and other fast stars from escaping, the Milky Way must have a large amount of dark matter.

But the nature of the dark matter pervading the outer Galaxy is unknown. Bright stars like the Sun could hardly exist in sufficiently large numbers to account for the Galaxy's total mass, simply because astronomers could easily see such luminous objects. Faint stars, such as red dwarfs, have high mass-to-light ratios and are better bets; but if there were enough of them to add up to all of the dark matter in the Galaxy, they probably would have been detected by now. For this reason, many astronomers favor hypothetical objects called brown dwarfs, stars whose centers never got hot enough to ignite their hydrogen fuel and so are dimmer than red dwarfs. If a typical brown dwarf has about 5 percent of the Sun's mass, it would take 20 trillion of them to add up to the trillion solar masses of dark matter in the Galaxy. Other candidates for the Milky Way's dark matter include white dwarfs that have cooled and faded, neutron stars, and black holes.

Yet even black holes are not the most exotic possibility. Every star, even one that has collapsed to form a black hole, consists of normal matter and is considered "baryonic"—that is, composed of protons and neutrons, which physicists call baryons. Baryonic matter makes up nearly everything we know, from a star like the Sun to a black hole like Cygnus X-1 to every human being on Earth.

But some astronomers have proposed that dark matter is nonbaryonic, consisting of subatomic particles that bear no resemblance to the stuff of which normal matter is made. These subatomic particles, which rarely interact with normal matter, fall into one of two categories. One type moves slowly compared with the speed of light and is called cold dark matter. Such subatomic particles are named WIMPs, for weakly interacting massive particles. The other type of particle moved at almost the speed of light when the universe was young and is called hot dark matter. The best candidate for hot dark matter is the neutrino, if it has mass.

In 1982, the dwarf galaxies astonished astronomers when the tiny objects narrowed down the nature of dark matter. Marc Aaronson at the

University of Arizona obtained spectra of stars in Ursa Minor and Draco, the nearest of the dwarfs. Their stars were faint, only seventeenth magnitude, but Aaronson managed to obtain radial velocities for one star in Ursa Minor and three in Draco. To his surprise, Aaronson found that the three Draco stars moved quickly relative to one another, with a velocity dispersion of 6 kilometers per second. This may not seem fast, since the velocity dispersion of stars in the solar neighborhood is several times higher. But for a tiny galaxy like Draco, it is enormous. To hold on to such speedy stars, dim Draco must have a large mass and an enormous quantity of dark matter.

"The Aaronson work stimulated a huge amount of interest," said Carlton Pryor, "and that's driven much of the observational effort. If you look at the central density—the mass per unit volume—of the dark matter in these dwarfs, it's shockingly high." Tiny galaxies like Ursa Minor and Draco are huge reservoirs of dark matter. As John Kormendy first pointed out in 1985, the dwarfs hoard even more dark matter, per unit volume, than large galaxies like the Milky Way.

Back in 1982, however, many astronomers distrusted Aaronson's result. Deducing a galaxy's mass from three stars, said these astronomers, was like predicting the outcome of an election from three voters. Aaronson began to measure other velocities that confirmed his original finding, but in 1987 he was killed during a freak accident at Kitt Peak. He was only thirty-seven years old.

Since Aaronson's death, additional work has strengthened his finding. The stars in Ursa Minor and Draco have even higher velocity dispersions than he first reported, about 12 kilometers per second. When astronomers work out the numbers, they estimate that both Ursa Minor and Draco have mass-to-light ratios of around 100, and recent observations suggest that Sextans and Carina also have large mass-to-light ratios, though not as high as those of Ursa Minor and Draco.

"There are fashions in astronomy, just as there are in all other fields," said Pryor. "The dwarfs are currently very popular and are in fashion, partly because there are some interesting questions that we think we can answer." One question the dwarfs answer is what dark matter is—or, at least, what it is not. "It's hard to cram hot dark matter, like neutrinos, into the dwarfs and get them to be as dense as they are," said Pryor. Hot dark matter zips around too fast for a small galaxy, like Ursa Minor or Draco, to retain. Thus, the dark matter in these galaxies, and presumably throughout the Milky Way's dark halo, must be something else: baryonic matter, like brown dwarfs or black holes; or cold dark matter, which means some type of slow-moving subatomic particle—WIMPs.

VICTIMS OF THE GALACTIC TIDE?

But all this talk of dark matter in dwarf galaxies is nonsense to Jeffrey Kuhn, an astronomer at National Solar Observatory in New Mexico who began investigating the dwarfs in 1988. He believes they contain little if any dark matter. "What caught my eye," he said, "was that they had been recognized as the smallest objects in the universe that were apparently dominated by dark matter. At least, that had been the story up until a few years ago." In general, Kuhn explained, the larger a system, the greater its proportion of dark matter. A small object, such as a globular cluster, contains no dark matter, whereas a large system, such as the Galaxy, has lots of dark matter, and clusters of galaxies harbor even more. "That's true," said Kuhn, "except for the dwarfs. They are a point well above the line." In other words, the dwarfs seem to have far more dark matter than they should, given their small size.

That suggests to Kuhn that the high speeds of the dwarfs' stars result not from dark matter but from the Milky Way's tide, which has shaken those stars the way the lunar tide stirs the terrestrial oceans. Tides are the way gravity distorts an extended object. The Milky Way's gravity pulls more strongly on the near side of an object than on its far side, so if the object is loosely bound, it splits apart. Everything in the Milky Way feels the Galactic tide. For example, in a globular cluster, stars nearest the Galactic center feel a stronger gravitational tug than other cluster stars and may get peeled away. But most globular clusters are too dense for the Galactic tide to tear apart.

Since their stars are so spread out, dwarf galaxies are much more fragile than the clusters—if the dwarfs have no dark matter to restrain their stars. Admittedly, the dwarfs do lie farther from the Milky Way's center and so feel only a weak tide. But Kuhn believes that the nearest dwarfs, Ursa Minor and Draco, are presently being torn apart, because their orbital motion around the Milky Way stimulates a resonance that amplifies the Galactic tide.

"Here's an analogy that demonstrates this," said Kuhn. "If I take a very stiff spring, I have to push very hard to make a change in the position of a weight on the end of the spring. On the other hand, if I very lightly bounce the weight at just the right frequency, then after a few bounces the weight can be moving with a very large amplitude even though the force that I used wasn't very strong. This is just the idea of a resonance: if you work with a very small tide that operates at the right frequency, it can have a very large effect on the system over several orbits." It is like pushing a child on a swing every time he swings toward you: you don't have to push him much before he swings quite high. In the case of a dwarf galaxy, Kuhn's resonance depends on a match between the dwarf's density and its distance, and Kuhn calculates that both Ursa Minor and Draco have the right densities and distances for a resonance to occur.

If so, then the high velocities of the stars in these galaxies say nothing about the dwarfs' masses or mass-to-light ratios. "The galaxies are oscillating," Kuhn said. "They're being pumped up by their interaction with the Milky Way, because the tide of the Milky Way takes orbital energy and puts it into the motion of the dwarfs' stars. So when you look out at a dwarf, what you're seeing is excess energy in the motions of its stars, which you observe as a large velocity dispersion that doesn't reflect the gravity of the system."

If Kuhn is right, and the dwarfs contain no dark matter to hold themselves together, the nearest ones are doomed. "Ursa Minor is probably completely unbound," said Kuhn. "In my opinion, we are definitely seeing Ursa Minor in its last stages, and Draco is close behind. This is probably Draco's last passage before it starts to look like Ursa Minor." In Kuhn's view, a huge "barrier reef" that tears dwarfs apart surrounds the Milky Way at a distance of 200,000 to 300,000 light-years. If a dwarf galaxy trespasses into this zone, the galaxy suffers a resonance, and the Milky Way pumps energy into the galaxy's stars. The stars then wander away and the dwarf disintegrates. Ursa Minor and Draco are passing through this treacherous zone, says Kuhn, and of all the dwarfs their stars have the highest velocity dispersions. Kuhn believes his theory also explains why Ursa Minor and Draco are the nearest dwarfs. Any dwarf that came closer got torn to shreds during its journey across the Galactic barrier reef. In fact, Kuhn thinks that the Milky Way may once have had as many dwarf galaxies as globular clusters, which would put the number of disrupted dwarfs at well over a hundred.

Opponents of this tidal idea, including Carlton Pryor, have argued that a dwarf caught in this resonance would quickly disperse, making it unlikely for astronomers to see at least two dwarfs that are now breaking up. The stars of Ursa Minor and Draco have velocity dispersions of around 12 kilometers per second. If these stars are not gravitationally bound to their parent galaxies, a velocity of 12 kilometers per second will carry the stars 4000 light-years from their homes in only 100 million years, meaning that a dwarf would vanish soon after it fell victim to the Galactic tide.

But Kuhn considers this calculation too simple, because his models indicate that stars in a galaxy orbiting the Milky Way can require over a billion years to disperse. "It's surprising how long the stars will appear to be in roughly the same place," said Kuhn. "It takes a long time for the stars to spread out along the orbit, so it's perfectly reasonable that we should look up at something like Ursa Minor and think of it as a bound object, when in fact it isn't. Its stars are just moving in a common orbit that is a relic of the original orbit that the dwarf had."

Pryor doesn't buy Kuhn's theory. "Jeff and I have had discussions on this topic," said Pryor, laughing. "Though I don't favor his theory, it's been popular among a lot of people, because it's hard to believe that these wimpy little galaxies have so much dark matter. But I argue that the problem with these tides is that once you pull things apart, the dwarfs disappear pretty quickly, so it's surprising to see so many dwarfs caught in the act of dispersing. Jeff claims that there are ways to slow that down, and I agree, but I think they're contrived. His resonance picture is very cute—where things spiral in, hit a resonance, and blow up—but I think that once they blow up, they'll disappear pretty fast."

Nor does Pryor worry that the two nearest dwarfs have the highest velocity dispersions. "That's certainly one of the basic arguments for tides," said Pryor. "But these dwarfs also have the lowest luminosities, so they're special in another way." It makes sense, he says, that the dimmer a galaxy, the greater its proportion of dark matter, which gives its stars their velocity dispersion and the galaxy its high mass-to-light ratio. To support his point, Pryor notes that the brightest dwarf, Fornax, has the lowest mass-to-light ratio, similar to that in the solar neighborhood. "You can imagine a picture in which the more luminous matter you have, the harder it is to detect the dark matter, just because the luminous matter tends to dominate more," said Pryor. Kuhn counters by arguing that the nearest dwarfs, Ursa Minor and Draco, are the dimmest precisely because the Milky Way has already started to rob them of stars.

Because the nearest dwarfs are the least luminous, the dispute goes on, since it is not clear whether their large velocity dispersions result from their nearness and the effects of the Galactic tide, as Kuhn argues, or from their low luminosity and large dark-matter content, as Pryor contends. One test that may help in the future would be to observe the most distant dwarfs, Leo I and Leo II, which are so remote that they are safe from the Galactic tide. According to Kuhn, the stars in these galaxies should have low velocity dispersions. According to Pryor, at least Leo II should have a large velocity dispersion, because this galaxy is almost as faint as Ursa Minor and Draco and therefore should contain a high proportion of dark matter. Since these galaxies are so distant, though, no data on their velocity dispersions have yet been published.

It would also help to discover additional dwarf galaxies, but that will be a challenge. Said Kuhn, "The last one, Sextans, was detected by computer, and you can imagine that any new dwarfs will only be detected by computerized techniques." Nevertheless, Kuhn believes several more are out there, and he even thinks he knows where they are. "I made a three-dimensional rendering of the dwarfs," he said, "and it was startling to me to notice that all of the dwarfs we know about are on the same side of the Milky Way as the Sun." The chance that eight galaxies happen to lie on the same side of the Galaxy as the Sun is the same as flipping a coin and getting heads eight times in a row—only 1 in 256—which suggests that several more dwarfs are hiding on the other side of the Galaxy.

Pryor is less optimistic about discovering more dwarfs. "Mike Irwin [who discovered the Sextans dwarf] has now surveyed most of the sky," said Pryor, "and he doesn't see any more. If there are any others out there, they have even lower surface brightness and lower luminosity than the systems we see. My guess would be that we have found all of the ones we're likely to find for quite a while." Several dwarfs, however, could lie buried among the stars in the Galactic plane. "If you get too close to the Galactic plane, all bets are off," said Pryor.

"Those would be hard to dig out." Of the eight known dwarfs, Carina is closest to the Galactic plane, lying 22 degrees south of it.

In 1975, though, a possible newcomer to the Galactic empire appeared just two degrees north of the plane, in the constellation Gemini, when S. Christian Simonson III detected 21-centimeter radio radiation from what he said was a small satellite galaxy. Someone dubbed the reputed galaxy Snickers, because it was peanuts compared with the Milky Way, but the object has never been seen at optical or infrared wavelengths, its reality is in question, and it does not appear on lists of known Galactic satellites.

More recently, in 1994, three astronomers in England—Rodrigo Ibata, Mike Irwin, and Gerard Gilmore—discovered a possible dwarf galaxy in the constellation Sagittarius. If that dwarf is confirmed, it will be unique, for it lies on the opposite side of the Galactic center from the Sun. But investigating the Sagittarius galaxy, if that is what it is, will be difficult, since the constellation is the home of the Galactic center and abounds with thick gas and dust that block the view of even the Milky Way's own stars, let alone those in other galaxies.

Meanwhile, both Pryor and Kuhn are trying to collect more data to see whose picture for the dwarfs is right. Pryor plans to measure velocities of additional stars in Ursa Minor and Draco. "In the tidal picture," said Pryor, "you should see one side of the galaxy going one direction and one side going the other. In other words, the mean radial velocity should vary pretty strongly across the galaxy, because it's being pulled apart, stretched like taffy." Pryor thinks the additional data will show that idea to be wrong, which will indicate that the dwarfs must harbor large amounts of dark matter. Since Kuhn believes the dwarfs have no dark matter and that the nearest ones are unbound, he is searching for stars that have escaped from Ursa Minor and Draco. And he also hopes to find traces of former dwarf galaxies that the Milky Way has destroyed. "Their stars haven't had time to completely diffuse," said Kuhn, "so there may be a very nice hint as to what the early dynamical properties of the Milky Way look like, when we find these remnant trails of stars in the outer Galaxy." If Kuhn is right, these star streams may wrap around the Galaxy the way jet contrails interlock in a clear sky.

CLOSE TO THE EDGE

No matter how the debate over the dwarfs' dark matter is resolved, astronomers have gained other knowledge from the galaxies. It used to be thought that the dwarfs resembled retirement homes, full of nothing but old stars, those formed 10 to 15 billion years ago. Every dwarf does have old stars, and Ursa Minor and Draco have nothing but old stars. Other dwarfs, however, break the pattern.

"Carina is the standout case," said Pryor. "For some reason, Carina sat around doing almost nothing for billions of years, and then 7 billion years ago it turned all its gas into stars. How you can get a galaxy to do something like that is very puzzling." Astronomers can trace Carina's history by examining its H-R diagram, which reveals that most of Carina's stars are about 7 billion years old. Fornax, the most luminous dwarf, has some stars that are even younger, only 2 or 3 billion years old.

Generally, though, the small sizes of the dwarfs have stymied their evolution. "If you're a very low-luminosity galaxy," said Pryor, "you have trouble when you first form. It's hard for a small galaxy—it doesn't have enough gravity to hold on to its gas, and the gas sails on out, so the galaxy can lose lots of its luminous mass." The gas ejected by supernovae was captured by the Milky Way, so the dwarf could not use this debris to form stars. And that may explain a trend that the dwarfs exhibit: the fainter the dwarf, the more metal-poor it is. The least luminous ones lost most of the metals their stars created and never recycled the metals into more stars. The faintest dwarf, Draco, has a metallicity around 1 percent of the solar value, whereas Fornax, the brightest, has a metallicity of 4 percent of the Sun's.

In contrast, the Magellanic Clouds have retained their star-forming material because they are much more massive. But even they lose gas. In 1973, astronomers discovered a strand of hydrogen gas extending from the Magellanic Clouds and stretching through the Milky Way's dark halo for some 300,000 light-years. This Magellanic Stream, as it is called, probably resulted from a recent near-collision between the Large and Small Clouds; the two galaxies fought over the hydrogen gas and lost it to the Milky Way. The galaxies may also have forfeited a star cluster. In 1992, Douglas Lin and Harvey Richer proposed that one of the Milky Way's globulars, Ruprecht 106, once belonged to the Magellanic Clouds, because it lies in the right direction and is younger than other globular clusters in the Milky Way.

Bright though they now are, the ultimate fate of the Magellanic Clouds is bleak, because they lie within the Milky Way's dark halo. Whatever form the dark matter takes, it exerts not only a gravitational force but also a drag force that will eventually snare the nearest of the Milky Way's satellites, including both of the Magellanic Clouds. When a large mass, such as a satellite galaxy, moves through a sea of smaller objects, it is slowed down by their gravity, just as celebrities are slowed down when they walk through a throng of fans. Because the satellite galaxy slows down, it loses energy, falls toward the Milky Way, and is swallowed by it. This drag force is called dynamical friction. In a few billion years dynamical friction will pull the Magellanic Clouds into the Milky Way, and our Galaxy will acquire all of the Clouds' stars for itself. Dynamical friction also operates on the nearest dwarf galaxies, slowly drawing them inward.

In order to learn more about both the dark halo and the stellar halo, astronomers have searched for the few individual stars that shine in the Milky Way's most remote regions. Most studies of the stellar halo have relied on the few halo stars passing through the disk near the Sun. Because these stars are nearby, they are bright and easier to study. But observations of nearby stars give only a partial view of the halo, because astronomers must select nearby halo stars from the overwhelming sea of disk stars by choosing those with either high velocities or low metallicities. Each presents problems: a high-velocity sample will miss any halo stars that have low velocities, and a metal-poor sample will miss any halo stars with high metallicities.

A more direct way to probe the halo is to look for stars within it, at distances exceeding 20,000 light-years above or below the Galactic plane. These stars will constitute an unbiased sample, selected for neither their kinematics nor their metallicity. Unfortunately, *in situ* studies of the halo are difficult, because the stars are distant and faint, and the first such searches did not begin until the 1980s. The most distant individual stars known—that is, stars that do not belong to any cluster, satellite galaxy, or group—are about 160,000 light-years away. They are

RR Lyrae stars and so have well-established distances. In the view of most astronomers, these few stars lie in a vast sea of dark matter surrounding the Galactic disk.

But not all astronomers accept this. "I have to admit," said Jeffrey Kuhn, "I have rather heretical views about the dark matter question. I keep thinking about the luminiferous ether from years past." Over a century ago, physicists convinced themselves that space was pervaded by an invisible substance called the ether, through which light waves propagated. Perhaps, said Kuhn, dark matter is today's equivalent of the ether and does not exist. Since dark matter cannot be seen, its presence is inferred solely from its gravitational effect on objects that astronomers can see. To quantify the gravitational force and the amount of dark matter required, astronomers use Newton's law of gravity. But if Newtonian gravity is wrong, dark matter may not exist. "I have real problems with dark matter," Kuhn said. "Any statement about what the dark matter is made of is really speculation, and it should be coupled with the speculation that there may be problems in understanding long-range interactions, like gravity."

In 1993, however, astronomers may finally have detected some of the dark matter in the dark halo. The discovery resulted from a clever proposal offered seven years earlier by Princeton astronomer Bohdan Paczyński. He suggested that if dark stars, such as brown dwarfs and black holes, make up the Milky Way's dark halo, their gravity should deflect the light of the visible stars they pass in front of. These hypothetical dark objects, which consist of baryonic matter, were later named MACHOs, for massive compact halo objects—to contrast with WIMPs, the hypothetical subatomic particles that constitute cold dark matter.

Paczyński recommended that astronomers search for MACHOs by monitoring stars in the Large Magellanic Cloud. Normally, the light from any of these stars diverges, spreading out in all directions from the star. But if a MACHO in the Milky Way's dark halo passed between us and the star, the MACHO's gravity would bend separate light rays from the star and blend them together. As a consequence, Earth-based astronomers would see the star gradually brighten as the MACHO passed in front of the star and then slowly fade back to normal brightness as the MACHO moved on. If, though, WIMPs make up the dark halo, astronomers would not see these brightenings, because an individual WIMP has far too little mass to deflect light.

The search for MACHOs began in the early 1990s. Although Paczyński's concept was simple, carrying it out was not. Even if MACHOs make up all the dark halo's mass, the chance of one passing in front of any star in the Large Magellanic Cloud was so minuscule that astronomers had to observe millions of stars in order to see just a few MACHO-induced brightenings. Worse yet, because many stars vary in brightness themselves, astronomers had to distinguish between a MACHO-induced variation and one intrinsic to the stars being observed.

Nevertheless, despite these obstacles, in 1993 two teams of astronomers observing the Large Magellanic Cloud reported the possible detection of MACHOs in the halo; and a third team, observing stars in the Galactic bulge, detected MACHOs in the Galactic disk. Although these discoveries are a promising start, years must pass to see how many events occur and whether enough dark stars exist to constitute all of the dark halo. For now, the question of exactly what makes up the dark halo—MACHOs or WIMPs—remains unanswered.

THE ORDER OF THE UNIVERSE

Whatever its form, most astronomers believe that dark matter dominates the Milky Way. But dark matter affects more than just the Galaxy. It dictates the fate of the entire universe. The universe is presently expanding from the big bang, the fiery explosion that occurred 10 to 15 billion years ago. If the universe does not have enough mass, it will continue to expand forever. But if instead the universe is more massive, the expansion will stop. In that case, the gravitational attraction of the mass will slow the expansion and eventually halt it, just as the Earth's gravity slows a stone you throw upward. Then the universe will reverse its expansion and collapse, just as that stone will fall back to Earth. The universe will become smaller, other galaxies will smash together, and the universe will implode in a dense fireball similar to the big bang—the aptly named "big crunch."

Whether the universe expands forever or ultimately collapses depends on its mass density. To quantify that density, which controls the universe's destiny, cosmologists employ the final letter of the Greek alphabet, omega (Ω). The greater the universe's density, the greater Ω. If Ω is less than 1.0, the universe will expand forever, because the gravitational attraction of its mass will be too weak to reverse the expansion. If instead Ω exceeds 1.0, the universe will someday collapse.

All the luminous material in the Milky Way and other galaxies adds up to little. If this were all the mass the universe had, Ω would be between 0.005 and 0.01, so the universe would expand forever. But the dark halo of the Milky Way, which has about ten times the Milky Way's luminous mass, and the dark halos of other galaxies imply that large amounts of dark matter exist throughout the universe and that Ω is some ten times greater, putting it between 0.05 and 0.10. Though larger, this value for Ω still means the universe will expand forever. Studies of the motions of galaxies in clusters imply that the clusters themselves must contain large amounts of dark matter to prevent their galaxies from escaping. These studies suggest that Ω is around 0.20, a number that many observational astronomers believe to be close to the truth.

Some theorists, however, believe that Ω is precisely 1.0. Few observations support this value, but many cosmologists find it aesthetically pleasing. Such a universe is in a delicate balance, for it will someday stop expanding, but it will not collapse. One basis for thinking Ω has this value is a popular but highly speculative theory called inflation, which claims that the universe expanded enormously just a fraction of a second after it was born. It turns out that this rapid expansion produces a universe in which Ω is exactly 1.0.

In exploring the fringes of the Milky Way, then, astronomers have unearthed large quantities of dark matter and confronted the destiny of the entire universe. But the Milky Way also takes astronomers back to the universe's very beginning, for the Galaxy's oldest stars preserve elements forged just minutes after the big bang. In fact, the primordial elements in the most ancient stars offer vivid clues into the nature of the big bang, the mass density of the universe (Ω), and the nature of dark matter itself. These ancient stars are the fossils of creation.

16

FOSSILS OF CREATION

THE MILKY WAY'S ancient stars trace not only the earliest years of the Galaxy's life but also the origin of the entire universe. These stars formed before many others had a chance to explode and contaminate the Galaxy with their ejecta. For this reason, the oldest stars in the Galactic halo are fossils of the universe's creation, still harboring the light elements—hydrogen, helium, and lithium—that were forged in the fiery aftermath of the big bang.

"The light-element abundances give us the best direct observational probe of the earliest times in the universe," said David Schramm, a cosmologist at the University of Chicago. "They really give us a lot of confidence that the basic picture of the big bang is in good shape." Schramm even advertises his enthusiasm on his car's license plates. "I bought the car the same year that Illinois allowed seven-digit vanity plates," he said. "So I immediately applied for BIG BANG and got it."

To most cosmologists, including Schramm, the light elements constitute the last of three pillars upholding big bang cosmology. The first pillar is the expansion of the universe, which Edwin Hubble announced in 1929. The second is the microwave background, which Arno Penzias and Robert Wilson discovered in 1965. It is the afterglow of the big bang fireball—radiation from that primordial event which still permeates all of space.

"Of those three pillars," said Schramm, "the one that's probing to the earliest moments and to the most exotic conditions is the light-element abundance probe." In contrast, the Hubble expansion reflects the present state of the universe, and the microwave background dates back to when the universe was hundreds of thousands of years old, because before then the universe was so small and dense that the big bang's radiation was completely absorbed by matter.

A UNIVERSAL DRAMA: PRIMORDIAL NUCLEOSYNTHESIS

Astronomers investigating the early history of the universe usually observe galaxies and quasars billions of light-years away. Because the light has traveled for billions of years, it carries information about conditions that prevailed long ago, when the universe was young. But other clues to the early universe lie much closer to Earth, in the halo stars of the Galaxy. The nearest, Kapteyn's Star, is only thirteen light-years away. These stars open a unique window on the chemical elements the universe created during the first minutes of its life. Moreover, these stars were untouched by the supernova explosions that later enriched the Galaxy's gas, so the oldest stars preserve a faithful record of the big bang itself.

According to standard cosmology, the big bang produced only five stable nuclei, all lightweight: hydrogen-1, helium-4, deuterium (or hydrogen-2), helium-3, and lithium-7. The quantities of these nuclei depended on the universe's density of normal, or baryonic, matter, because that density determined the speed of the nuclear reactions. Observations of the light elements in old stars therefore provide information about Ω, the mass density of the universe, or at least about that part of Ω which consists of baryonic matter.

To see how this works, pretend that the early universe was an epic play and that each nucleus created right after the big bang was an actor. The starring roles will be taken by hydrogen-1 and helium-4, for these two nuclei will wind up with over 99.9 percent of the mass of the universe—76 percent in hydrogen-1, 24 percent in helium-4. Hydrogen-1 is the simplest nucleus, consisting of a single proton, and helium-4 is the most tightly bound light nucleus, made of two protons and two neutrons. (Electrons also existed in the early universe but played only a minor role in the nuclear reactions, since electrons lie outside an atom's nucleus.)

In addition to hydrogen-1 and helium-4, three other nuclei in the big bang drama will play vital roles. Deuterium, or hydrogen-2, is an isotope of hydrogen, made of a proton and neutron. Another isotope, helium-3, is the lighter cousin of helium-4 and consists of two protons but only one neutron. Hydrogen-2 will emerge with about 0.006 percent of the mass of the universe and helium-3 with about 0.001 percent. Lithium-7, with three protons and four neutrons, is the heaviest stable actor in the play, but it will also end up the loser, coming away with just 0.00000001 percent of the universe's mass.

Two other nuclei will make fleeting appearances in the early universe: hydrogen-3, or tritium, which consists of one proton and two neutrons and so is an even heavier hydrogen isotope than deuterium; and beryllium-7, which has four protons and three neutrons. Both tritium and beryllium-7 are radioactive, so any that formed during the big bang vanished long ago.

Let the play begin, with a big bang. The universe, smooth and homogeneous, explodes into being and expands rapidly. One second after the big bang, the temperature is 10 billion degrees Kelvin, or 18 billion degrees Fahrenheit. Because of the intense heat, photons—particles of light—greatly outnumber protons and neutrons, both of which are only trace components in a sea of light. The protons, in turn, outnumber the neutrons by a factor of five or so, but the protons are really hydrogen-1 nuclei, so hydrogen-1 starts with a big lead over the other nuclei.

The size of the neutron-to-proton ratio, it turns out, depends on Ω: the greater Ω, the larger the neutron-to-proton ratio. Nearly all the neutrons will be incorporated into helium-4, since hydrogen-1, the most common nucleus, has no neutrons, and the other nuclei, which do contain neutrons, are rare. So the greater Ω, the greater the amount of helium-4 that will emerge from the big bang.

But neither helium-4 nor any other nucleus heavier than hydrogen-1 has yet formed. It is now ten seconds after the big bang, and as the universe has expanded, the temperature has fallen to 3 billion degrees Kelvin. Protons and neutrons swirl about furiously, but not much happens. Two protons cannot join, because like charges repel each other. Nor can two neutrons join, for the resulting product is unstable and splits apart. Instead, the only possible reaction occurs when a proton and neutron meet to make deuterium (hydrogen-2),

$$p + n \rightarrow {}^2H.$$

This is the first step in primordial nucleosynthesis—the creation of elements in the early universe. But the universe will not allow this reaction to proceed, for the universe is hot and filled with high-energy photons, which tear apart the deuterium as soon as it forms. Thus, the play enters an ironic moment: the universe is hot enough to promote this first nuclear reaction, but the same heat that fosters the reaction annihilates the very thing it produces.

The universe grows larger and cools further. Finally, about a hundred seconds after the big bang, the temperature drops to a billion degrees Kelvin, the high-energy photons diminish, and the deuterium can finally survive. As soon as this happens, a cascade follows in which protons, neutrons, and other deuterium nuclei hit the newly formed deuterium to produce isotopes with mass numbers of three, radioactive hydrogen-3,

$$^2H + n \rightarrow {}^3H$$
$$^2H + {}^2H \rightarrow {}^3H + p,$$

and stable helium-3,

$$^2H + p \rightarrow {}^3He$$
$$^2H + {}^2H \rightarrow {}^3He + n.$$

Because it is radioactive, the hydrogen-3 will decay, but even before it gets the chance to do so, protons and deuterium nuclei convert most of it into rock-solid helium-4,

$$^3H + p \rightarrow {}^4He$$
$$^3H + {}^2H \rightarrow {}^4He + n.$$

Likewise, neutrons, deuterium, and other helium-3 nuclei transform most of the helium-3 into helium-4,

$$^3He + n \rightarrow {}^4He$$
$$^3He + {}^2H \rightarrow {}^4He + p$$
$$^3He + {}^3He \rightarrow {}^4He + p + p.$$

The result is that nearly all deuterium, hydrogen-3, and helium-3 become helium-4, a prime actor that now steps into the spotlight. The denser the universe and the greater Ω, the more collisions occur and the faster these reactions proceed. So the greater Ω, the less deuterium and helium-3 will emerge from the big bang.

At this point, however, the nucleosynthesis runs out of steam, because no stable nuclei have masses of five or eight. If helium-4 collides with a neutron or proton to produce helium-5 or lithium-5, the product decays into lighter nuclei. And because no stable nuclei have a mass of eight, when two helium-4 particles—now the most abundant nuclei after hydrogen-1—join to create beryllium-8, the beryllium immediately splits back into two helium-4 nuclei.

Nevertheless, a little lithium-7 does manage to appear. The small amount of hydrogen-3 that did not become helium-4 can collide with helium-4 to create lithium-7,

$$^4He + {}^3H \rightarrow {}^7Li.$$

But protons smash into lithium-7 and destroy it,

$$^7Li + p \rightarrow {}^4He + {}^4He.$$

Protons are common, for they are simply hydrogen-1 nuclei, the dominant actor. The denser the universe and the greater the baryonic mass density Ω, the more protons there are to destroy the lithium-7, so less lithium-7 results. Also, the denser the universe, the less hydrogen-3 there is to create lithium-7, because a dense universe has converted more of its hydrogen-3 into helium-4.

But for larger values of Ω, lithium-7 can arise in another way. The denser the universe, the more beryllium-7 forms from a collision between helium-4 and helium-3,

$$^4\text{He} + {}^3\text{He} \rightarrow {}^7\text{Be}.$$

Beryllium-7 is radioactive, for it captures an electron and decays into lithium-7,

$$^7\text{Be} + e^- \rightarrow {}^7\text{Li}.$$

This decay occurs long after the universe has cooled, so the protons that might have torn the lithium apart no longer have the energy to do so, and the lithium therefore survives.

Consequently, the behavior of lithium-7 as a function of Ω is complicated. At first, its abundance decreases with increasing Ω, due to the greater destruction by protons, but then it increases with increasing Ω, because of the greater amount of beryllium-7 that a denser universe produces.

So the play ends, about fifteen minutes after it began. The abundances of deuterium, helium-3, helium-4, and lithium-7 relative to hydrogen-1 all depend on Ω. And all the abundances that astronomers have measured point toward a single value for Ω.

"Each of those abundances has come out right on the money," said David Schramm. "It's a quantitative prediction ranging over ten orders of magnitude and down to levels of accuracy of one part in 10 billion." That latter figure is for lithium-7, the rarest of the five nuclei and the last to be found, in 1981. "The lithium discovery in halo stars was critical," he said, "because it gave us lithium at this level of one part in 10 billion. That was the keystone that tied it all together." But as astronomer Virginia Trimble has written, "I find it slightly unsatisfactory that anything as scarce and unromantic as lithium should be so important."

THE DENSITY—AND DESTINY—
OF THE UNIVERSE

Because the primordial nuclear reactions occurred at a rate that depended on the universe's density, the products of those nuclear reactions constrain the baryonic mass density of the universe (Ω) and thus offer clues to the destiny of the universe—whether it will expand forever or ultimately collapse. Before the discovery of lithium in old stars, astronomers had determined the primordial abundances of the other light nuclei, finding the combined number of deuterium and helium-3 nuclei to be about one ten thousandth that of hydrogen-1, and the mass of helium-4 to be 23 percent that of the primordial universe. These numbers suggested a value of Ω that implied a low abundance of lithium, only one ten billionth that of hydrogen. But the lithium level in the Galactic disk was ten times higher, so if this were really the primordial value, the big bang prediction would have been wrong.

However, since the old halo stars better probe the early universe, in 1981 French astronomers Monique and François Spite searched for lithium in the Galactic halo. Using what was then the new 3.6-meter Canada-France-Hawaii telescope atop Mauna Kea, the Spites took aim at the halo stars and hunted for a spectral line at 6707 angstroms, a red wavelength that lithium absorbs.

"When François started the work," said Schramm, "he thought he was going to be able to prove the model wrong, because he was assuming that he was going to get numbers similar to what they got with disk stars, that is, at the level of about one billionth. So he was expecting that he was going to cause us a real problem, but to his surprise and our joy it came out right at the one ten billionth level." In all but one of the halo stars, the Spites detected lithium at the abundance predicted by the big bang model. Furthermore, though the stars had differing iron abundances, each star with lithium had the same amount. "That plateau proves," said Schramm, "that the lithium you're seeing in those stars did not get made by the Galaxy along with the iron but in fact had another source, the big bang." In contrast, elements produced during Galactic evolution, such as oxygen, become more abundant as the iron abundance increases.

One possible problem remained, though. Temperatures of only a few million degrees destroy lithium, so the present lithium abundance on a star's surface may be much less than the quantity the star was born with. For this reason, some halo stars—the ones that drag their surface material, which is cool, into their hot interiors—have no lithium at all. This meant that the lithium abundance the Spites found, which agreed with big bang predictions, might be only a fraction of the abundance these stars had started their lives with.

In 1991, however, three astronomers at the University of Texas at Austin—Verne Smith, David Lambert, and Poul Nissen—made an intriguing observation. While studying the halo star HD 84937 in the constellation Leo, the astronomers discovered a small amount of lithium-6, the lighter and rarer isotope of lithium. Lithium-6 is even more fragile than lithium-7 and was produced by the Galaxy rather than by the big bang. In HD 84937 the lithium-6 abundance is only a twentieth that of lithium-7, but the presence of the former means that the star has not destroyed any of the latter. If the star had destroyed lithium-7, there would have been no lithium-6. So the star's abundance of lithium-7 reflects the amount that the big bang produced.

The primordial lithium-7 abundance, along with the abundances of deuterium, helium-3, and helium-4, reveal the mass density of the universe, Ω. By considering the abundances of all four light nuclei relative to hydrogen-1, cosmologists calculate that Ω must be low—between 0.01 and 0.10—which suggests that the universe has little mass and will expand forever. This number far exceeds the universe's amount of visible matter, which gives an Ω of only 0.005 to 0.01. Primordial nucleosynthesis therefore provides independent evidence for dark matter, and agrees with studies of the Milky Way's dark halo.

But the Ω derived from primordial nucleosynthesis falls far short of 1.0, the number that inflationary cosmologists embrace. This does not, however, stop such cosmologists—like Schramm—from believing in inflation, because the low Ω implied by primordial nucleosynthesis applies only to normal, baryonic matter. Nonbaryonic matter did not participate in nor affect the primordial nuclear reactions, so its abundance is unconstrained by the light nuclei.

"That tells you," said Schramm, "that if the total Ω of the universe is near 1.0, as most of us believe, then the bulk of the matter in the universe has to be something other than baryons. So you need some kind of exotic stuff to do that." If inflation is correct, 90 to 99 percent of the universe's mass must be nonbaryonic.

A BUMPY BIG BANG?

In 1991, however, possible trouble for the standard model of primordial nucleosynthesis appeared when astronomers reported large quantities of the element beryllium in the halo star HD 140283. Beryllium has atomic number four, and the standard model says that essentially none of it should have arisen during the big bang. Yet HD 140283 had thousands of times more beryllium than the prediction. The beryllium hinted that the early universe might not have been smooth and homogeneous, as the standard model assumes, but inhomogeneous, with some regions denser than others. It turns out that an inhomogeneous early universe could have produced different amounts of some of the light elements, including beryllium.

One of the astronomers who detected beryllium in HD 140283 was Gerard Gilmore, the champion of the Milky Way's thick-disk population. Gilmore is no cosmologist; he got into cosmology through his research on the Galaxy. "It was partly by having worked for some time on the chemical evolution of the Galactic halo and therefore being aware of the problem in that context," said Gilmore. "And in fact I remember chatting over a glass of wine in Martin Rees's garden with Willy Fowler, who was enthusing wildly over this very idea that the early universe might have been inhomogeneous. And it really went on from there."

Beryllium might signify an inhomogeneous early universe, but detecting the element was difficult. The spectral lines Gilmore and his colleagues were looking for, at 3130 and 3131 angstroms, lie in the ultraviolet, which the Earth's ozone layer absorbs. In addition, beryllium is rare; its abundance on the Sun's surface relative to that of hydrogen is only 0.000000000014. Yet in 1991 Gilmore's team reported the discovery of beryllium in HD 140283 and later in several other halo stars as well.

At the time, the discovery did hint that the early universe was inhomogeneous, an idea suggested in 1982 by Matt Crawford and Schramm and then worked out in more detail two years later by Edward Witten, a physicist at the Institute for Advanced Study in Princeton. When the universe was only a fraction of a second old, it had no protons or neutrons but instead was full of quarks, sub-subatomic particles that make up protons and neutrons. As the universe expanded and cooled, the quarks joined to create protons and neutrons, a "phase transition" that took place when the universe was one hundred thousandth of a second old. This phase transition might have left the universe lumpy and inhomogeneous, just as a lake can freeze irregularly.

In 1985, James Applegate and Craig Hogan explored this idea to see how an inhomogeneous early universe might affect primordial nucleosynthesis. As in the homogeneous model, the inhomogeneous universe starts off with more protons than neutrons, both of which cluster inside the high-density regions. The neutrons, though, diffuse into the low-density regions. The protons try to do likewise but do not succeed. Protons are positively charged, so the attractive electrical force of the negatively charged electrons keeps the protons from escaping. The neutrons, which have no charge, face no such barrier.

As a consequence, two different areas emerge: high-density regions, which are proton-rich, and low-density regions, which are neutron-rich. In the proton-rich regions, the nucleosynthesis mimics that of the standard homogeneous model. But in the neutron-rich regions, nuclear reactions occur that are otherwise impossible. Take, for example, lithium-7, the heaviest stable nucleus that the standard big bang produces in measurable amounts. If lithium-7 collides with a proton, it splits into two helium-4 nuclei. In the homogeneous model, such collisions are likely, because protons outnumber neutrons everywhere. But in the neutron-rich regions of an inhomogeneous universe, lithium-7 will probably meet a neutron, which creates lithium-8:

$$^7\text{Li} + \text{n} \rightarrow {}^8\text{Li}.$$

Like all nuclei with mass eight, lithium-8 is radioactive. It has a half-life of just 0.8 seconds. But before it decays, it may encounter the common nucleus helium-4 and create boron-11,

$$^8\text{Li} + {}^4\text{He} \rightarrow {}^{11}\text{B} + \text{n}.$$

Boron has atomic number five, and boron-11—with five protons and six neutrons—is its most common isotope.

Another product of an inhomogeneous early universe would be beryllium. The neutron-rich regions have a fair amount of hydrogen-3, the hydrogen isotope that contains two neutrons. If hydrogen-3 hits lithium-7, the result is beryllium-9, the only stable isotope of beryllium,

$$^7Li + {}^3H \rightarrow {}^9Be + n.$$

This beryllium might be exactly what Gilmore team's had discovered in the halo stars. After finding beryllium in the halo stars, Gilmore and other astronomers discovered boron as well, the other element that suggests an inhomogeneous early universe.

But David Schramm does not believe the early universe was inhomogeneous, because he thinks these elements did not come from the big bang. "Beryllium and boron are telling us something about early Galactic evolution," he said, "but not cosmology. The only thing that would have me change that statement would be if somebody would decisively show a plateau in the beryllium and boron abundances as you go to low metallicity, and so far the data doesn't show anything like that." In contrast with lithium, whose abundance is the same throughout most of the halo, the beryllium and boron abundances in the halo increase as the metallicity does. This suggests that the Galaxy, and not the big bang, generated the beryllium and boron. In Schramm's view, the elements formed when cosmic rays smashed into and split heavier elements that were floating in space. The resulting beryllium and boron were then inherited by stars that formed from this material.

Furthermore, said Schramm, early calculations showing that inhomogeneous models could create beryllium violated the data for the other light elements. "It's true you can take an inhomogeneous model and you can make a lot of beryllium," he said. "But when you do that, it doesn't make anything else right. So it's not a consistent model."

At first, cosmologists also thought the inhomogeneous model might increase their estimate for Ω in baryons. Although the primordial abundances of deuterium, helium-3, helium-4, and lithium-7 imply that the Ω in baryons lies between 0.01 and 0.10, the calculations assume that the universe was homogeneous when those nuclei originated. But if the early universe was actually inhomogeneous, the same abundances could have resulted with a different density and therefore a different value of Ω. Some scientists even thought that an inhomogeneous early universe could raise the value of Ω in baryons to the magic number of 1.0 demanded by inflationary cosmologists, which would mean that an inflationary universe could consist entirely of normal, baryonic matter.

But later calculations invalidated the idea, for even in an inhomogeneous early universe, Ω in baryons cannot be 1.0. "Instead of being a loophole," said Schramm, "it's ended up showing how robust the basic calculation really is, and that's been one of the things that's given people a lot more confidence with the conclusion that Ω in baryons really does have to be low."

This does not necessarily mean, though, that the early universe was smooth and homogeneous. If the beryllium and boron in the halo stars were manufactured by cosmic rays, as Schramm believes, rather than in an inhomogeneous big bang, the boron abundance should be about ten times higher than the beryllium, because cosmic rays make more boron than beryllium. Unfortunately, present observations of halo stars are ambiguous. The measured boron-to-beryllium ratio agrees with the cosmic-ray theory, but the observational errors are large, and Gilmore thinks ruling out an inhomogeneous early universe would be premature.

"In Scotland," said Gilmore, "they have this delightful legal system which instead of saying innocent or guilty says innocent, guilty, or not proven. Not proven means that we're not actually convinced enough that you're innocent to say you're innocent, but we can't say you're guilty. It has a hint of guilt about it without being conclusive. So 'not proven' is definitely the right verdict at this stage." Future observations should pin down the abundances of beryllium and boron and determine whether or not they might have originated in an inhomogeneous early universe, but for now the issue remains unresolved.

Whatever the outcome, the abundances of the other light elements in the oldest Galactic stars bolster big bang cosmology. Yet the same stars also pose its greatest challenge. Astronomers who estimate stellar ages have reached an unsettling conclusion: the Galaxy's oldest stars look older than many cosmologists believe the universe itself to be, a logical impossibility that could torpedo conventional cosmology altogether.

17

OLDER THAN THE UNIVERSE?

No ONE EVER thought it would come to this. No one ever thought the Milky Way Galaxy might topple the big bang. Large though the Milky Way is, it is just one galaxy in a universe tens of billions of light-years wide. But today, a disturbing specter haunts astronomy: the Galaxy may be older than the universe, a logical contradiction that could demolish standard cosmology.

According to best estimates, the Galaxy's most ancient stars, such as those in metal-poor globular clusters, are 15 billion years old, a number derived from the clusters' H-R diagrams. The universe, then, must be at least this old, and it is probably older, because the Milky Way did not form instantly after the big bang. Yet when they apply the equations of big bang cosmology, most astronomers find the universe to be younger than 15 billion years.

To derive the age of the universe, cosmologists use the Hubble constant, the rate at which the universe is presently expanding from the big bang. By running the universe's expansion backward, cosmologists can compute at what point in the past the universe had zero size. The time to this cosmic ground zero is the age of the universe. The faster the universe's present expansion—the larger the Hubble constant—the less time that has elapsed since the big bang and the younger the universe must be. In the last decade, many astronomers have measured values for the Hubble constant so high that they suggest the universe is only 11 or 12 billion years old.

Although the Hubble constant is pivotal to the determination of the universe's age, astronomers have yet to agree on its numerical value. In fact, the measurement of the Hubble constant is so contentious that the participants more resemble knights doing battle than scientists conducting research. For twenty years, astronomers have repeatedly attacked one another, some finding the Hubble constant twice as large as others claim. So persistent is the Hubble war that it resembles the Hundred Years' War between England and France—whose duration, had it been known beforehand, might have deterred both sides from entering into it.

"Both sides are so deeply entrenched," said Allan Sandage, stalwart champion of a low Hubble constant and an old universe, "and amazingly, answers that seem positive are gotten on both sides. I don't believe that the debate will end in my lifetime. Only when all of us die will you young people be able to look at the thing dispassionately."

For now, though, the Hubble war is anything but dispassionate. A few months after one side reports its favored value, the other side counterattacks and offers support for its preferred value. During a 1993 talk in Berkeley, one of Sandage's opponents said the struggle looks to outsiders like a soccer game, with two competing teams trying to kick the ball in opposite directions. But that perception is wrong, he said; to those involved, the Hubble war resembles mud wrestling.

"I find contention, controversy, terrible," said Sandage. "I now don't read anything that will disturb me, if I have a suspicion that it will. There's a lot of propaganda out there. I always thought science would be its own best advertisement for ferreting out truth and falseness, but that's not true. It's a political game."

The Hubble war is bitter because so much is at stake. On a personal level, the careers of several astronomers hang in the balance, and on a scientific level, the age of the universe hinges on the outcome. And though few acknowledge it, big bang cosmology itself is in jeopardy, for if the Hubble constant is too high, standard cosmology will claim the impossible: that the universe is younger than the Milky Way.

THE ELUSIVE HUBBLE CONSTANT

The Hubble constant was first determined, in the loosest sense of the word, by Edwin Hubble himself, after he discovered that the universe is expanding. But it was actually another astronomer, Vesto Slipher at Lowell Observatory, who first discerned clues of the universe's expansion. During the 1910s Slipher obtained spectra of what are now known to be other galaxies and found that most had redshifts, a sign that the galaxies were receding from the Milky Way. Slipher also found that fainter and presumably farther galaxies tended to have larger redshifts and so were moving away faster.

But during the 1920s, Hubble employed larger telescopes than Slipher did and was able to do what Slipher could not: measure the galaxies' distances. He did so by examining nearby galaxies, such as Andromeda and M33, and detecting individual stars in them. Comparing the stars' intrinsic brightnesses with their apparent brightnesses, he deduced the distances to the stars and therefore to their galaxies.

In 1929, after determining the distances of different galaxies, Hubble reported that the farther the galaxy, the greater the redshift Slipher had measured. This correlation exists because the universe is expanding, not because the Milky Way inhabits a special, central point in the universe from which all other galaxies recede. To see this, imagine an expanding balloon. No point on the balloon's surface is at the center, but every point on the balloon's surface moves away from every other point on the surface. And the farther two points are from each other, the faster they move apart. In the universe, the farther a galaxy is from us, the more expanding space there is between us and the galaxy, so the faster the galaxy is carried away.

But the redshift that most galaxies exhibit is not a Doppler shift. A Doppler shift occurs when an object moves through space, either toward us, so that the light waves get scrunched up, producing a blueshift, or away from us, so that the light waves get stretched, producing a redshift. Most galaxies move away from us, but it is not this movement which causes the redshift. Instead, it is the expansion of space between the galaxy and the Milky Way that stretches the light waves and produces the redshift. Imagine a distant galaxy that emits light with a single wavelength. As the light journeys toward us, the space it travels through expands, so the light's wavelength gets stretched and a redshift results. Astronomers call this redshift, which results from the expansion of space, the cosmological redshift. The longer the galaxy's light has traveled through space—that is, the farther the galaxy is from Earth—the greater the redshift, just as Hubble discovered.

The Hubble constant states how fast the universe is presently expanding. In order to determine the Hubble constant, astronomers measure a galaxy's redshift and divide it by the galaxy's distance; the resulting number shows how much the redshift increases with distance. Suppose an astronomer observes other galaxies and finds that, on average, for every megaparsec farther a galaxy is, the redshift increases by 40 kilometers per second. (One megaparsec—the unit of distance that cosmologists use—equals 3.26 million light-years, somewhat greater than the distance to the Andromeda Galaxy.) Then the Hubble constant would be 40 kilometers per second divided by 1 megaparsec, or 40 kilometers per second per megaparsec. But now suppose each galaxy is only half as far as the astronomer thought. Then the redshifts grow twice as fast per megaparsec, so the universe is expanding twice as fast and the Hubble constant is twice as high. Numerically, one would divide 40 kilometers per second by *half* a megaparsec, which yields a Hubble constant of 80 kilometers per second per megaparsec.

Redshifts themselves are easy to measure, and everyone agrees on their observed values. What astronomers fight over is the measurement of *distances* to galaxies, which are extremely difficult to establish. The only galaxies whose distances astronomers know well are the few that lie near the Milky Way, such as the Large Magellanic Cloud (160,000 light-years away), Andromeda (2.4 million light-years away), M33 (2.6 million light-years away), and M81 (11 million light-years away). These distances are agreed on because astronomers can see the galaxies' Cepheids, the pulsating yellow supergiants that reveal the distances of the galaxies in which they reside.

Unfortunately, these nearby galaxies do not yield the Hubble constant. Because they feel the gravitational attraction of the Milky Way and other galaxies near them, their overall motion does not reflect the pure motion caused by the universe's expansion. In fact, some galaxies, such as Andromeda, actually show a *blue*shift and are moving toward the Milky Way. This is because the strong gravitational attraction between Andromeda and the Milky Way overwhelms the expansion of space between them and makes them approach each other, despite the universe's expansion.

To probe the expansion of the universe, astronomers must therefore observe galaxies that are hundreds of millions of light-years away. Their redshifts are caused almost entirely by the expansion of space between them and the Milky Way. The problem, though, is that these galaxies are so distant that their Cepheids cannot be seen, so their distances are unknown. Hubble himself severely underestimated the distances to other galaxies and obtained an enormous Hubble constant of around 500 kilometers per second per megaparsec, which implied that the universe was less than 2 billion years old—younger than the Earth.

Later work increased the estimated distances to galaxies and thereby lowered the value of the Hubble constant, and during the 1960s and 1970s Sandage and his long-time collaborator Swiss astronomer Gustav Tammann refined the distances further. Sandage and Tammann found that the Hubble constant was around 50 kilometers per second per megaparsec. This was a low value that meant the universe was old, probably older than the Milky Way. At the time, other astronomers believed the result, and harmony prevailed.

THE BATTLE BEGINS

In 1976, however, the opening shot in the Hubble war was fired when a French-born astronomer at the University of Texas at Austin, Gérard de Vaucouleurs, attacked Sandage and Tammann's result. De Vaucouleurs used elaborate methods to determine the distances of distant galaxies, and found that the galaxies were only half as far as Sandage had claimed. This led to a Hubble constant of around 100, twice the Sandage value, and implied that the universe was only half as old.

After de Vaucouleurs's challenge, other astronomers entered the fray. Some found low values for the Hubble constant, in support of

Sandage, while others found high values, in support of his arch-rival. During the two decades of warfare, the fronts on the battlefield have shifted, but a factor of two still separates the opponents. Today the low side often finds values for the Hubble constant in the 40s, whereas the high side now favors a value around 80. Such wildly different numbers result because each side uses different methods to measure the distances of galaxies. According to Sandage, all reliable methods give a low value for the Hubble constant, while his opponents claim that all reliable methods give a high value. In the Hubble war, it seems, a method is considered "reliable" only if it yields the answer the astronomer already believes.

In recent years, many astronomers have found a high value, and some have even stated that the Hubble constant is finally determined, with a value around 80. But as one long-time observer of the Hubble war remarked, claiming that the Hubble constant is known is one of the great lies in astronomy. Indeed, the first two papers in the September 20, 1993, issue of *The Astrophysical Journal* demonstrate that the dispute remains as unresolved as ever. In the first paper, Sandage and Tammann report that the universe is expanding slowly and that the Hubble constant is 47 ± 5. In the second paper, de Vaucouleurs says the universe is expanding nearly twice as fast and that the Hubble constant is 87 ± 1. Little wonder, then, that a recent review attempting to sort out the mess began with a passage from Mark Twain: "The researches of many commentators have already thrown much darkness on this subject, and it is probable that, if they continue, we shall soon know nothing at all about it."

Every war makes famous the sites of its greatest battles. Just as World War II gave us Pearl Harbor, the Hubble war has given us IC 4182, an obscure galaxy in the constellation Canes Venatici. A cross between a small spiral and an irregular, IC 4182 illustrates the ferocity of this war, for both sides have attacked and tried to seize the dim galaxy.

At first sight, it hardly seems worth fighting for, since IC 4182 is a ragtag galaxy comparable in size to the Small Magellanic Cloud. In 1937, however, IC 4182 gave birth to a type Ia supernova, an object that some soldiers in the Hubble war hold sacred. Type Ia supernovae are so luminous they can be seen exploding in galaxies billions of light-years away. Moreover, many astronomers believe these supernovae are

"standard candles," all peaking at the same intrinsic brightness because all are thought to arise from the same kind of star, an exploding
white dwarf that exceeds the Chandrasekhar limit of 1.4 solar masses.
By comparing a type Ia supernova's intrinsic brightness with its apparent brightness, astronomers can determine the distance of both the
supernova and the galaxy that spawned it.

Unfortunately, no one knows the intrinsic brightness of a type Ia
supernova, because none has recently exploded in a galaxy with a
known distance. But IC 4182 is fairly nearby, and if one knew its
distance, one would also know the intrinsic brightness of the 1937
supernova. If other type Ia supernovae have the same intrinsic brightness, astronomers could determine the distances to the distant galaxies
in which those supernovae exploded. These distances, along with the
galaxies' redshifts, would give the Hubble constant. (The distance of
IC 4182 alone does not give the Hubble constant, since the galaxy is
too near the Milky Way for its motion to reveal the expansion rate of
the universe.)

The best way to determine the distance to IC 4182 is by observing
its Cepheids, but during the 1980s none could be seen. Instead, in
1982, Sandage and Tammann reported on IC 4182's red supergiants.
While red supergiants are not as reliable as Cepheids, they outshine
Cepheids and so are easier to see. And like Cepheids, red supergiants
yield a distance estimate, because one galaxy's brightest red supergiants are about as luminous as those in another galaxy. Using the red
supergiants, Sandage and Tammann placed IC 4182's distance at 14
million light-years. This distance gave the 1937 supernova's intrinsic
brightness, which Sandage and Tammann then applied to type Ia
supernovae in more distant galaxies to reach a Hubble constant of
50 ± 7.

In 1992, however, a team led by Michael Pierce of Kitt Peak National Observatory stormed IC 4182 and seized it for themselves.
Pierce and his colleagues armed themselves with CCDs—charge-cou-
pled devices—electronic light detectors that are more sensitive and
accurate than the photographic plates Sandage and Tammann had
used. Pierce's team found that the red supergiants had brighter apparent magnitudes than Sandage and Tammann had said and that the
galaxy lay only 8 million light-years away. The smaller distance meant
that the 1937 supernova was less luminous, which in turn indicated

that the distant galaxies were less distant and that the Hubble constant was 86 ± 12.

But later that year, Sandage's side counterattacked and regained possession of the galaxy. Using the Hubble Space Telescope, Sandage and his colleagues succeeded in detecting IC 4182's Cepheids. These indicated that the galaxy is 16 million light-years away, close to Sandage's original 1982 estimate. The Cepheid distance gave a Hubble constant of 45 ± 9.

Still, the battle for IC 4182 is far from over. As soon as Sandage announced his result, the other side speculated that dust in IC 4182 dims its Cepheids, making them look farther away than they really are. Furthermore, some astronomers now suspect that type Ia supernovae are not even reliable standard candles but instead peak at different luminosities.

Meanwhile, several new techniques for measuring galaxy distances have given high values for the Hubble constant. For example, John Tonry of the Massachusetts Institute of Technology has invented a clever method for establishing the distance of a galaxy by observing how smooth it looks, just as a newspaper photograph of a blue sky looks smooth from a distance but breaks up into little dots if viewed close up with a magnifying glass. Through this method, Tonry finds a Hubble constant around 80. Another technique, developed by Kitt Peak's George Jacoby, uses planetary nebulae in a galaxy to measure its distance. He also finds a Hubble constant around 80.

THE AGE OF THE UNIVERSE

If correct, high values like these for the Hubble constant imply a young universe. In standard cosmology, the age of the universe depends on the Hubble constant and also on Ω, the mass density of the universe, because mass exerts a gravitational attraction that slows the expansion. To see how this works, first imagine a universe with space but no mass ($\Omega = 0$). The expansion of such a universe is represented by the straight line marked A in Figure 20. Because there is no mass to slow the expansion, this universe always expands at the same speed, and the age of the universe—the time to the big bang—is just the inverse of the Hubble constant. This time is called the Hubble time.

But the real universe has mass, and the gravitational pull of that mass slows the expansion, as shown by the curved line marked B in Figure 20. This means that the universe expanded faster in the past than it does today. The Hubble constant measures only the present expansion, so if the universe once expanded faster, it must be younger than if it had no mass. In other words, a universe with mass is younger than the Hubble time, and the more mass the universe has, the younger it is, because the faster it expanded in the past.

The age of the universe therefore depends on Ω. But the value of Ω is even more uncertain than that of the Hubble constant. Most observations suggest that Ω is low, around 0.1 or 0.2. But inflationary cosmology, which postulates that the universe inflated enormously a fraction of a second after its birth, demands that Ω be 1.0.

Using a complicated equation, cosmologists can calculate how the age of the universe depends on Ω. For example, if Ω is 0.1, this equation indicates that the universe's age is 90 percent of the Hubble time —that is, 90 percent of the age the universe would have if Ω were 0. If the universe is more massive and Ω is 0.2, then this equation reveals the age to be 85 percent of the Hubble time. And if Ω is 1.0, as inflation demands, then the universe's age is only 67 percent of the Hubble time.

Suppose the Hubble constant is only 50, as Sandage believes. Then the Hubble time is 19.6 billion years. If Ω is 0.1, the universe's age would be 90 percent of this—17.6 billion years old. If Ω is 0.2, the universe would be 85 percent as old as the Hubble time, or 16.6 billion years old. Both ages exceed and are therefore in harmony with the Milky Way's oldest globular clusters, which are about 15 billion years old. But if Ω is 1.0, the universe's age would be only 67 percent of the Hubble time—13.0 billion years old, which is younger than the oldest globulars.

Now suppose that the Hubble constant is 80, as favored by those on the high side of the Hubble war. Then the Hubble time is only 12.2 billion years, already younger than the globulars. But the real universe is even younger. If Ω is 0.1, the universe would be 11.0 billion years old; if Ω is 0.2, the universe would be 10.3 billion years old; and if Ω is 1.0, as inflation claims, the universe would be only 8.1 billion years old. All these ages are younger than the ancient stars residing in the Milky

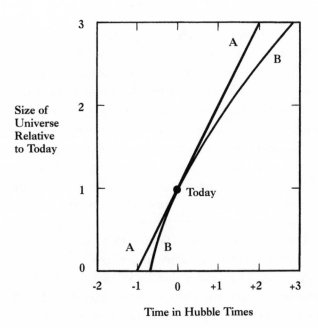

Figure 20. This figure shows the expansion of two model universes, each with the same present expansion rate. Universe A has no mass ($\Omega = 0$) and universe B has considerable mass ($\Omega = 1.0$), so universe A is older than universe B.

Way's globular clusters and pose serious trouble for cosmologists—if the ages of those globulars are right.

THE PROBLEMS OF DATING AN OLDER STAR CLUSTER

High values for the Hubble constant obviously press on those astronomers—both observers and theorists—who find old ages for the Milky Way's globular clusters. "Nobody leaves threatening messages on my answering machine," said Michael Bolte at Lick Observatory, "but people who find 75 or 80 for the Hubble constant just assume we're doing something wrong and would like us to find out what it is. People have been scratching their heads to find ways to make globular clusters look younger. But I think it actually makes it all more interesting to have this conflict."

Surprisingly, Bolte himself believes the Hubble constant is high, around 75, and thinks methods giving a low value—such as that favored by Sandage, type Ia supernovae—are wrong. "I think it's a bad idea," said Bolte. "In fact, I think it's a demonstrably bad idea. Type Ia supernovae are bad standard candles. As people have looked at more and more type Ia's, they've started throwing more and more of them out as being either overluminous or underluminous."

But Bolte also admits that the ages of globulars are more uncertain than many Hubble warriors realize. These ages are based on the clusters' H-R diagrams. As a cluster ages, its brightest main-sequence stars die, then somewhat less bright main-sequence stars die, and so on. Thus, the luminosity of the brightest surviving main-sequence stars indicates the cluster's age: the fainter they are, the older it is. In Bolte's view, the chief uncertainty in dating a globular cluster is the cluster's distance. The farther one thinks a particular cluster is, the more luminous its stars will seem, so the younger the cluster will look. A distance error of only 10 percent translates into an age error of about 1.8 billion years.

Distances to globulars are difficult to establish, for even the nearest lie thousands of light-years away. To estimate globular distances, astronomers can use the clusters' RR Lyrae stars, but the intrinsic brightness of RR Lyrae stars is somewhat uncertain, because none is close enough to the Sun to have a reliable parallax and thus a reliable distance. Astronomers can also use the main-sequence stars in globulars, but to do so they need to know the absolute magnitudes of subdwarfs—metal-poor main-sequence stars—because the oldest globulars are metal-poor. Unfortunately, subdwarfs are rare and only two are close enough to have good parallaxes, reliable distances, and accurate absolute magnitudes. One is Kapteyn's Star, 13 light-years away, but it is red and does not help, because red subdwarfs glow too dimly to be seen in globulars. The other is Groombridge 1830, which is 28 light-years away and a yellow star similar to the ones astronomers can observe in globulars. The cluster distances—and ages—therefore revolve primarily around Groombridge 1830, plus six other subdwarfs with less accurate parallaxes. Such a strong reliance on such a small sample of stars hardly inspires great confidence. Fortunately, the situation will soon improve, because from 1989 to 1993 the European satellite Hipparcos measured accurate parallaxes for dozens of subdwarfs.

Still, even after astronomers analyze the Hipparcos data, Bolte believes his estimates will indicate that the globulars are old. "I think that when everything gets said and done, which is going to be in about five years, globular clusters are going to be older than 13 billion years, and they'll probably be around 15 billion years." Nevertheless, with present data, he can't rule out somewhat younger ages. "We could probably live with 13 [billion years] right now," he said. "It pushes everything, but we could live with that. I think 12 is going to be hard; 11 is almost out of the question; and 10 is definitely out of the question."

But Bolte does not think the globulars' old age necessarily conflicts with a high Hubble constant of 75. "It completely depends on what you believe Ω is," said Bolte. "If you just accept that Ω equals 1.0—if you're an inflation sort—then there's a big conflict. But if Ω is 0.1, which I think is more consistent with the data, then you can just barely squeeze the cluster ages in there. So all that really forces you into is that globular clusters formed very early in the universe."

In order to calculate a globular cluster's age, astronomers must use theoretical models that describe a star's evolution. These models yield a theoretical H-R diagram that Bolte and other astronomers then compare with the observed H-R diagrams to establish the cluster's age. Although Bolte believes that these theoretical models are sound, one astronomer who has spent his career computing them has recently grown concerned. New observations, including some by Bolte, hint that the metal-poor globulars have a different luminosity function—that is, number of stars of a given luminosity—than the models predict.

"Until the recent work that showed the discrepancy with the luminosity functions," said Don VandenBerg of the University of Victoria in British Columbia, "I myself would have said that the ages were 14 ± 2 or 3 billion years, in that I think it's highly unlikely that the globulars are younger than 11 or 12 billion years or very much older than 17 or 18 billion years. Mike Bolte's work, especially, has given me real concern. I'm nearly at the point of believing that there's a problem with standard models of very metal-poor stars."

Before a yellow main-sequence star like the Sun becomes a red giant, it expands into a subgiant. But the metal-poor globulars appear to have more subgiants than VandenBerg's models predict. The relative number of subgiants depends on how fast stars pass through this stage. Because the models seem to underestimate the number of subgiants, the models say that the stars speed through this stage faster than they really do.

"If you can't reproduce the relative lifetimes of stars in different evolutionary phases," said VandenBerg, "then that's a deep stellar structure problem that you're looking at. The only way in which we can have real confidence in the derived ages is if the model predictions match the observed luminosity functions." If the discrepancy persists after new observations are obtained, then the models must be missing something, but VandenBerg doesn't know what that is, or whether it will make his estimates of globular ages rise or fall. Despite these problems, though, he suspects that the estimated ages of the globulars may well remain high. "Somehow," he said, "cosmology may have to accommodate ages of globular clusters of 14 or 15 billion years."

THE COSMOLOGICAL CONSTANT: BLUNDER OR BREAKTHROUGH?

So desperate are some astronomers that they have recently pulled a wild card out of their hat: the notorious cosmological constant, introduced but then abandoned by Albert Einstein, who called it the biggest blunder of his life. In 1916 Einstein published his greatest triumph, the general theory of relativity, which describes gravity. But in 1917, when Einstein applied his equations to the universe, he found that the gravitational force of its mass would have made the universe collapse. To solve the problem, he postulated that empty space exerted a repulsive force, a sort of antigravity that held up the universe. This repulsive force was represented by a term in his equations he called the cosmological constant. It kept the universe static—that is, unchanging in size—as astronomers then believed it was.

Of course, they now know the universe is expanding, not static. In fact, by the time Einstein introduced the cosmological constant, Vesto Slipher had already found that most of what are now known to be other galaxies had redshifts, but Einstein never heard the news. Only after Hubble discovered the expansion of the universe did Einstein throw out the cosmological constant, and it has lived in infamy ever since.

Lately, though, the cosmological constant has returned, for it can make the universe older than the globular clusters. Because it represents a repulsive force, one that opposes gravity, the cosmological constant behaves in the opposite way from Ω. The more massive the universe and the greater Ω, the faster the universe expanded in the past and the younger it now is; but the greater the cosmological constant, the more slowly the universe expanded in the past and the older it now is. This is because as the universe gets larger, it has more space and so expands faster and faster, since the cosmological constant represents the repulsive force of empty space. Therefore, in the past, when the universe was smaller and had less space, it must have expanded more slowly, if the cosmological constant exists. This is illustrated by model C in Figure 21.

Astronomers designate the size of the cosmological constant with the Greek letter lambda, λ. If there is no cosmological constant, λ is zero. But if the cosmological constant exists, empty space has energy, and that energy can be expressed as mass, since as Einstein showed,

Figure 21. A cosmological constant increases the age of the universe. Universe C has little mass ($\Omega = 0.01$) but a large cosmological constant ($\lambda = 0.99$); it is therefore older than universes A ($\Omega = 0$) and B ($\Omega = 1.0$), which have no cosmological constant.

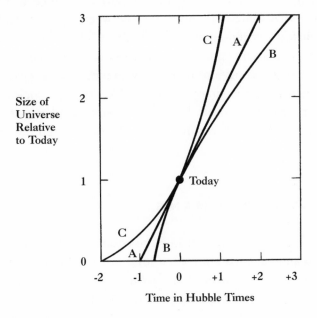

mass and energy are equivalent. If, for example, Ω is 0.1 and λ is also 0.1, the equivalent mass represented by the cosmological constant would equal the real mass of the universe.

But the cosmological constant counteracts and can even defeat Ω. For example, if there is no cosmological constant, and if Ω is 2.0, the universe will someday collapse, because the gravitational pull of its mass would reverse the expansion. But if λ is 1.0, the repulsive force of the cosmological constant would overwhelm even an Ω of 2.0 and keep the universe expanding forever.

The key feature of the cosmological constant, though, is that it increases the estimated age of the universe. Suppose the Hubble constant is high—75—and Ω is 0.15. If the cosmological constant does not exist, the universe's age would be only 11.3 billion years. But if the cosmological constant does exist and λ is 0.85, the universe would be 15.1 billion years old, about as old as the globular clusters.

Powerful though it is, most cosmologists do not like the cosmological constant, deeming it an ugly fudge factor. But in 1990 they realized that they could actually measure its value. All other things being equal, the larger the cosmological constant, the farther away a distant object with a particular redshift should be. This is because the repulsive force of the cosmological constant has accelerated the object away from us in the time since it emitted its light. The farther the object, the greater the chance that some other object lies between it and us and gravitationally distorts, or lenses, the original object's light. So the larger the cosmological constant, the more gravitational lenses there should be. But recent searches have turned up few gravitationally lensed quasars, which suggests, in the simplest model, that the cosmological constant is small or zero. If this result holds up, a cosmological constant will not resolve the conflict between the age of the Galaxy and the age of the universe.

DUMP THE BIG BANG?

Although cosmologists do not like the cosmological constant, most of them like another possibility even less: that the big bang never happened. Deriving the universe's age by rewinding the universe's expansion to the time of the big bang presupposes that the universe actually began with a big bang. But if the big bang never happened, that approach is obviously invalid.

During the late 1940s and early 1950s, a similar conflict between the ages of two objects bolstered the big bang's rival, the steady state theory, which Fred Hoyle, Hermann Bondi, and Thomas Gold had proposed in 1948. At the time, estimates of the Hubble constant were so high that the big bang theory made the universe even younger than the Earth. The steady state cosmology ran into no such problems, because it maintained that the universe had existed forever. Fortunately for big bang proponents, later work lowered the Hubble constant and erased the discrepancy between the age of the Earth and the age of the universe. But the present discrepancy may not vanish so easily.

In 1993, Hoyle, Geoffrey Burbidge, and Jayant Narlikar revived the steady state cosmology, noting the conflict between the ages of the Galaxy and the universe. According to the steady state cosmology, the universe is expanding, just as in the big bang cosmology. However, if the universe is indeed infinitely old, it should have expanded so much that the density of matter would have fallen to zero unless new matter entered the universe. To deal with this problem, the original steady state theory proposed that particles of matter were spontaneously created in intergalactic space at just the right rate to compensate for the thinning effect of the universe's expansion. The new steady state theory, which Hoyle and his colleagues dub the quasi-steady state, says that matter enters the universe through titanic explosions called "little big bangs" that occur throughout the universe. Quasars may be one example of these explosive events. The little big bangs cause space to expand and also create the light elements—hydrogen, helium, and lithium—in the amounts observed in the Galaxy's oldest stars. "This paper is not intended to give a finished view of cosmology," wrote Hoyle and his colleagues in 1993. "It is intended rather to open the door to a new view which at present is blocked by a fixation with big bang cosmology."

But the new steady state theory has attracted few admirers. "For whatever reason," said Burbidge, "people have never liked the steady state cosmology. They've always attacked it, and I think part of the reason is religious." Since the steady state universe had no beginning, the cosmology contradicts those religions, like Christianity, which hold that the universe had a definite origin. During the 1950s, when the original steady state battled the big bang, the Catholic Church officially sided with the latter cosmology.

Burbidge believes that the Hubble constant is low—around 50—so its value does not threaten big bang cosmology, but he nevertheless considers the new steady state theory attractive. He does not claim that it is necessarily right, only that it deserves a fair hearing, which he says it has not yet received. Critics maintain that the new steady state, like the old one, cannot explain the microwave background, which most cosmologists believe is the big bang's afterglow. Hoyle, Burbidge, and Narlikar counter that intergalactic space is filled with small metallic needles that get heated by starlight and radiate at microwave frequencies.

Burbidge is displeased with the scientific establishment's attitude toward the new steady state theory. "There now appears to be blatant hostility from some quarters any time a minority opinion in a popular subject is brought up," he said. "That says essentially that a minority point of view is guilty until proven innocent. You have to ask yourself what that says about new ideas in science. It's *damning* for the younger people, because they know that they can't get to first base by suggesting anything new. They've got to fight their way over this barrier, which has got higher and higher since I was a young man. In the old days, it was much easier to publish a paper in *Nature* about something new and different; nowadays, it gets refereed to death.

"This peer review business has gone mad," said Burbidge. "It's something to do with people's need to be right, and in science, most of the time we're wrong. The history of science shows you that. If you read what people thought two hundred years ago, they were quite convinced they were right, but they were wrong." Without pausing, Burbidge added, "And the same thing will happen again."

Although few cosmologists are as eager as Burbidge to question orthodox cosmology, most of them acknowledge the dilemma posed by the Milky Way's apparent old age and the universe's apparent youth. There are four possible solutions: the Hubble constant is low and the universe old; the globular clusters are younger than astronomers now estimate; the cosmological constant is large; or the big bang is wrong. Most astronomers today would opt for one of the first two, a few for the third, and a tiny minority for the fourth. But no one can predict the outcome.

1 8

An Intelligent Galaxy

THE ALARM CLOCK rang at three in the morning. On April 8, 1960, Frank Drake walked through the cold fog to the 85-foot radio telescope in Green Bank, West Virginia. His mission was no ordinary one, for he was about to search for radio signals from extraterrestrial life.

Drake's first target was Tau Ceti, a yellow G-type star eleven light-years away that resembles the Sun. Drake and his colleagues pointed the telescope at the star and began to monitor it for unusual radio emissions. A mechanical chart recorder received the signals and wrote them out on paper, providing a continuous record of the star's output. All morning long, as the astronomers observed the star, the chart recorder's pen rose and fell, but the signals were just random noise. Tau Ceti was quiet.

At noon, Drake swung the telescope toward his second target, Epsilon Eridani, an orange dwarf that also lies eleven light-years away. At first, it was also quiet. A few minutes later, though, the pen on the chart recorder went wild, jumping off scale and banging the top of the recorder eight times a second.

Now what? Drake had never thought finding extraterrestrial life would be so easy. To test the apparent discovery, Drake moved the radio telescope away from Epsilon Eridani, and the signal vanished, suggesting that the star itself, rather than terrestrial interference, was the source. But when he pointed the telescope back at the star, the signal did not resume. Puzzled, he observed Epsilon Eridani into the afternoon, until the star set. It remained silent.

In the following days, Drake and his colleagues continued to track the star, but the Epsilon Eridanians refused to stage an encore. In order to test the extraterrestrial nature of any additional signals, the scientists had set up a small radio antenna that picked up terrestrial noise. A true extraterrestrial signal would therefore arrive via only the main receiver.

A week and a half after the first detection, while the astronomers were observing Epsilon Eridani, the strong signal returned, once again pulsing eight times a second. This time, though, the scientists noted with disappointment that the pulsations came through both receivers, which meant that the signal originated nearby, probably from a passing plane. The scientists had indeed found proof of intelligent life—on Earth rather than around Epsilon Eridani.

Searching for Life Among the Stars

Although searches for extraterrestrial intelligence have so far failed, some astronomers believe the Milky Way teems with intelligent life, both because the Galaxy has hundreds of billions of stars and because, unlike most galaxies, it is metal-rich, containing the heavy elements, like carbon, nitrogen, and oxygen, that life requires. In addition, the Galaxy has already proven its power by producing at least one intelligent species: us.

For all its grandeur, however, the Milky Way probably does not harbor life around most of its stars. After all, a good star is hard to find. Life exists on Earth because our planet lies the right distance from a bright yellow star that pours out generous amounts of light and heat. Most of the Galaxy's stars, though, are fainter and cooler than the Sun and probably cannot support life.

Of particular interest to those who seek life in the Milky Way are the stars nearest the Sun, like Tau Ceti and Epsilon Eridani. Such stars are close enough for a radio telescope on Earth to detect weak signals emanating from them and to issue a reply that would take only a few years to reach the recipients. Furthermore, if scientists ever launch an

TABLE 18-1: THE NEAREST STARS

Star	Distance (Light-Years)	Spectral Type	Magnitudes Apparent Visual	Absolute Visual	Velocities (Km/Sec.) U	V	W
Sun	0.00	G2	−26.74	4.83	−9	+12	+7
Alpha Centauri A	4.35	G2	0.00	4.38	+20	+13	+21
B	4.35	K1	1.36	5.74	+20	+13	+21
C	4.25	M5	11.09	15.52	+16	+9	+20
Barnard's Star	5.96	M3.5	9.55	13.24	+132	+17	+25
Wolf 359	7.8	M6	13.45	16.56	+19	−36	−5
Lalande 21185	8.25	M2	7.50	10.48	−55	−41	−67
Sirius A	8.6	A1	−1.46	1.43	−24	+12	−4
B	8.6	WD	8.39	11.28	−24	+12	−4
Luyten 726-8 A	8.8	M5.5	12.52	15.38	+35	−8	−12
B	8.8	M5.5	13.02	15.88	+35	−8	−12
Ross 154	9.5	M3.5	10.47	13.14	+7	+10	0
Ross 248	10.3	M5.5	12.29	14.79	−42	−62	+7
Epsilon Eridani	10.7	K2	3.73	6.15	−6	+20	−14
Ross 128	10.9	M4	11.12	13.50	−27	+18	−26
Luyten 789-6 A	11.1	M5	12.33*	14.66*	+60	+11	+48
B	11.1	—			+60	+11	+48
Epsilon Indi	11.2	K4	4.69	7.00	+68	−26	+11
Struve 2398 A	11.3	M3	8.90	11.20	+15	0	+33
B	11.3	M3.5	9.68	11.98	+15	0	+33
Tau Ceti	11.4	G8	3.50	5.79	−27	+41	+19
Procyon A	11.4	F5	0.38	2.66	−14	+3	−11
B	11.4	WD	10.8	13.1	−14	+3	−11
61 Cygni A	11.4	K5	5.21	7.50	+84	−41	−1
B	11.4	K7	6.03	8.32	+84	−41	−1
Lacaille 9352	11.5	M1	7.34	9.61	+91	−2	−49
Groombridge 34 A	11.6	M1	8.08	10.33	+40	0	+3
B	11.6	M3	11.07	13.32	+40	0	+3
Giclas 51-15	11.8	M6.5	14.79	16.99	−3	+6	−14

* These magnitudes are the combined output of Luyten 789-6 A and B.

interstellar spacecraft, its destination will be a nearby star. Because these stars lie so close, their parallaxes are well known and hence the stars' distances and luminosities are well determined.

By human standards, though, even the nearest star to the Sun is extremely distant. If the Galaxy shrank so that the space between the Sun and Earth became one inch, then Neptune, the farthest planet visited by spacecraft, would lie only thirty inches from the Sun. On the same scale, the nearest star system would be over four *miles* away. The

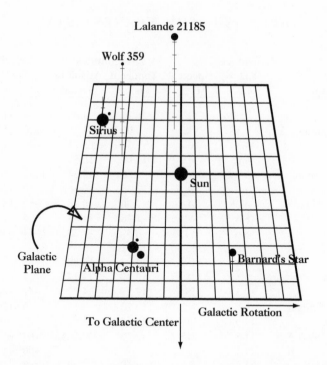

Figure 22. The very nearest stars to the Sun include the bright stars Alpha Centauri and Sirius as well as the faint red dwarfs Barnard's Star, Wolf 359, and Lalande 21185. In this figure, each side of a square equals one light-year.

fastest spaceships ever built would require tens of thousands of years to reach Alpha Centauri.

Nevertheless, within twelve light-years of Earth are twenty star systems, including the Sun. Seven of these systems are double, and one, Alpha Centauri, is triple, yielding twenty-nine individual stars. Three nearby systems, Alpha Centauri, Sirius, and Procyon, are among the brightest stars visible from Earth, but most of the Sun's neighbors are red dwarfs too faint for the naked eye to see. The abundance of red dwarfs in the Sun's vicinity reflects the Galaxy's preponderance of these stars.

Several stars on this list may warm Earthlike planets that have oceans, an atmosphere, and some form of life, while other nearby stars are probably circled only by dead worlds. To separate stars that might support life from those which do not, astronomers subject them to five tests. These tests are admittedly an extrapolation, based on the one intelligent civilization we know, but they are the best astronomers can do.

The first test a star must pass is that it be on the main sequence, as the Sun is. For most stars, this stage provides a long period of stability during which the star's light is fairly constant, giving life time to arise and evolve. In contrast, stars that have left the main sequence to become red giants and white dwarfs have wracked their planets with drastic changes. Since 90 percent of the Galaxy's stars are on the main sequence, most pass this test, and of the twenty-nine stars within twelve light-years, only three do not fit the bill. Sirius B and Procyon B are white dwarfs, and Procyon A is making the transition from the main sequence to the subgiant stage. But the other twenty-six stars remain candidates.

The second test is much tougher, and most stars fail. To support intelligent life, a main-sequence star must have the right spectral type, because this controls how long the star remains on the main sequence and how much light it generates during this stage. Most stars, of course, fall into one of the seven main spectral types—O, B, A, F, G, K, M—but astronomers divide these for further precision. For example, spectral type G, which includes the Sun, breaks into ten subtypes from hot to cool: G0, G1, G2, G3, G4, G5, G6, G7, G8, and G9. The Sun is spectral type G2. All G stars are yellow, with temperatures like the Sun's, but a G1 star is slightly hotter and a G3 star slightly cooler.

To foster intelligent life, a main-sequence star must live long enough to give intelligence time to evolve. On Earth, intelligent life took 4.6 billion years to evolve. This is a long time, between half and a third the age of the Galaxy itself, but neither astronomers nor biologists know whether this period is typical. Perhaps on other Earthlike planets, intelligence emerges much faster or much more slowly. Since no one knows, the safest assumption is that intelligence, if it evolves, usually takes several billion years, which means that high-mass main-sequence stars die too soon to foster intelligent life. For example, the blue O and B main-sequence stars live no more than a few hundred million years. Such stars are rare, however, and none lies within twelve light-years. The white A stars are also troublesome, since most die within a billion years of their birth, probably too soon for intelligent life to develop. This rules out the one A-type main-sequence star within a dozen light-years, Sirius A. The yellow-white F stars may or may not live long enough. The hotter F-type stars—with spectral types from F0 to F5—probably leave the main-sequence too fast, but the cooler F stars—with spectral types between F6 and F9—may be acceptable. Procyon A has a spectral type of F5, but it has already been eliminated because it has started to leave the main sequence.

In contrast, the cooler stars have lower masses and longer life-times. The yellow G stars live for many billions of years. The orange K and red M stars live even longer, but many of them face another problem because they are faint. The dimmest star within twelve light-years, Giclas 51-15, is a red dwarf discovered in 1972 with an absolute magnitude of 16.99, which means that the star produces less visible light in a century than the Sun does in a day. Since terrestrial life derives nearly all its energy from sunlight, the only way a planet circling a red dwarf could receive enough warmth would be to lie close to the star. But then the star's gravity would force one planetary hemisphere to face the star and the other to turn away, frying one side and freezing the other. For this reason, most astronomers believe that no red dwarf can sustain life. Some of the orange dwarfs, especially the cooler ones, have the same problem and are also unlikely prospects.

After this pruning, astronomers are left with only the cooler F stars, all the G stars, and the warmer K stars. Of the twenty-nine near-est stars, few survive the cut. Three G stars are on our list, the Sun, Alpha Centauri A, and Tau Ceti. Alpha Centauri A's spectral type is G2, the same as the Sun's, so this star is perfect. Tau Ceti, whose spectral type is G8, is somewhat cooler and fainter, with a luminosity 40 percent of the Sun's, but it too is a possibility, which is why Drake chose it as his first target. Five K stars also lie within a dozen light-years, Alpha Centauri B, Epsilon Eridani, Epsilon Indi, and the two stars of 61 Cygni. Alpha Centauri B and Epsilon Eridani have spectral types of K1 and K2, so they are probably warm enough, but the other three stars are cooler, fainter, and less likely to foster life. Thus, out of twenty-nine stars, between five and eight remain, one of which is the Sun.

The third test a star must pass is stability. If a star erupts, it will harm and possibly destroy any life on its planets. Red dwarfs often spew out huge flares that, for a few minutes, outshine the rest of the star. The first red dwarf flare star discovered, in 1948, was Luyten

726-8 B, which lies 8.8 light-years away and also bears the name UV Ceti. Astronomers have since found that many other red dwarfs, including Alpha Centauri C, Wolf 359, and Ross 154, emit powerful flares, but these stars have already been eliminated from consideration because they are red dwarfs. In contrast, most F, G, and K main-sequence stars are stable and provide a steady source of energy for their planets.

The fourth test is age. Even if a star is on the main sequence, of the right spectral type, and stable, it may not support intelligent life if insufficient time has passed for intelligence to evolve. A billion years ago, even the Sun had yet to deliver any intelligent life to the Galaxy. Therefore, the older the star, the better. Determining the ages of individual stars is difficult, however. If the star is a member of a cluster, the cluster's H-R diagram would give an estimate of its age; but nearby stars do not belong to any star cluster, so astronomers cannot apply this dating technique. Nevertheless, they can sometimes infer the ages of the nearby stars in other ways. In particular, Alpha Centauri A and B appear to be middle-aged stars somewhat older than the Sun, making them ideal candidates for intelligent life. This conclusion follows from Alpha Centauri A's luminosity, because as a main-sequence star ages, it gradually brightens. The Sun is 40 percent brighter today than it was at birth. Alpha Centauri A has the same G2 spectral type but is 50 percent more luminous than the present-day Sun. When astronomers work out the numbers, and account for the star's somewhat greater mass, they estimate Alpha Centauri A to be 5 to 6 billion years old, about a billion years older than the Sun. This age also applies to Alpha Centauri B, which formed at the same time as its mate.

Another otherwise promising neighbor, the orange dwarf Epsilon Eridani, is much too young to have intelligent life. Since Drake examined it for radio signals in 1960, astronomers have discovered that Epsilon Eridani rotates fast, once every eleven days, which is only a third of the Sun's rotation period. Rapid rotation signifies youth, because G, K, and M stars usually spin more slowly as they age, so astronomers estimate that Epsilon Eridani is less than a billion years old. It may nourish intelligent life in the future, but it probably does not do so today.

The two other nearby star systems containing orange dwarfs, Epsilon Indi and 61 Cygni, look more promising, because they are probably older than the Sun. Neither Epsilon Indi nor the two stars of 61 Cygni rotate rapidly, and both systems have moderately high U, V, W velocities, a sign of old age. Unfortunately, these stars are cool K dwarfs and may be too faint to support life. The G star Tau Ceti, Drake's first target, has the most uncertain age. It rotates slowly, which suggests that it is not young, but its space velocity is not particularly high. This star could be somewhat younger than the Sun or somewhat older.

The fifth and final test is based on metallicity. If a star is metal-poor, then small, rocky, Earthlike planets may not have formed around it, because such worlds consist mostly of heavy elements like iron, silicon, and oxygen. Moreover, life itself requires heavy elements, especially carbon, nitrogen, and oxygen. Alpha Centauri passes the metallicity test with flying colors, for it is even more metal-rich than the Sun. Epsilon Eridani has a metallicity 70 percent of the solar value, which may be enough to support life, but it has already been eliminated because of its youth. Tau Ceti's still lower metallicity, about 30 percent solar, is so low that it may have stunted the formation of planets and any life on them.

So only two or three nearby stars survive these five tests. Alpha Centauri is the best candidate, for both Alpha Centauri A and B are stable, metal-rich main-sequence stars somewhat older than the Sun. The brighter star, Alpha Centauri A, has a spectral type of G2, the same as the Sun's, and Alpha Centauri B has a spectral type of K1, probably warm enough for life. Tau Ceti, a single star with a G8 spectral type, remains a possibility, with its low metallicity the most serious liability.

THE QUEST FOR EARTHLIKE PLANETS

However, life needs more than just a good star. It also requires a good planet. Indeed, if ours is the only solar system in the Galaxy, then ours is also the only intelligent civilization in the Galaxy. Unfortunately, because planets emit no light of their own, astronomers have never seen one around any normal star besides the Sun. For decades, the nearby red dwarf Barnard's Star has exhibited an apparent wobble in its proper motion that some astronomers thought was caused by the gravitational tug of orbiting planets. More recent work, though, suggests that the wobble lay in the telescope, not the star. In 1991, American radio astronomers Aleksander Wolsczan and Dale Frail startled their colleagues by discovering at least two planets around a most unlikely star, PSR B1257+12, which is a pulsar in the constellation Virgo. Wolsczan and Frail achieved this feat by finding small changes in the pulsar's pulses that resulted from the pull of orbiting worlds. Planets in such a hostile environment could hardly support life, but that they exist around a pulsar suggests that they may also circle more ordinary stars. In fact, most stars probably do have planets, because the formation of planets is a natural consequence of star birth. Many young stars have disks of material orbiting them that are probably solar systems in the making.

The Sun has nine planets, but only one has intelligent life. The Earth is at the right distance from the Sun to have liquid water, the prime requirement for life. Terrestrial life originated in the oceans, and water makes up over half the human body. Venus lies closer to the Sun and is too hot for water or life, and Mars lies farther away and is too cold, with all of its water frozen.

A planet's size is also important. Earth has enough mass and gravity to retain a thick atmosphere. But the Moon, which is at the same distance from the Sun, is so small that it has almost no air at all. Mars, half Earth's size, might have been able to support life if it had been larger, for then it would have been warmer. One reason Earth is warm is that it has atmospheric carbon dioxide, which traps solar heat. This and other greenhouse gases keep Earth 60 degrees Fahrenheit warmer than it would otherwise be; without these gases, Earth would be an ice-covered world hostile to life. Most of the carbon dioxide comes from Earth's volcanoes, and though Mars also has volcanoes, they stopped erupting because the red planet is smaller and its interior cooled faster, in the same way that a small roll cools faster than a loaf of bread. If Mars had been larger, its interior would still be warm and its volcanoes would erupt today, injecting carbon dioxide into its atmosphere and giving Mars a warm climate despite its greater distance from the Sun.

Beyond Mars are the solar system's giant planets, Jupiter, Saturn, Uranus, and Neptune, which differ greatly from small rocky worlds like Earth and Mars. So huge are the giants that their atmospheres alone outweigh the entire Earth. These atmospheres are made of hydrogen and helium, which means the planets probably could not support life even if they were closer to the Sun. The solar system is therefore fortunate: the rocky planets, which can support life, lie near enough to the Sun for one of them to do so; the giant planets, which probably could not support life, are too far away to do so anyway.

Planetary scientists suspect that other solar systems follow the same pattern, a positive sign for those searching for extraterrestrial intelligence. The Sun's planets condensed out of a spinning disk of material that surrounded the newborn Sun. The inner part of the disk was hot and gave birth to Mercury, Venus, Earth, and Mars—planets of rock and iron, substances with high melting points. In the outer disk, the temperatures were lower, and ices also became part of the planets. These planets—Jupiter, Saturn, Uranus, and Neptune—grew so massive that their gravity grabbed the hydrogen and helium gas near them and produced thick atmospheres. If other solar systems formed the same way, similar circumstances probably prevail there. The inner planets, born from hot material, should be small and rocky, whereas the outer planets, born from cool material, should be large and gaseous.

But this does not mean that every solar system has a planet in the so-called life zone, the region around the star where temperatures are right for liquid water and the development of life. If the life zone in our solar system were narrow and the inner planets had different distances from the Sun, our Earth might not have fallen into that zone and the solar system would be uninhabited. The wider the life zone, the greater the chance that a planet will be in it. Astronomers usually measure solar system distances in astronomical units, or AU, where one AU is the mean distance between the Sun and Earth—93 million miles. Mercury is 0.4 AU from the Sun, Venus 0.7 AU, Earth 1.0 AU, and Mars 1.5 AU, so the rocky planets are spaced about 0.4 AU apart. If the life zone is 0.4 AU wide, then one planet will lie within it.

Even in our solar system, however, scientists do not know exactly how wide the life zone is. One estimate, made many years ago, said it was narrow, extending only from 0.95 to 1.01 AU, which suggested that the Earth was lucky to be within it. If the life zones around other F, G, and K stars were this narrow, and if planets were spaced like those orbiting the Sun, most Sunlike stars would have lifeless solar systems. More recent work, though, indicates that the Sun's life zone is wider, stretching from 0.95 to 1.5 AU. If so, a typical Sunlike star should have at least one planet in the life zone.

Of the stars nearest the Sun, Alpha Centauri and Tau Ceti stood out as the best candidates for intelligent life. But the necessity for planets places an additional burden on a star. In particular, Alpha Centauri faces trouble because it is triple. The two brightest members, Alpha Centauri A and B, lie 23 AU apart and orbit each other every 80 years. (The system's third member, Alpha Centauri C, or Proxima Centauri, is a red dwarf 12,000 AU from the others and does not affect them.)

Double stars like Alpha Centauri A and B do not bode well for planets, because a planet orbiting one star may be yanked away by the other long before even primitive life forms. If two stars in a binary are close together—for example, 0.1 AU apart—then planets could orbit both stars, and residents of such a world would see a double sun in their sky. Conversely, if the two stars lie far apart, separated by hundreds of AU, each star could have its own solar system. Residents of such a solar system would call the star they orbit their sun and see the other as an extremely bright point of light. Unfortunately, Alpha Centauri falls between these extremes. A and B have a mean separation of 23 AU, a little greater than the distance between the Sun and Uranus, and their orbit is eccentric, so the separation varies and the situation is even more unstable. When closest, Alpha Centauri A and B lie 11 AU apart, a little more than the distance between the Sun and Saturn, and when farthest, they are 35 AU apart, a bit greater than the distance between the Sun and Neptune.

Nevertheless, each star could have planets, provided the planets always orbit close to one star or the other. The most difficult time for an orbiting planet occurs when the two stars come closest together, so astronomers express the condition for a planet's survival in terms of the stars' minimum separation from each other. A planet can remain in orbit if its distance from its star is less than a fifth of the star's minimum separation from the other star. Since Alpha Centauri A and B never get closer than 11 AU, planets could exist around either star out to 2.2 AU. Each star could therefore have four planets like Mercury, Venus, Earth, and Mars, giving the system a total of eight planets.

But such planets may never have formed in the first place. When the stars were developing, the gravity of one may have perturbed the disk of material circling the other so that neither star gave birth to planets. All that may exist around these otherwise ideal stars are billions of worthless asteroids. In fact, the asteroids in our solar system—most of which lie between the orbits of Mars and Jupiter—may be the remains of a world that never formed. When the solar system was young, the gravity of Jupiter, the Sun's most massive planet, probably disturbed the material inside its orbit so that a planet never arose there. Solar-type stars, such as Alpha Centauri A and B, have a thousand times more mass than Jupiter, and their gravity could have been fatal.

Unlike Alpha Centauri, the G-type star Tau Ceti is single, so it may have planets. The same applies to the orange dwarfs Epsilon Eridani and Epsilon Indi, which are also single. Each star in the 61 Cygni system could have planets as well, because the two stars lie some 80 AU apart, which means that each has room for a solar system. Even if Alpha Centauri does not harbor a warm, watery, Earthlike planet, perhaps one of these other nearby stars does.

LIFE AT LAST?

That does not mean, however, that such a world has life, let alone intelligent life. Some scientists believe that life, even intelligent life, develops easily and will inevitably arise on any Earthlike planet circling a Sunlike star, which would mean that the Galaxy abounds with extraterrestrial intelligence. Other astronomers have argued that if intelligent life did exist elsewhere, some of it would have visited or contacted Earth by now. In their view, the continuing failure to find extraterrestrial intelligence looks more and more like the Galaxy's failure to produce any.

The truth hinges not on astronomy, where the arguments and observations are reasonably secure, but on biology, where they are not. Even for Earth, biologists do not know how life originated. If, even under ideal circumstances, life rarely develops, then the Galaxy could be full of Earthlike worlds that have no life. And if primitive life does form, there is no guarantee that it will evolve into more complex organisms, like plants and animals, as it did on Earth.

The major obstacle, though, may be the difficulty with which intelligence develops. Because intelligence did appear on Earth, scientists often expect the same thing will happen on other life-bearing planets. But Earth may be a special planet, one that has rewarded intelligence more than most. That may be why we are now its dominant species, so dominant that many of the problems Earth now faces, such as overpopulation and the depletion of the ozone layer, have resulted from our own success.

Intelligence is an asset, but it is not the only one that benefits a species. On other planets, strength or speed may count for more than intelligence, so even if a species like us arose, it might be killed off by stronger, faster, dumber competitors. As one example, consider the ice ages that have periodically gripped Earth. Although they were not conducive to the overall development of life, they may have been pivotal to the emergence of intelligent life, because intelligence gave human beings fire, and fire kept humans warm through the cold. On planets that lack ice ages, intelligence may be less of an advantage; and still other planets may suffer such harsh climatic changes that even intelligent life cannot survive them.

In 1992, George Wetherill, a planetary scientist at the Carnegie Institution of Washington, discovered another possible barrier to the emergence of intelligent life. According to his findings, intelligent life might never have developed on Earth if the solar system did not have Jupiter and Saturn. Wetherill was conducting computer simulations of the solar system's formation. In the real solar system, large numbers of comets once roamed interplanetary space, but the gravity of Jupiter and Saturn, the most massive planets, ejected the comets from the solar system, leaving few to strike the Earth today. Wetherill tried a simulation in which the solar system had no planet as large as Jupiter or Saturn. He found that many comets remained and that they struck Earth a thousand times more often. The last catastrophic impact on the real Earth occurred 65 million years ago, when the dinosaurs went extinct, and since their demise there has been time for intelligence to emerge. But in a Jupiter-less solar system, Wetherill found, large comets would hit Earth once every hundred thousand years, allowing insufficient time for intelligent life to evolve. Therefore, Jupiter and Saturn may be responsible, in part, for the existence of intelligent life on Earth. Furthermore, Wetherill's simulations suggest that the formation of planets as massive as Jupiter and Saturn is difficult, which in turn implies that such immense worlds are rare in other solar systems. If so, Earthlike planets elsewhere may have no intelligent life.

Of course, this is all speculation, and claims that intelligent life is rare are hardly popular, especially among science fiction writers who like to depict Earth's inhabitants as backward and barbaric in a Galaxy full of wise and sophisticated aliens. But just as the Sun is not an average star and the Milky Way is not an average galaxy, the Earth may not be an average life-bearing planet. Earth may be one of the few planets on which intelligent life has developed. If so, ours could be the most advanced race in the Galaxy, giving us a responsibility that falls to no other species in the Milky Way. Moreover, if Earth has allowed us to succeed where, on other planets, we would have failed, we should feel greater gratitude toward our planet, and greater urgency toward saving it from the problems we ourselves have created.

Whatever the case—whether the Milky Way abounds with intelligent life, or whether we are its sole example—the Galaxy has already accomplished a major feat: it has given birth to intelligent life. Since the Galaxy's formation from lifeless hydrogen and helium gas, the Milky Way has created star after star, seen them live and seen some die, and for billions of years has witnessed the great alchemy that these stars have performed in transforming the light elements of the big bang into the life-giving heavy elements of which we are now made. We, rather than the Galaxy's countless stars, immense size, or vast empire of satellites, may be the greatest testimony to the Milky Way's elegance and power.

STATISTICAL SUMMARY OF
THE MILKY WAY GALAXY

Number of Stars:	Hundreds of billions
Number of Known Solar Systems:	2
Number of Known Civilizations:	1
Number of Satellite Galaxies:	10
Type:	Spiral
Star Formation Rate:	10 stars per year
Absolute Visual Magnitude:	About −20.6
Total Visual Luminosity:	About 15 billion Suns
Total Mass:	Roughly a trillion Suns
Mass-to-Light Ratio:	Roughly 70
Location of Galactic Center:	Sagittarius
Distance of Sun from Galactic Center:	27,000 light-years
Diameter of Disk:	About 130,000 light-years
Diameter of Dark Halo:	Larger, but unknown
Farthest Satellite Galaxy:	890,000 light-years
Sun's Orbital Period:	230 million years
Rotational Velocity at Solar Position:	220 kilometers per second
Solar U Velocity:	−9 kilometers per second
Solar V Velocity:	+12 kilometers per second
Solar W Velocity:	+7 kilometers per second
Member in Good Standing of:	Local Group
	Local Supercluster
Noteworthy Accomplishments:	Second largest galaxy in Local Group
	Far larger than most other galaxies in the universe
	Rules ten other galaxies
	Supports only known civilization in the universe

GLOSSARY

A. Spectral type for white stars, such as Sirius, Vega, Altair, Deneb, and Fomalhaut.

Absolute Magnitude. A measure of the intrinsic brightness of a star or galaxy. Absolute magnitude is defined as the apparent magnitude the star or galaxy would have if it were 32.6 light-years (10 parsecs) from Earth. The lower an object's absolute magnitude, the *greater* its intrinsic brightness. For example, the Sun has an absolute magnitude of +4.83, while Sirius, whose intrinsic brightness is greater, has an absolute magnitude of +1.43. A star that is one absolute magnitude brighter than another (e.g., +4 versus +5) is 2.5 times intrinsically brighter; a star that is 5 absolute magnitudes brighter is 100 times intrinsically brighter; and a star that is 10 absolute magnitudes brighter is 10,000 times intrinsically brighter.

Absorption Line. Atoms and molecules in a star's atmosphere swallow light of particular wavelengths, so the star appears darker at these wavelengths than at others near it; this darkness appears as an absorption line in the star's spectrum. For example, strong absorption lines appear in the Sun's spectrum at 3934 and 3968 angstroms because singly ionized calcium atoms absorb light at these wavelengths. A star's absorption lines let astronomers determine the star's spectral type and metallicity.

Aldebaran. The brightest star in the constellation Taurus, Aldebaran is an orange K-type giant that lies 60 light-years away.

Algol. The most famous eclipsing binary, Algol was probably the first variable star discovered. It lies in the constellation Perseus and consists of two stars that orbit each other every 2.87 days. When one star passes in front of the other, the light of the system dims.

Alpha Centauri. The nearest star system to the Sun and the third brightest star in the night sky. Unfortunately, Alpha Centauri is so far south that it is visible only from latitudes below 25 degrees north. The system consists of three stars: Alpha Centauri A, the brightest, which is a yellow G-type main-sequence star like the Sun; Alpha Centauri B, the second brightest, which is an orange dwarf; and Alpha Centauri C, by far the faintest, which is a red dwarf. Alpha Centauri A and B both lie 4.35 light-years from Earth and orbit each other every 80 years; Alpha Centauri C lies far from its mates and 4.25 light-years from Earth. Because it is closer to Earth than are A and B, Alpha Centauri C is usually called Proxima Centauri.

Altair. The brightest star in the constellation Aquila, Altair is a white A-type main-sequence star that lies 16 light-years away.

Andromeda. 1. A constellation near Perseus and Pegasus. 2. The nearest giant galaxy to the Milky Way. Also designated M31, Andromeda lies 2.4 million light-years from us and is the largest member of the Local Group. Like the Milky Way, it is a spiral galaxy. Together the Andromeda Galaxy and the Milky Way rule the Local Group.

Angstrom. A unit that measures the wavelength of light and equals 0.00000001 of a centimeter. Blue light has a wavelength of about 4400 angstroms, yellow light 5500 angstroms, and red light 6500 angstroms.

Antares. A red supergiant star in the constellation Scorpius. Antares is the brightest star in Scorpius and lies about 500 light-years from Earth, on the inner edge of the Orion spiral arm.

Antennae. A famous pair of interacting galaxies in the constellation Corvus. Each galaxy's tidal force has drawn out a long tail of stars from the other. The Antennae are also known as NGC 4038 and NGC 4039.

Anticenter. See **Galactic Anticenter.**

Apogalacticon. The point in a star's orbit farthest from the Galactic center.

Apparent Brightness. See **Brightness.**

Apparent Magnitude. A measure of how bright a star looks in the sky. The brighter the star, the smaller the apparent magnitude. A star that is one magnitude brighter than another (e.g., +1 versus +2) looks 2.5 times brighter. The brightest star of all, of course, is the Sun, whose apparent magnitude is −26.74, followed by Sirius, whose apparent magnitude is −1.46, Canopus (−0.72), Alpha Centauri (−0.27), Arcturus (−0.04), and Vega (+0.03). Stars of the Big Dipper are fainter, most of them around magnitude +2. On a clear, dark night, the unaided eye can see stars as faint as apparent magnitude +6, and the largest telescopes penetrate to apparent magnitude +30.

Arcminute. One sixtieth of a degree of angular measure. The Moon is 31 arcminutes across.

Arcsecond. One sixtieth of an arcminute, or 1/3600 of a degree. Jupiter is 40 arcseconds across.

Arcturus. A beautiful orange star that is the brightest in the constellation Bootes and the fourth brightest in the night sky. It lies 34 light-years away and is a member of the thick-disk population. Historically, Arcturus is famous because it was one of the first stars to have its proper motion measured.

Asteroid. A small rocky body that orbits a star. In the solar system, most asteroids lie between the orbits of Mars and Jupiter. The largest asteroid is Ceres, about 900 kilometers in diameter.

Astrometry. The branch of astronomy that deals with measuring the positions of celestial objects, especially stars. Astrometrists measure parallaxes and proper motions, which allow astronomers to determine the distances and velocities of the stars.

Astronomical Unit (AU). The mean distance from the Sun to Earth, about 93 million miles or 150 million kilometers. The AU is the preferred unit for distances within the solar system. Mercury, the innermost planet, lies on average 0.39 AU from the Sun; Pluto, normally the farthest planet, lies on average 39.5 AU from the Sun.

Asymmetric Drift. The negative of the mean V velocity of a stellar population. In general, the older the stellar population, the more negative the V velocity and therefore the greater the asymmetric drift. The young thin disk has an asymmetric drift of 0 kilometers per second, whereas the halo has an asymmetric drift of 200 kilometers per second.

Atom. The building block of matter. The nucleus of an atom consists of one or more protons and may contain neutrons as well; any electrons surround the nucleus. The number of protons in the atom—the atomic number—determines the element.

Atomic Hydrogen. Individual hydrogen atoms that do not belong to molecules. In its neutral form (H I), atomic hydrogen consists of a proton and an electron and generates radio waves that are 21 centimeters long. In its ionized form (H II), atomic hydrogen is simply a proton. H II regions look red because a few of the protons capture electrons, which can radiate red light as they settle into position around the protons.

Atomic Mass Number. The total number of protons and neutrons in an atom's nucleus. For example, oxygen-16 has a mass number of sixteen, because it has eight protons and eight neutrons.

Atomic Number. The number of protons in an atom's nucleus. This determines the type of element. For example, hydrogen has an atomic number of one, so all hydrogen atoms have one proton; helium has an atomic number of two, so all helium atoms have two protons; and oxygen has an atomic number of eight, so all oxygen atoms have eight protons.

A-Type. Having a spectral type of A—that is, hot and white, like Sirius and Vega.

AU. See **Astronomical Unit.**

B. Spectral type for blue stars, such as Rigel, Spica, and Regulus. B-type stars are hot, but even hotter blue stars are designated spectral type O.

Baade's Window. A clearing in the dust clouds of the constellation Sagittarius where astronomers can view stars in the Galactic bulge. Baade's window lies four degrees south of the Galactic center, so an observer's line of sight passes within 1800 light-years of the Milky Way's center.

Bahcall-Soneira Model. A model for the Galaxy first published by John Bahcall and Raymond Soneira in 1980. In its original form, it sought to reproduce star counts in different parts of the sky by employing only a (thin) disk and a halo; it had no thick disk.

Barnard's Star. Discovered in 1916 by Edward Emerson Barnard, this red dwarf lies 5.96 light-years away and is the second nearest star system to the Sun. Barnard's Star has the largest proper motion of any star, 10.3 arcseconds per year, which means that the star moves the equivalent of a lunar diameter every 180 years.

Barred Spiral Galaxy. A spiral galaxy whose bulge is oval instead of round. The Milky Way may be a barred spiral.

Baryonic. Consisting of baryons—protons and neutrons—baryonic matter is "normal" matter. The Sun and the Earth are made of baryonic matter.

Beryllium. Element with atomic number four. Beryllium is rare and fragile, and nuclear reactions in stars destroy it. Most and possibly all beryllium originated when cosmic rays smashed into heavier atoms in space and split them into lighter ones, such as beryllium.

Betelgeuse. A red supergiant star in the constellation Orion and the brightest red supergiant in Earth's sky.

B²FH. An epic paper, published in 1957, by Margaret Burbidge, Geoffrey Burbidge, William Fowler, and Fred Hoyle, who described in detail how the stars had created nearly every element in the universe.

Big Bang. According to standard cosmology, the explosion that started the universe expanding 10 to 15 billion years ago.

Binary Star. A star system having two stars that revolve around each other.

Black Hole. An object with such strong gravity that nothing—not even light, the fastest thing in the universe—can escape it. The most famous black hole candidates are Cygnus X-1, discovered in 1971, and A0620-00, discovered in 1986. A massive black hole, containing a million solar masses, probably lies at the Milky Way's center.

Blue Giant. A giant star with spectral type O or B.

Blueshift. The shift to the blue of an object's spectrum. A blueshift arises when an object moves toward us: its light waves get compressed and reduced in wavelength, so that the entire spectrum is shifted to shorter, or bluer, wavelengths. The greater a star's blueshift, the faster the star is moving toward us. A few galaxies also show blueshifts, the most famous being Andromeda, but most show redshifts, due to the expansion of the universe.

Blue Supergiant. A supergiant star with spectral type O or B. All blue supergiants are hot and young. Rigel, in the constellation Orion, is the best example.

Boron. Element with atomic number five. It is rare and fragile, and nuclear reactions in stars destroy it. Most boron is created in space, by cosmic rays that smash into heavier atoms and split them.

Brightness. Refers to the amount of light coming from an object. Apparent brightness is the light we *see;* intrinsic brightness, which has more importance but is more difficult to measure, is the light an object *actually* gives off—also known as luminosity. By itself, the word "bright" can mean either apparent or intrinsic brightness, depending on the context.

Brown Dwarf. A star with too little mass to ignite its hydrogen-1 fuel. If brown dwarfs exist, they shine faint red for a time, as they convert gravitational energy into heat, and then fade and cool. Their masses range from 1 to 8 percent of the Sun's mass.

B-Type. Having a spectral type of B—that is, hot and blue, such as Rigel and Regulus.

Bulge. The stellar population that lies within several thousand light-years of the Galactic center. The bulge is old, dense, and metal-rich.

Canopus. The brightest star in the constellation Carina and the second brightest star in the night sky. It is spectral type F and shines yellow-white.

Capella. The brightest star in the constellation Auriga and the sixth brightest star in the night sky, Capella lies 42 light-years away and consists of two yellow giants.

Carbon. Element with atomic number six and the basis of all terrestrial life. Carbon is produced during helium burning in red giants and is ejected into the Galaxy when these stars form planetary nebulae. Some carbon also comes from high-mass stars that explode as supernovae.

Carbon Burning. The stage when a star fuses carbon into heavier elements, making neon and magnesium. Carbon burning eventually occurs in all stars born with more than eight solar masses.

Carbon Dioxide. A molecule consisting of one carbon atom and two oxygen atoms (CO_2). It is a gas in Earth's atmosphere that helps to keep the planet warm by trapping solar heat.

Carbon Monoxide. A molecule consisting of one carbon atom and one oxygen atom (CO). It is the most abundant interstellar molecule after molecular hydrogen and is especially useful because it radiates at radio wavelengths, so astronomers can use it to map the distribution of molecular hydrogen.

Carbon Star. A cool star that has a large amount of carbon on its surface.

Carina. 1. A constellation in the southern sky and home of the bright star Canopus. 2. A dwarf galaxy that orbits the Milky Way. Discovered in 1977, Carina lies 350,000 light-years from the Galactic center.

Cepheid. A yellow supergiant that pulsates, alternately brightening and dimming. Cepheids allow astronomers to measure distances, because the longer a Cepheid's period of variation, the greater the Cepheid's mean intrinsic brightness. To determine a Cepheid's distance, all an astronomer has to do is measure the Cepheid's period; comparing the star's mean intrinsic brightness with the star's mean apparent brightness then yields the distance. Cepheids are so bright that we can see them in other galaxies, allowing us to establish distances to entire galaxies beyond the Milky Way.

Chandrasekhar Limit. The most mass a white dwarf can have, about 1.4 solar masses. If a white dwarf receives material from a companion star and exceeds the Chandrasekhar limit, the white dwarf explodes as a type Ia supernova.

Cluster. 1. A gathering of hundreds, thousands, or even a million stars. Star clusters come in two varieties: open clusters and globular clusters. 2. A gathering of hundreds or thousands of galaxies. The nearest large galaxy cluster is the Virgo cluster.

CNO Cycle. One way that a star converts hydrogen into helium. During the CNO cycle, carbon, nitrogen, and oxygen catalyze the nuclear reaction, so the total number of carbon, nitrogen, and oxygen nuclei remains the same.

However, carbon and oxygen gradually get converted into nitrogen. The CNO cycle powers the hydrogen burning that occurs in main-sequence stars with more than 1.5 solar masses and in giants and supergiants of all masses.

CO. See **Carbon Monoxide.**

CO₂. See **Carbon Dioxide.**

Cold Dark Matter. Hypothetical subatomic particles that move slowly compared with the speed of light.

Color. Color indicates temperature: blue stars are hot, yellow stars are warm, and red stars are cool.

Comet. A small icy body that orbits a star. The best-known comet is Halley's. Most comets in our solar system spend most of their time far beyond Neptune and Pluto.

Constellation. 1. As used by astronomers, constellation refers to a particular region of the sky. There are 88 official constellations that blanket the entire sky, so every star, known or unknown, is in one constellation or another. Constellations are like states, and stars are like cities within those states. 2. As used by many people, constellation means a pattern of stars. By analogy, if you draw lines between the cities of a particular state, you would have the equivalent of this type of constellation.

Cosmic Rays. High-energy particles that travel through space. They can smash into atoms and split them apart, creating lighter elements, such as lithium, beryllium, and boron.

Cosmological Constant. A possible third parameter in cosmology, in addition to the Hubble constant and omega (Ω). Most cosmologists believe the cosmological constant is zero, but if it is not, it would make the universe older than astronomers calculate from the Hubble constant and Ω. The size of the cosmological constant is designated by the Greek letter lambda (λ).

Cosmological Redshift. The redshift produced by the expansion of the universe and the reason most galaxies in the universe have redshifts. Contrary to popular belief, this is *not* a Doppler shift. A Doppler redshift arises when an object moves away from us. Most galaxies move away from us, but this is *not* the cause of their redshifts. Instead, as a light wave travels through the fabric of space, the universe expands and the light wave gets stretched and therefore redshifted. It's a subtle difference, but a difference it is. The farther a galaxy, the longer its light waves have traveled through space and the more redshifted they have become.

Cosmology. The branch of astronomy that deals with the universe as a whole, especially its origin, structure, and evolution.

Crab Nebula. A supernova remnant in the constellation Taurus. The star that produced it exploded in A.D. 1054.

Cygnus X-1. A black hole candidate in the constellation Cygnus and a source of x-rays.

Dark Halo. The massive outer region of the Milky Way that surrounds the disk and stellar halo. The dark halo consists mostly of dark matter, whose form is unknown. Though it emits almost no light, the dark halo outweighs the rest of the Galaxy.

Dark Matter. Material astronomers cannot see but whose presence they believe in either because they detect its gravitational influence or because certain theories predict its existence. For example, astronomers believe that the outer part of the Galaxy harbors dark matter, because they notice its gravitational influence on the stars they can see; and inflationary cosmologists believe that the universe is full of dark matter, because inflation predicts that the universe has a large density.

Delta Cephei. A pulsating star in the constellation Cepheus. It was the second Cepheid discovered and lent its name to the entire class of stars.

Deneb. The brightest star in the constellation Cygnus, Deneb is a white A-type supergiant that generates more light in a single night than the Sun produces during an entire century. Deneb lies 1500 light-years away and is the most distant first-magnitude star.

Density. An object's mass divided by its volume. Cotton has a low density; lead has a high density. Red giants have a low density; white dwarfs have a high density.

Density Wave Theory. One possible explanation for spiral arms. According to this theory, the spiral arms represent regions of somewhat enhanced density (density waves) that rotate more slowly than the galaxy's stars and gas. As gas enters a density wave, it gets squeezed and makes new stars, some of which are short-lived blue stars that light the arms.

Deuterium. Hydrogen-2, the rare heavy isotope of hydrogen. Deuterium has one proton and one neutron, whereas normal hydrogen has one proton and no neutrons.

Differential Rotation. As an object rotates, different parts of it may move at different rates. The Galaxy rotates differentially.

Dissipation. When a galaxy forms and gas clouds start crashing into and impeding one another. The Galaxy's disk probably formed through dissipation.

Doppler Shift. The blueshift or redshift produced by an object's motion toward or away from us. If a star moves toward us, its light waves get compressed and its spectrum is blueshifted; if a star moves away from us, its light waves get stretched and its spectrum is redshifted. The Doppler shift allows astronomers to measure the radial velocities of stars. The Doppler shift is *not* responsible for the redshifts that most galaxies exhibit; that is a cosmological redshift.

Double Star. A star system having two stars that revolve around each other.

Draco. 1. A constellation in the northern sky. 2. A dwarf galaxy that orbits the Milky Way and lies about 250,000 light-years from the Galactic center. Draco is the least luminous galaxy known, with an absolute magnitude of −8.6.

Dwarf. 1. A star that is on the main sequence—that is, a star fusing hydrogen into helium at its core. 2. A small, faint galaxy, exemplified by those that orbit the Milky Way: Ursa Minor, Draco, Sculptor, Sextans, Carina, Fornax, Leo II, and Leo I.

Dynamical Friction. The process by which a large mass gets slowed down as it moves through a sea of smaller objects and feels their gravitational pull. For example, a satellite galaxy that moves through the Milky Way's dark halo gets slowed down by the dynamical friction with the dark matter and will spiral into the Milky Way as a result.

Eccentric. An orbit that has a high eccentricity—that is, highly elliptical.

Eccentricity. A measure of how round or elliptical an orbit is. A perfect circle has an eccentricity of 0 percent, and an extremely elliptical orbit has an eccentricity of just under 100 percent. The Sun has an orbital eccentricity of 6 percent, which means that at perigalacticon the Sun is 6 percent closer to the Galactic center than its mean distance and at apogalacticon the Sun is 6 percent farther from the Galactic center than its mean distance.

Eclipsing Binary. A double star in which at least one of the two stars passes in front of and/or behind the other so that the system's total light periodically fades. The most famous eclipsing binary is Algol.

Electromagnetic Radiation. Visible light, radio waves, infrared radiation, ultraviolet radiation, x-rays, and gamma rays. In a vacuum, all electromagnetic radiation travels at the speed of light. The shorter its wavelength, the more energetic it is.

Electron. A small, negatively charged particle that appears in every neutral atom, surrounding the positively charged nucleus like bees around honey.

Element. Different elements are distinguished by the number of protons in their nuclei. All hydrogen atoms have one proton; all helium atoms have two protons; all oxygen atoms have eight protons.

Elliptical Galaxy. A galaxy that looks round or elliptical. One example is M87, in the constellation Virgo.

ELS. A classic paper published in 1962 by Olin Eggen, Donald Lynden-Bell, and Allan Sandage, who argued that the Galaxy formed from a single huge cloud of gas that rapidly collapsed.

Epsilon Eridani. A young orange dwarf star in the constellation Eridanus that is visible to the naked eye and lies just 10.7 light-years away.

Epsilon Indi. An old orange dwarf star in the southern constellation Indus that lies 11.2 light-years away.

Equipartition of Energy. If all stars have the same kinetic energy, equipartition of energy prevails. Because kinetic energy depends on both a star's mass and its velocity, high-mass stars must move more slowly than low-mass stars, if equipartition of energy prevails.

Eta Aquilae. A pulsating star in the constellation Aquila. It was the first Cepheid discovered, in 1784.

Extragalactic Astronomy. The field that deals with objects beyond the Milky Way, especially galaxies and quasars.

F. Spectral type for yellow-white stars, which are slightly hotter than the Sun. The brightest F-type stars in Earth's sky are Canopus and Procyon.

Field Star. A star that is not part of any star cluster. Most stars, including the Sun, are field stars.

Flare Star. A star that emits flares, which can outshine the entire star. Many red dwarfs are flare stars.

Fomalhaut. The brightest star in the constellation Piscis Austrinus, Fomalhaut is a white A-type main-sequence star 21 light-years away.

Fornax. 1. A faint constellation in the southern sky. 2. A dwarf galaxy that orbits the Milky Way and lies 440,000 light-years from the Galactic center. It was discovered in 1938.

47 Tucanae. A globular cluster in the southern constellation Tucana and a member of the thick-disk population.

Free-Fall. A collapse in which gas clouds do not hit or impede one another. According to ELS, the Galactic halo formed in a free-fall collapse.

F-Type. Having a spectral type of F—that is, yellow-white, like Canopus and Procyon.

G. Spectral type for yellow stars, such as the Sun, Alpha Centauri A, and Capella.

Galactic. 1. When capitalized, the word refers to our Galaxy. 2. When not capitalized, it refers to a galaxy.

Galactic Anticenter. The point in the Galactic plane that lies directly opposite the Galactic center. Here we gaze toward the edge of the Galactic disk. The nearest bright star to the anticenter is El Nath, in the constellation Taurus.

Galactic Astronomy. The study of the Milky Way.

Galactic Latitude. The angle between the line of sight to a star and the Galactic plane. Galactic latitude ranges from +90 degrees to −90 degrees; the Galactic plane has a Galactic latitude of 0 degrees. Regions north of the Galactic plane have positive Galactic latitude; regions south have negative Galactic latitude. The point with a Galactic latitude of +90 degrees is called the north Galactic pole, and the point with a Galactic latitude of −90 degrees is called the south Galactic pole.

Galactic Longitude. A measure of a star's position with respect to the Sun and Galactic center. Galactic longitude ranges from 0 degrees to 360 degrees. Imagine the Sun at the center of a giant clock, with the Galactic center located in the direction of six o'clock. A Galactic longitude of 0 degrees would correspond to the direction of six o'clock, a Galactic longitude of 90 degrees to the direction of three o'clock, a Galactic longitude of 180 degrees to the direction of twelve o'clock, and a Galactic longitude of 270 degrees to the direction of nine o'clock.

Galactic Plane. The plane that contains the disk of the Milky Way. By definition, one direction perpendicular to this plane is called "above" or "north," and the opposite direction, also perpendicular to the Galactic plane, is called "below" or "south." From Earth, due Galactic north is marked by the north Galactic pole, which lies near the bright star Arcturus, and due Galactic south is marked by the south Galactic pole, which lies in the faint constellation Sculptor.

Galactic Pole. Either of the two points in the sky where we look perpendicular to the disk of the Milky Way. The north Galactic pole is the Galactic pole located above the disk; the south Galactic pole is the Galactic pole located below the disk.

Galactic Tide. See Tide.

Galactocentric Distance. A star's distance from the Galactic center. The Sun's Galactocentric distance is about 27,000 light-years.

Galaxy. A huge collection of millions, billions, or trillions of stars. When referring to the Milky Way, "galaxy" is capitalized, otherwise not; thus: "Andromeda is the nearest giant galaxy to the Galaxy."

Galaxy Cluster. A conglomeration of hundreds or thousands of galaxies. The nearest large galaxy cluster is the Virgo cluster.

Giant. 1. A star that has evolved off the main sequence and is roughly a hundred times as luminous as the Sun. Giants can be of any color, but yellow, orange, and red giants are the most common. 2. A planet much more massive than Earth. The solar system has four giant planets, all far from the Sun: Jupiter, Saturn, Uranus, and Neptune.

Giant Molecular Cloud. A huge complex of interstellar gas and dust, consisting mostly of molecular hydrogen, that typically stretches over 150 light-years and contains 200,000 solar masses. Giant molecular clouds give birth to new stars.

Globular Cluster. A star cluster that packs hundreds of thousands of stars into a region only about a hundred light-years across. Bright globular clusters include Omega Centauri, 47 Tucanae, and M13. In the Milky Way, all globular clusters are old. Most are members of the Galactic halo, and the rest are members of the thick disk.

Gold. Element with atomic number 79. It is produced entirely by the r-process, in supernovae.

Gravitational Lensing. The bending of light caused by the gravity of an object lying between us and the light source. This may cause the light source to look brighter than it normally does.

Groombridge 1830. A famous halo star that lies 28 light-years away in the constellation Ursa Major. Its proper motion, discovered in 1841, was then the largest known, displacing that of 61 Cygni.

Group. A small gathering of galaxies, smaller than a cluster. The Milky Way is part of the Local Group, which contains about thirty galaxies.

G-Type. Having a spectral type of G—that is, yellow, like the Sun.

H I. Neutral hydrogen gas. It emits radio waves that are 21 centimeters long.

H II. Ionized hydrogen—that is, hydrogen with its electron missing.

H II Region. An area of ionized hydrogen. Most H II regions are red and arise from hot blue O and B stars, whose ultraviolet light can ionize all the hydrogen for dozens or even hundreds of light-years in every direction. The most famous H II region is the Orion Nebula.

Half-Life. The length of time it takes for half the number of radioactive atoms to decay.

Halo. The somewhat round population of old, metal-poor stars in the Milky Way. Also, the huge entity that surrounds the disk and contains most of the Galaxy's dark matter. To distinguish between the two, astronomers call the former the stellar halo and the latter the dark halo. Most of the stellar halo lies closer to the Galactic center than the Sun, while most of the dark halo lies farther from the Galactic center than the Sun.

HD. *Henry Draper Catalogue,* which lists over 200,000 stars. It was published in nine volumes between 1918 and 1924.

HD 19445. A subdwarf in the constellation Aries that was used, along with HD 140283, to establish that some stars have much lower metallicities than the Sun.

HD 122563. A yellow giant star in the constellation Bootes whose metallicity is only 0.2 percent of the Sun's.

HD 140283. A subdwarf or subgiant in the constellation Libra that was used, along with HD 19445, to establish that some stars have much lower metallicities than the Sun.

Helium. The second lightest (atomic number two) and second most common element in the universe. Most of it was produced by the big bang, with main-sequence stars making an additional contribution. It has two stable isotopes, helium-3 (two protons and one neutron) and helium-4 (two protons and two neutrons). The latter isotope is by far the more common; it is also the most stable and tightly bound of the light nuclei.

Helium Burning. The stage when a star fuses helium into carbon and oxygen. All stars born with more than half a solar mass eventually burn helium.

Hertzsprung-Russell Diagram. A plot of stellar color, temperature, or spectral type versus stellar luminosity. The H-R diagram segregates three principal types of stars: the main sequence, which forms a diagonal band from bright blue stars to faint red ones; red giants and supergiants, which appear in the upper right of the diagram; and white dwarfs, which lie below the main sequence.

High-Velocity Star. A star whose U and/or V and/or W velocities are much greater or much less than zero. Such stars usually have eccentric orbits around the Galaxy.

Hipparcos. A European satellite that from 1989 to 1993 measured the parallaxes of stars.

Horizontal-Branch Star. A metal-poor star, similar in mass to the Sun, that fuses helium into carbon and oxygen at its core. Such stars range in color from blue to yellow. RR Lyrae stars are horizontal-branch stars. Stars bluer than RR Lyraes are called blue horizontal-branch stars; stars redder are called red horizontal-branch stars, even though they are actually yellow. All other things being equal, the more metal-poor a globular cluster, the bluer its horizontal branch; the older a globular cluster, the bluer its horizontal branch, too.

Hot Dark Matter. Subatomic particles that moved almost as fast as light when the universe was young.

H-R Diagram. See **Hertzsprung-Russell Diagram.**

Hubble Constant. The present expansion rate of the universe, in units of kilometers per second per megaparsec. The larger the Hubble constant, the younger the universe.

Hubble Time. The inverse of the Hubble constant and a crude measure of the universe's age. For a Hubble constant of 50, one can calculate that the

Hubble time is 19.6 billion years; for a Hubble constant of 80, the Hubble time is 12.2 billion years. If there is no cosmological constant, the universe is younger than the Hubble time. In particular, if the mass density of the universe (designated Ω) is 0.1, the universe's age is 90 percent of the Hubble time; if Ω is 1.0, the universe's age is 67 percent of the Hubble time.

Hyades. The nearest star cluster, 150 light-years from Earth. The cluster lies in the constellation Taurus and is 600 million years old. It is an open star cluster.

Hydrogen. The lightest and most common element in the universe. It has atomic number one and was produced by the big bang. Hydrogen-1 (one proton and no neutrons) is the most common isotope; hydrogen-2 (one proton and one neutron), or deuterium, is rarer; and hydrogen-3 (one proton and two neutrons), or tritium, is radioactive.

Hydrogen Burning. The fusion of hydrogen into helium and the process by which all main-sequence stars generate energy. Every star born with more than 0.08 solar masses burns hydrogen.

IC 4182. A nearby galaxy in which a type Ia supernova exploded in 1937.

Inflation. The idea that, when it was a fraction of a second old, the universe expanded dramatically. If inflation is correct, then the mass density of the universe (Ω) should be 1.0, if there is no cosmological constant; if there is a cosmological constant and inflation is correct, the sum of Ω and the cosmological constant (λ) should be 1.0.

Infrared. Radiation that has a somewhat longer wavelength than visible light. It penetrates dust much more readily than visible light does.

Inhomogeneous Early Universe. The idea that during the first few minutes after the big bang, the universe had regions of different density. An inhomogeneous early universe can produce elements different from those of the standard homogeneous early universe.

Interarm Region. The area between a spiral galaxy's spiral arms. These areas look dark, not because they lack stars, but because they contain none of the young, luminous stars that light the arms.

Interstellar Cloud. A collection of gas and dust that lies between the stars.

Interstellar Medium. The space between the stars.

Intrinsic Brightness. The amount of light an object actually emits, as opposed to how bright the object looks from Earth. An apparently bright star can be intrinsically bright and far away or intrinsically faint and nearby.

Ionized Hydrogen. A hydrogen atom that has lost its electron. Hydrogen gets ionized by hot O and B stars in H II regions. The most famous H II region is the Orion Nebula.

Iron. Element with atomic number 26, created mostly by type Ia supernovae, with an additional contribution from type Ib, Ic, and II supernovae. It is the most stable element.

IRS. Infrared source.

IRS 7. A red supergiant that lies less than one light-year from the Galactic center. It resembles a comet, because a wind hits its atmosphere and creates a tail that points away from the Galactic center.

IRS 16. A star cluster near the Galactic center.

Isotope. An element with different numbers of neutrons. For example, hydrogen-1 and hydrogen-2 are both isotopes of hydrogen; each has one proton in its nucleus, but the former has no neutrons and the latter has one.

Jupiter. The largest planet in the solar system, having about 0.001 solar masses. The planet has more mass than all of the other planets put together.

K. Spectral type for orange stars, such as Arcturus, Aldebaran, and Alpha Centauri B. K-type stars are somewhat cooler than the Sun.

Kapteyn's Star. A nearby star discovered in 1897 by Jacobus Kapteyn. It is a red subdwarf that lies 13 light-years away and is both the nearest halo star to the Sun and the nearest star that orbits the Galaxy backward.

Kapteyn Universe. An incorrect model for the Galaxy proposed by Jacobus Kapteyn in which the Milky Way was small and the Sun located at or near the Galaxy's center.

Kelvin. The temperature scale that astronomers usually use. On the Kelvin scale, the coldest possible temperature is 0 degrees. This corresponds to −273° Celsius or −460° Fahrenheit. Room temperature is about 295 degrees Kelvin.

Kilometer Per Second. The unit of speed in astronomy. One kilometer per second is 2237 miles per hour—five times the speed of an airplane.

Kinematics. The motions of stars, especially as these motions pertain to the stars' orbits around the Galaxy.

K-Type. Having spectral type K—that is, orange, like Arcturus and Aldebaran.

Lambda (λ). See **Cosmological Constant.**

Large Magellanic Cloud. The nearest and largest of the many galaxies that orbit the Milky Way. It is 160,000 light-years away.

Lensing. See **Gravitational Lensing.**

Leo I. The most distant galaxy that orbits the Milky Way. A dwarf galaxy, Leo I was discovered in 1950 and lies 890,000 light-years from the Galactic center. It is close to the bright star Regulus, whose glare interferes with the study of the galaxy.

Leo II. The second most distant galaxy that orbits the Milky Way, lying 720,000 light-years from the Galactic center. Like Leo I, Leo II is a dwarf galaxy that was discovered in 1950.

Life Zone. The region around a star where a planet can have liquid water and so may support life.

Light Elements. Usually, hydrogen, helium, and lithium, which have atomic numbers of one, two, and three; sometimes also beryllium and boron, which have atomic numbers of four and five.

Light-Year. The distance light travels in one year: 5.88 trillion miles, or 9.46 trillion kilometers. The nearest star system to the Sun is 4.3 light-years away.

Lithium. Element with atomic number three. Some lithium formed in the big bang, along with huge amounts of hydrogen and helium. Lithium has two stable isotopes: the rarer, lithium-6 (three protons and three neutrons); and the more common, lithium-7 (three protons and four neutrons).

Local Bubble. The region of the Galaxy near the Sun which has little neutral hydrogen gas. It extends about a hundred light-years in most directions but up to a thousand in some. The local bubble may have been produced by supernovae.

Local Group. The gravitationally bound collection of nearby galaxies ruled by the Andromeda Galaxy and the Milky Way, which are its largest members. The Local Group has about 30 known galaxies.

Local Standard of Rest. An imaginary point, located at the Sun's distance from the Galactic center, that revolves clockwise around the Galaxy on a circular orbit. Astronomers measure a star's U, V, and W velocities with respect to the local standard of rest rather than with respect to the Sun, because the Sun has a slightly noncircular orbit. The orbital velocity of the local standard of rest around the Galaxy is about 220 kilometers per second.

Local Supercluster. The supercluster to which the Local Group belongs. It is shaped like a cigar, with the Virgo cluster of galaxies at its center and the Local Group near one end.

Low-Velocity Star. A star whose U, V, and W velocities are all near zero. Such stars have nearly circular orbits around the Galaxy.

Luminosity. The total amount of energy radiated by a star—that is, its true, or intrinsic, brightness.

Luminosity Class. A measure of a star's intrinsic brightness, as determined from the star's spectrum. Supergiants have luminosity class I, bright giants have luminosity class II, giants have luminosity class III, subgiants have luminosity class IV, and main-sequence stars have luminosity class V.

Luminosity Function. 1. The number of stars in the Galaxy with a particular absolute magnitude. The luminosity function reveals that luminous stars are rare and intrinsically faint stars common. 2. The distribution of galaxies by absolute magnitude. Luminous galaxies are rare and intrinsically faint ones are common.

Luminous. Intrinsically bright, as opposed to being just apparently bright.

M. Spectral type for red stars, such as Betelgeuse, Antares, and Proxima Centauri.

M13. A great globular cluster in the constellation Hercules.

M31. The Andromeda Galaxy, the largest member of the Local Group. It is a giant spiral galaxy that lies 2.4 million light-years away.

M32. An elliptical galaxy that orbits the Andromeda Galaxy.

M33. The Pinwheel Galaxy, the third largest member of the Local

Group, after Andromeda and the Milky Way. It is a spiral galaxy that lies 2.6 million light-years away.

M42. The Orion Nebula, a star-forming region in the constellation Orion.

M45. The Pleiades, a beautiful open star cluster in the constellation Taurus. It is 410 light-years away.

M51. The Whirlpool Galaxy, a stunning spiral in the constellation Canes Venatici.

M81. A giant spiral galaxy 11 million light-years away in the constellation Ursa Major. It rules the M81 group, the second nearest galaxy group to the Local Group.

M87. A giant elliptical galaxy in the Virgo cluster.

M104. The Sombrero Galaxy, in which galactic rotation was first detected. It lies in the constellation Virgo.

MACHO. Massive compact halo object. MACHOs are dark stars or planets that may make up the Milky Way's dark halo.

Magellanic Clouds. The Large and Small Magellanic Clouds, the two nearest and largest of the galaxies that orbit the Milky Way. The Magellanic Clouds lie in the southern sky and cannot be seen from the United States.

Magellanic Stream. A strand of gas spanning 300,000 light-years that the Milky Way has ripped out of the Magellanic Clouds.

Magnesium. Element with atomic number twelve. It is the fifth most common metal in the universe and was produced by high-mass stars that exploded.

Magnitude. A measure of a star's brightness. Apparent magnitude measures a star's apparent brightness—that is, how bright a star looks from Earth. Absolute magnitude measures a star's intrinsic brightness—that is, how much light the star actually emits.

Main-Sequence Star. A star, like the Sun, that fuses hydrogen into helium at its core. Ninety percent of all stars are main-sequence stars; examples are Sirius, Vega, Altair, and Alpha Centauri A, B, and C.

Main-Sequence Turnoff. The point on the H-R diagram of a star cluster where main-sequence stars are beginning to leave the main sequence. The main-sequence turnoff measures age: all other things being equal, the older a star cluster, the fainter the main-sequence turnoff.

Mass Number. See **Atomic Mass Number.**

Mass-to-Light Ratio. The amount of mass in an object divided by its luminosity, both measured in solar units. The Sun has a mass-to-light ratio of one, because it has one solar mass and one solar luminosity. Stars brighter than the Sun, such as upper main-sequence stars, giants, and supergiants, have low mass-to-light ratios, because most have somewhat more mass than the Sun but much more luminosity. Stars fainter than the Sun, such as red, orange, and white dwarfs, have high mass-to-light ratios, because most have smaller masses than the Sun but much smaller luminosities. Dark matter, by definition, has a high mass-to-light ratio: it has much mass but radiates little or no light.

Megaparsec. A unit of distance equal to a million parsecs, or 3.2616 million light-years.

Merger. The formation of a galaxy from the collision of two or more separate galaxies.

Metal. To an astronomer, a metal is any element heavier than hydrogen and helium; thus, not only are iron and copper metals, but so are elements like oxygen and neon.

Metallicity. An object's abundance of metals. In practice, this usually means the abundance of iron, which is easy to measure.

Metallicity Gradient. The progressive change in metallicity from the center of a galaxy to its edge. A galaxy exhibiting a metallicity gradient is more metal-rich at its center than at its edges.

Metal-Poor. Having a low metallicity.

Metal-Rich. Having a high metallicity.

Microwave Background. The 2.7-degree Kelvin radiation that pervades the universe and is believed to be the afterglow of the big bang.

Milky Way. 1. Our Galaxy. 2. The band of light that stretches across the sky during summer and winter, produced by innumerable stars in the plane of the Galaxy.

Mintaka. One of the three stars in Orion's belt, and the star along whose line of sight interstellar gas was first spectroscopically detected.

Mira. A red giant in the constellation Cetus that varies in brightness as it pulsates. When brightest, Mira is visible to the naked eye; when dimmest, Mira can be viewed only with optical aid. Mira is the prototype of all pulsating red giants, which are called Miras in its honor.

Molecular Cloud. A cloud of interstellar gas and dust that consists mostly of molecular hydrogen.

Molecular Hydrogen. A molecule consisting of two hydrogen atoms (H_2) and the most common molecule in space.

M-Type. Having a spectral type of M—that is, red like Betelgeuse and Antares.

Nebula. A region of gas and dust in space, like the Orion Nebula, the Ring Nebula, and the Lagoon Nebula. Galaxies were once called "nebulae" as well, before astronomers knew what these objects really were.

Neon. Element with atomic number ten and the third most common metal in the universe. It is produced by carbon burning in high-mass stars and released into the Galaxy when they explode.

Neon Burning. The stage in which a star burns neon into oxygen and magnesium.

Neutral Hydrogen. A hydrogen atom that has a proton and an electron and so is electrically neutral. Neutral hydrogen produces radio waves that are 21 centimeters long.

Neutrino. A neutral subatomic particle with little or no mass that travels at or near the speed of light. Neutrinos hardly ever interact with matter.

Neutrino-Process. Nucleosynthesis induced by neutrinos. It may create fluorine and boron.

Neutron. A neutral, massive subatomic particle that occurs in all nuclei except hydrogen-1. For example, helium-4 has two neutrons, and oxygen-16 has eight.

Neutron Star. A dead, collapsed star that consists mostly of neutrons and is only about 20 kilometers across. Neutron stars are much denser than white dwarfs.

NGC. *New General Catalogue*, a catalogue of 7840 nebulae, star clusters, and galaxies that was published in 1888 by John Dreyer.

NGC 253. A large edge-on spiral galaxy and the largest member of the Sculptor Group, the group of galaxies nearest to the Local Group.

NGC 288. A globular cluster that played a key role in proving that not all globulars are the same age. In the late 1980s, NGC 288 was shown to be 3 billion years older than NGC 362.

NGC 362. A globular cluster. See **NGC 288.**

NGC 4565. An edge-on spiral galaxy in the constellation Coma Berenices that resembles the Milky Way.

Nitrogen. Element with atomic number seven and the fourth most common metal in the universe. It formed during hydrogen burning in main-sequence stars and red giants, via the CNO cycle.

Nonbaryonic Matter. Material that consists of exotic subatomic particles. These subatomic particles can move slowly (cold dark matter) or fast (hot dark matter). Cosmologists who believe in inflation—or anyone else who thinks the mass density of the universe (Ω) is 1.0—believe that most of the universe consists of nonbaryonic matter.

North Galactic Pole. A point in the constellation Coma Berenices where we look perpendicular to and above the Galactic plane. The nearest bright star to the north Galactic pole is Arcturus, in the neighboring constellation Bootes.

North Star. See **Polaris.**

Nova. An exploding star, but one that never attains the enormous luminosity of a supernova. A nova usually arises from a double-star system in which one member is a white dwarf. The other star dumps material onto the white dwarf, and the nova occurs when this material catches nuclear fire and explodes. The explosion does not destroy either star.

Nucleon. A proton or neutron. For example, oxygen-16, with eight protons and eight neutrons, has sixteen nucleons.

Nucleosynthesis. The transformation of one element or isotope into another. Nucleosynthesis occurred just after the big bang, but today most nucleosynthesis takes place in stars—for example, the Sun presently converts hydrogen into helium.

Nucleus. The central part of an atom, which contains the atom's protons and neutrons.

O. Spectral type for the hottest blue stars, even hotter than B-type stars. O-type stars are rare and short-lived.

OB. Spectral type O or B—that is, hot and blue.

OB Association. A loose gathering of O and B stars that typically stretches over hundreds of light-years and contains a few dozen OB stars.

Old Thin Disk. The older part of the thin-disk population, ranging in age from about 1 to 10 billion years. The Sun and most other nearby stars belong to the old thin disk. The scale height of the old thin disk is about 1000 light-years.

Omega (Ω). The mass density of the universe. If Ω is greater than 1.0, the universe is so dense that the gravitational attraction of all the mass will halt the expansion and cause the universe to collapse; if Ω is less than 1.0, the universe does not have enough mass to reverse the expansion and will expand forever. The larger Ω, the younger the universe.

Omega Centauri. A bright globular cluster in the constellation Centaurus.

Omicron^2 Eridani. A triple star that lies 16 light-years away and has the first white dwarf ever discovered.

Open Cluster. A small, loose cluster of stars that typically contains several hundred members. The best examples are the Hyades and the Pleiades, both in the constellation Taurus. Open clusters line the Galactic plane, in contrast with globular clusters, which are members of the Galaxy's halo or thick disk.

Orange Dwarf. A main-sequence star with spectral type K. These stars are somewhat fainter, cooler, and smaller than the Sun and account for 15 percent of the Galaxy's stars.

Orange Giant. A giant star with spectral type K. These stars are about a hundred times more luminous than the Sun but are cooler. The two brightest in the sky are Arcturus and Aldebaran.

Orion Arm. The spiral arm containing the Sun. It lies between the Sagittarius arm and the Perseus arm.

Orion Nebula. A large cloud of gas and dust giving birth to young stars in the constellation Orion and visible to the naked eye. It is an H II region 1500 light-years away.

O-Type. Having a spectral type of O—that is, hot and blue.

Oxygen. The most abundant metal in the universe, and the third most abundant element overall, after hydrogen and helium. Oxygen has atomic number eight and is produced by massive stars—those born with over eight solar masses—which eject the element into the Galaxy when they explode.

Oxygen Burning. The stage when a star fuses oxygen into silicon and sulfur. It occurs only in stars born with over eight solar masses.

Parallax. The tiny shift in a star's apparent position that occurs when the star is viewed from slightly different perspectives as the Earth revolves around the Sun. The larger a star's parallax, the closer the star is to Earth.

Parsec. A unit of distance equal to 3.261633 light-years. A star that is one parsec from the Sun has a parallax of one arcsecond.

Perigalacticon. The point in a star's orbit around the Galaxy when the star lies closest to the Galactic center. The Sun is near perigalacticon now.

Period-Luminosity Relation. Cepheids obey this relation: the longer the Cepheid's pulsation period, the more luminous the star. Since measuring a Cepheid's period is easy, the period-luminosity relation allows astronomers to determine the Cepheid's intrinsic brightness and hence distance. If the Cepheid is part of another galaxy, the Cepheid's distance gives the distance to the entire galaxy.

Perseus Arm. The spiral arm that lies next out from the arm containing the Sun. The most famous members of the Perseus arm are the young star clusters h and Chi Persei.

Photon. A particle of light.

Pinwheel Galaxy. M33, a spiral galaxy that lies 2.6 million light-years away and is the third largest member of the Local Group, after Andromeda and the Milky Way.

Planet. An object that formed in the disk surrounding a star. To be called a planet, an object must be more massive than Pluto (1/500 the Earth's mass) and less massive than ten times Jupiter's mass. Unlike stars, planets do not produce light of their own but merely reflect that of the star(s) they orbit.

Planetary Nebula. A bubble of gas surrounding a hot, dying star. The star is so hot that it makes the planetary nebula glow, which allows astronomers to see it. The star was once the core of a red giant, which ejected its outer atmosphere and created the planetary. A planetary nebula has nothing to do with a planet, but through a small telescope, it looks like a planet's disk, hence the misleading name.

Platinum. Element with atomic number 78. It is produced almost entirely by the r-process, in supernovae.

Pleiades. A nearby star cluster in the constellation Taurus that lies 410 light-years away. It is young, containing blue stars, and has an age of 70 million years.

Polaris. The North Star, a second-magnitude star in the constellation Ursa Minor. The star is a yellow-white F-type supergiant that lies 330 light-years away. Until recently, Polaris was also a Cepheid—in fact, it was the nearest Cepheid to Earth—but as this book goes to press there are signs that the star is ceasing to pulsate.

Population. See **Stellar Population.**

Primordial Nucleosynthesis. The creation of elements that occurred just minutes after the big bang. According to standard theory, primordial nucleosynthesis gave the universe only five nuclei, all lightweight: hydrogen-1, hydrogen-2 (or deuterium), helium-3, helium-4, and lithium-7.

Procyon. The brightest star in the constellation Canis Minor and one of the nearest stars, lying just 11.4 light-years from Earth. Procyon is the eighth brightest star in the night sky. It consists of two stars: Procyon A, a bright yellow-white F-type star that has just started to evolve off the main sequence; and Procyon B, a dim white dwarf.

Proper Motion. The apparent movement of a star, year after year, caused by the star's velocity across the line of sight. If the star's distance is known, this velocity—called the tangential velocity—can be computed. The star with the largest proper motion is Barnard's Star, whose proper motion is 10.3 arcseconds per year.

Protogalaxy. An object that becomes a galaxy.

Proton. A subatomic particle with positive electric charge. Every atom has at least one proton in its nucleus; the number of protons determines the element. For example, all atoms with one proton are hydrogen, all atoms with two protons are helium, and so on.

Proton-Proton Reaction. The nuclear sequence by which the Sun and all other main-sequence stars with less than 1.5 solar masses fuse hydrogen into helium.

Proxima Centauri. The faintest of the three stars that make up the Alpha Centauri star system. See **Alpha Centauri.**

PSR B1257+12. A pulsar in the constellation Virgo and the site of the first solar system to be discovered outside our own. The planets were detected in 1991.

Pulsar. A fast-spinning neutron star that emits radiation toward Earth every time it rotates.

Quasar. The brightest objects in the universe, quasars can generate over a trillion times as much light as the Sun from a region little larger than the solar system. Most are extremely distant, which means that they existed long ago.

Radial Velocity. The speed at which an object moves toward or away from us. It can be measured from a star's spectrum: a star moving toward us has a blueshifted spectrum, and a star moving away from us has a redshifted spectrum. The larger the blueshift or redshift, the larger the radial velocity. The present radial-velocity champion is a star in the constellation Lacerta named Giclas 233-27, which moves toward us at 583 kilometers per second.

Radio. Electromagnetic radiation with the lowest energy and longest wavelength. Unlike visible light, radio waves penetrate dust and can be detected from throughout the Galaxy.

Reddening. The scattering away of blue light that occurs when light passes through gas and dust, thereby leaving red (and infrared) light more dominant. This phenomenon occurs at sunset, when the Sun looks orange or red because its light passes through the thick air on the Earth's horizon. Reddening is unrelated to redshift.

Red Dwarf. A main-sequence star with spectral type M. Red dwarfs are much fainter, cooler, and smaller than the Sun but are the most common type of star in the Galaxy, accounting for 70 percent of all stars. The nearest red dwarf, Proxima Centauri, lies just 4.25 light-years away, but neither it nor any other is visible to the naked eye.

Red Giant. A giant star with spectral type M. Such stars are in a more advanced state of evolution than the Sun, for they do not burn hydrogen into helium at their cores. Instead, they may fuse hydrogen into helium in a layer surrounding their cores, or they may fuse helium into carbon and oxygen, or they may do both. Often, astronomers use "red giant" loosely, to include not only M giants but G and K giants, too.

Redshift. The shift to the red of a star's spectrum, caused by the star's movement away from us. This movement stretches the star's light waves and increases their wavelength. Since red has a longer wavelength than blue, this shift is called a redshift. The larger a star's redshift, the faster the star is moving away from us.

Most galaxies also show redshifts, not because of the galaxy's movement away from us (although the galaxy is moving away from us) but because of the expansion of the universe. As a galaxy emits a light wave toward us, the light wave travels through the fabric of space; en route to Earth, it is stretched by the expansion of space and exhibits a redshift. The farther the galaxy, the larger the redshift. To distinguish this type of redshift from one caused by movement, astronomers call it the "cosmological redshift."

Red Supergiant. A supergiant with spectral type M. Red supergiants are the largest stars in the universe: if put in place of the Sun, some would touch Saturn. The two brightest red supergiants in Earth's sky are Betelgeuse and Antares.

Resolution. The ability to see detail in an object.

Rigel. A blue supergiant some 900 light-years away in the constellation Orion. Rigel is the brightest star in Orion and the seventh brightest star in the night sky.

Ring Nebula. A famous planetary nebula in the constellation Lyra.

R-Process. The creation of elements heavier than zinc through the rapid bombardment of other elements by neutrons. The r-process occurs in supernovae. Examples of r-process elements are gold, iodine, and europium.

RR Lyrae Star. An old metal-poor white or yellow-white giant star that pulsates like a Cepheid and therefore varies in brightness. Most RR Lyrae stars have periods of under one day, which is shorter than periods for Cepheids. RR Lyrae stars are also fainter than Cepheids, with absolute magnitudes around +0.7, corresponding to a luminosity about 45 times the Sun's. RR Lyrae stars are excellent distance indicators because they all have nearly the same intrinsic brightness. They take their name from the star RR Lyrae, in the constellation Lyra.

Sagittarius A* (pronounced "ay star"). The very center of the Milky Way, Sagittarius A* is a strong source of radio waves and probably a massive black hole.

Sagittarius Arm. The spiral arm that lies next in from the arm containing the Sun.

Satellite Galaxy. A galaxy that orbits a larger one. The Milky Way has at least ten satellite galaxies: the Large Magellanic Cloud, the Small Magellanic Cloud, Ursa Minor, Draco, Sculptor, Sextans, Carina, Fornax, Leo II, and Leo I.

Scale Height. The mean distance of a group of stars from the Galactic plane. In general, old stars have larger scale heights than young ones.

Sculptor. 1. A faint constellation in the southern sky. 2. A dwarf galaxy that orbits the Milky Way in the constellation Sculptor. It is 255,000 light-years from the Galactic center.

Sculptor Group. The nearest group of galaxies to the Local Group, 4 to 10 million light-years away. Its brightest member is the beautiful edge-on spiral NGC 253.

Searle and Zinn. The idea, published in 1978 by Leonard Searle and Robert Zinn, that the outer part of the Galaxy's halo formed from the accretion of smaller galaxies.

Second Parameter. The color of a globular cluster's horizontal branch is determined largely by its metallicity: all other things being equal, the more metal-poor a cluster, the bluer its horizontal branch. However, all other things are not always equal, because globulars with similar metallicities sometimes have different horizontal-branch colors, so a second parameter must be responsible. Searle and Zinn speculated that the second parameter was age and said that all globulars had not formed at the same time.

Sextans. 1. A faint constellation south of Leo. 2. A dwarf galaxy that orbits the Milky Way. Discovered by computer in 1990, Sextans lies 295,000 light-years from the Galactic center.

Silicon. Element with atomic number fourteen and the sixth most common metal in the universe. It is produced by high-mass stars that explode.

Silicon Burning. The end of the line for a high-mass star, silicon burning creates iron and other elements of similar mass and presages a supernova.

Silver. Element with atomic number 47. It is produced by both the r-process and the s-process, but more by the former.

Sirius. The brightest star in the night sky. It is a white, A-type star that lies just 8.6 light-years from Earth in the constellation Canis Major. Orbiting the main star (officially called Sirius A) is a faint white dwarf, Sirius B. Sirius A is the nearest A-type main-sequence star to Earth; Sirius B is the nearest white dwarf to Earth.

61 Cygni. The first star other than the Sun to have its parallax, and hence distance, measured. The star is a double orange dwarf that lies in the constellation Cygnus 11.4 light-years away.

Small Magellanic Cloud. The second largest, and the second nearest, of the galaxies that orbit the Milky Way. It lies in the southern sky, 190,000 light-years away.

Snickers. A possible satellite galaxy of the Milky Way, reported in 1975. Its existence is in dispute.

Solar Mass. The amount of mass in the Sun, and the unit in which stellar and galactic masses are expressed.

Solar Motion. The velocity of the Sun through space, relative to the local standard of rest. The solar motion is $U = -9$ kilometers per second, $V = +12$ kilometers per second, and $W = +7$ kilometers per second.

Solar System. Objects that orbit a star—planets, asteroids, comets.

Sombrero Galaxy. A spiral galaxy in the constellation Virgo. It was the first galaxy whose rotation was detected.

South Galactic Pole. A point in the constellation Sculptor toward which our line of sight is perpendicular to and below the Galactic disk.

Space Velocity. A star's total velocity with respect to the local standard of rest. This is the combination of the star's U, V, and W velocities:

$$\text{space velocity} = \sqrt{U^2 + V^2 + W^2}.$$

For example, the Sun ($U = -9$, $V = +12$, $W = +7$) has a space velocity of 17 kilometers per second.

Spectral Type. Classification of a star's spectrum, which correlates with the star's temperature and color. There are seven main spectral types. From hot and blue to cool and red, they are O, B, A, F, G, K, and M. For further precision, astronomers divide each spectral type. For example, from warmest to coolest, spectral type G is G0, G1, G2, G3, and so on to G9. The Sun is spectral type G2.

Spectroscopy. Measuring the spectrum of an object.

Spectrum. The breakdown of light into a rainbow of colors. A good spectrum reveals a star's spectral type, radial velocity (from the spectrum's Doppler shift), and metallicity.

Spica. The brightest star in the constellation Virgo, Spica consists of two blue B-type stars about 220 light-years from Earth.

Spiral Galaxy. A galaxy that looks like a pinwheel. The Milky Way, the Andromeda Galaxy, M33, and M51 are all spiral galaxies.

Spitzer-Schwarzschild Scattering Mechanism. The process by which stars in the Milky Way's disk encounter interstellar clouds and are accelerated by them. Over time, this perturbs the stars, so that older disk stars have more elliptical orbits, larger velocity dispersions, and greater scale heights than younger disk stars. This mechanism cannot, however, explain the motions of halo stars.

S-Process. The process by which elements heavier than copper are formed through a slow flux of neutrons. The s-process operates in red giant stars; prominent s-process elements include barium, zirconium, yttrium, and lanthanum.

Standard Candle. An object—usually a star or a galaxy—of known intrinsic brightness. Measuring the apparent brightness of a standard candle yields its distance.

Star Cluster. A gathering of hundreds, thousands, or even a million stars. Star clusters are of two types: the less massive open clusters and the more massive globular clusters.

Star Count. Determination of the number of stars in a region of the sky as a function of apparent magnitude and sometimes color.

Star Stream. Discovered by Kapteyn in 1902, a star stream is a group of stars traveling in more or less the same direction. Kapteyn found what he thought were two oppositely directed star streams, but astronomers now recognize that these simply reflect the tendency of stars to have their largest velocities in the U direction.

Star System. A few stars that orbit each other. For example, a double star system consists of two stars; a triple star system consists of three stars; and so on.

Steady State. A cosmological model that proposes that the universe is eternal, with no beginning or end.

Stellar Evolution. How a star changes with time.

Stellar Halo. See Halo.

Stellar Parallax. See Parallax.

Stellar Population. A Galaxy-wide group of stars of all types that have similar ages, locations, kinematics, and metallicities. As astronomers presently know the Milky Way, they recognize four stellar populations: the thin disk; the thick disk; the stellar halo; and the bulge.

Subdwarf. A metal-poor main-sequence star. On the H-R diagram, subdwarfs lie slightly below the metal-rich main sequence, because they are fainter than metal-rich main-sequence stars of the same color.

Subgiant. A star making the transition from the main sequence to the giant stage.

Sulfur. Element with atomic number sixteen and the eighth most common metal in the universe. It was produced by oxygen burning in high-mass stars that exploded.

Sun. The star that Earth orbits. The Sun is a yellow main-sequence star that is spectral type G2, shines with apparent magnitude −26.74, and has an absolute magnitude of +4.83. The Sun is 4.6 billion years old. It lies 27,000 light-years from the Galactic center, or about 40 percent of the way from the center to the edge of the Galactic disk.

Supercluster. A huge assemblage of galaxies. The Milky Way and the entire Local Group are part of the Local Supercluster, which is centered on the Virgo cluster.

Supernova. A titanic explosion that destroys a star. Type Ia supernovae are explosions of white dwarfs that receive material from a companion star and exceed the Chandrasekhar limit, whereas type Ib, Ic, and II supernovae are explosions of high-mass stars—those born with more than eight solar masses—that run out of fuel. Type Ia, Ib, and Ic supernovae have no hydrogen in their spectra, whereas type II supernovae do have hydrogen in their spectra.

Surface Brightness. The measure of the amount of light that an object, especially a galaxy, emits per area of the sky. Even a luminous galaxy can be hard to see if it has a low surface brightness.

Tangential Velocity. A star's velocity across an observer's line of sight. To calculate a star's tangential velocity, one must know the star's distance and proper motion.

Tau Ceti. A G-type main-sequence star that lies in the constellation Cetus, 11.4 light-years away. It is a single star like the Sun and could support life.

Technetium. Radioactive element with atomic number 43. It was seen in red giants in 1952; because it is unstable, its presence indicated that the stars themselves had made it.

Temperature. The hotter a star, the bluer it is: blue stars are hot, yellow stars warm, and red stars cool.

Thick Disk. The stellar population that contains Arcturus and about 4 percent of the other stars near the Sun. It has a scale height of about 3500 light-years and consists of old stars.

Thin Disk. The stellar population that contains the Sun and most other nearby stars. Most of its stars have a scale height of 1000 light-years and orbit the Galaxy on fairly circular orbits. The stars of the thin disk range in age from 0 to about 10 billion years. The thin disk breaks into two subpopulations, the young thin disk (ranging in age from 0 to 1 billion years) and the old thin disk (ranging in age from 1 to about 10 billion years). The young thin disk has a smaller scale height than the old thin disk, and the former's stars have more circular orbits.

Tide. A differential gravitational force. In the Galaxy, a tide results because the Milky Way's gravity pulls more strongly on the side of an object facing the Galactic center than on the object's other side, so the object may get torn apart.

Triple Star. A star system having three stars that revolve around one another.

Tritium. Hydrogen-3, the heaviest isotope of hydrogen, which contains one proton and two neutrons. It is radioactive.

Turnoff. See **Main-Sequence Turnoff.**

21-Centimeter Radiation. Emission of radio waves from neutral hydrogen gas.

U. See **U Velocity.**

Ultraviolet. Electromagnetic radiation that has a somewhat shorter wavelength than visible light. In general, the hotter and bluer a star, the more ultraviolet radiation it produces.

Ultraviolet Excess. Property of a star that emits more ultraviolet radiation than one would have expected, based on its visual color. In general, the greater the ultraviolet excess, the lower the star's metallicity, because metals in a star's atmosphere absorb ultraviolet radiation.

Ursa Minor. 1. The Little Bear (or Little Dipper), a constellation in the northern sky that contains Polaris, the North Star. 2. A dwarf galaxy in the constellation Ursa Minor that orbits the Milky Way and lies 215,000 light-years from its center.

U Velocity. The component of a star's motion away from the Galactic center. If a star moves away from the Galactic center, the star's U velocity is positive; if a star moves toward the Galactic center, the U velocity is negative; and if the star moves neither toward nor away from the Galactic center, the U velocity is zero. The Sun has a U velocity of –9 kilometers per second, so the Sun is moving toward the Galactic center at 9 kilometers per second.

V. See **V Velocity.**

Variable Star. A star whose light varies. Some variables vary simply because they consist of two stars, one of which eclipses the other; Algol is the most famous example. Other variables, however, vary because the stars themselves actually change in brightness; the most famous are the Cepheids, RR Lyraes, and Miras, all of which pulsate.

Vega. The brightest star in the constellation Lyra and the fifth brightest star in the night sky. Vega is a white A-type main-sequence star 25 light-years away.

Velocity Dispersion. The spread of a velocity distribution—that is, how stars move relative to one another. Technically, the velocity dispersion is the standard deviation of the velocity distribution. Stars with similar velocities have a small velocity dispersion, whereas stars with wildly different velocities have a large velocity dispersion.

V Velocity. A star's velocity in the direction of Galactic rotation, as measured relative to a nearby star that has a circular orbit. If a star revolves faster than such a star, the V velocity is positive; if it revolves more slowly, the V velocity is negative; and if both revolve at the same rate, the V velocity is zero. The Sun has a V velocity of +12 kilometers per second, so it revolves 12 kilometers per second faster than it would if it had a circular orbit.

Since a star on a circular orbit revolves around the Galaxy at 220 kilometers per second, a star with a V velocity of 0 is not stationary; rather, it revolves at 220 kilometers per second. The Sun therefore revolves around the Galaxy at $220 + 12 = 232$ kilometers per second.

W. See **W Velocity.**

Warp. The deviation from flatness in the outer Galactic disk. Some parts of the outer disk lie above the Galactic plane; others lie below it.

Whirlpool Galaxy. M51, the most beautiful spiral galaxy in the sky. It lies in the constellation Canes Venatici.

White Dwarf. A small, faint, dense, dying star that has used up its nuclear fuel and is slowly fading from view. A typical white dwarf has 60 percent of the Sun's mass but is little larger than the Earth. White dwarfs are common, accounting for 10 percent of all stars in the Galaxy; the nearest is Sirius B, just 8.6 light-years away. But no white dwarf is visible to the naked eye.

White Giant. A giant star of spectral type A. Some RR Lyrae stars are white giants.

White Supergiant. A supergiant star with a spectral type of A. White supergiants are rare; the nearest is Deneb, which lies 1500 light-years away.

WIMP. Weakly interacting massive particle. Some astronomers believe that these exotic subatomic particles make up most of the mass of the universe.

Wolf-Rayet Star. A hot blue star with a peculiar spectrum. Wolf-Rayet stars have ejected their outer layers, exposing some of the elements the stars have created.

W Velocity. A star's velocity perpendicular to the Galactic plane. If a star is moving up, its W velocity is positive; if a star is moving down, its W velocity is negative; and if a star does neither, its W velocity is zero. The Sun has a W velocity of +7 kilometers per second, so it is moving up at 7 kilometers per second. In general, the greater a star's W velocity when it crosses the Galactic plane, the farther above and below the plane the star will travel.

W Virginis Star. Also known as a population II Cepheid, a W Virginis star is a bright yellow star that pulsates like a Cepheid but is older and fainter. W Virginis stars appear in globular clusters.

X-Process. The unknown nucleosynthetic process that Burbidge, Burbidge, Fowler, and Hoyle said had formed the light nuclei deuterium, lithium, beryllium, and boron.

X-Ray. An energetic form of electromagnetic radiation that is more powerful than ultraviolet radiation but less powerful than gamma rays.

.   .   .

Yellow Giant. A giant star with a spectral type of G. The nearest and brightest yellow giants are the two composing the double star Capella.

Yellow Supergiant. A supergiant star with a spectral type of G.

Young Thin Disk. A subpopulation in the thin disk whose stars range in age from 0 to 1 billion years old. The stars of the young thin disk have a scale height of 350 light-years and have very circular orbits around the Galaxy.

Zirconium. Element with atomic number 40. It arises almost entirely from the s-process, in red giant stars.

Notes

Introduction: A New Galaxy

All quotes are from Irwin (1993). Irwin, Bunclark, Bridgeland, and McMahon (1990) reported the discovery of Sextans, and Da Costa, Hatzidimitriou, Irwin, and McMahon (1991) and Suntzeff et al. (1993) made more recent observations of the galaxy.

1: Welcome to the Milky Way

During the last forty years, three excellent books on the Milky Way have appeared that are more technical than *The Alchemy of the Heavens*. In historical order, they are Blaauw and Schmidt (1965), Mihalas and Binney (1981), and Gilmore, King, and van der Kruit (1990). Good accounts of many celestial objects appear in Burnham (1978).

2: River of Stars

The best accounts of many of the developments in this chapter are by Berendzen, Hart, and Seeley (1976) and Paul (1993). Whitney (1971) gives a broader and more popular overview. Jaki (1971, 1972) discusses early Milky Way research, including that before Galileo. Hoskin (1970) and Hoskin and Rochester (1992) discuss Wright; Jones (1971) discusses Kant. Bennett (1976) and Gingerich (1984) describe William Herschel's work.

Pages 5 through 11 of Hoskin (1982a) give the history of parallax determinations up to Bessel's success. Williams (1979) discusses Flamsteed's spurious parallax for Polaris. Halley (1718) discovered the proper motion of Arcturus and Sirius from the shift of their positions relative to those on ancient star maps; using the same method, he also claimed that he had detected the proper motions of Aldebaran and Betelgeuse, but his numbers were wrong. For Aldebaran, though, he was able to deduce its proper motion from a lunar occultation of the star that occurred in A.D. 509. Serio (1990) describes how Piazzi discovered 61 Cygni's proper motion. Bessel (1838) detected 61 Cygni's parallax. Hoskin (1987) discusses John Herschel's work. Hoskin (1982b) and Dewhirst and Hoskin (1991) describe Rosse's work, and de Vaucouleurs (1987) commemorated the hundredth anniversary of the discovery of Andromeda's spiral structure.

Anonymous (1921, 1922), Curtis (1913), Eddington (1922), Galloway (1902), Glaisher (1902), Jeans (1923), Paul (1993), Seares (1922), and van Maanen (1922) describe the life and work of Kapteyn. The first quote concerning Kapteyn is from Anonymous (1921); the Seares quote is from Seares (1922). The discovery of Kapteyn's Star is mentioned by Aitken (1898). Barnard (1916) discovered Barnard's Star; for the history of Groombridge 1830, see the delightful article by Ashbrook (1974). Kapteyn (1914a) reports on the selected areas. The final models of the Kapteyn universe appeared in Kapteyn and van Rhijn (1920, 1922a) and Kapteyn (1922). Hartmann (1904) reported interstellar calcium gas between us and Mintaka. Slipher (1909) later did similar work. Kapteyn (1909a,b, 1914b) investigated the question further, as did van Rhijn (1916). Shapley (1916) reported little if any interstellar absorption.

3: BIG GALAXY, BIG UNIVERSE

Berendzen, Hart, and Seeley (1976) and Paul (1993) discuss many of the developments in this chapter. Seeds (1980) describes Algol. Most books incorrectly say that Delta Cephei was the first Cepheid discovered; for the correct history, see the excellent article by Hoskin (1979). Accounts of Goodricke's life appear in Gilman (1978), Goodricke (1912), and Joy (1937). Duncan (1909) said Clerke was responsible for coining "Cepheid." Contrary to claims in most books, Leavitt (1907) first reported the period-luminosity relation in 1907; Leavitt (1912) strengthened it. Hertzsprung (1913) first used Cepheids to derive distances.

Shapley published a long series of papers on the globulars, the most important from a Galactic viewpoint being Shapley (1918a,b, 1919); see Shapley (1918c) for a summary. Before them, though, Shapley (1916) stated that the globulars were outside the Milky Way. Pickering (1901) reported the variability of RR Lyrae (also named +42° 3338), discovered by Williamina Fleming. Shapley's use of the Cepheids is described by Gingerich and Welther (1985).

Gingerich (1988), Hoskin (1976, 1988), and Smith (1983) describe the "Great Debate," and the first of these is the source of the quote concerning Shapley's negative debate performance. Baade's quote is from page 9 of Baade (1963), and the Mihalas-Binney quote is from page 9 of Mihalas and Binney (1981). Kapteyn and van Rhijn (1922b) questioned Shapley's results concerning the RR Lyrae stars; see also Luyten (1922). Gingerich (1990) gives Shapley's response to Hubble's discovery of Cepheids in Andromeda. Hubble (1929b) discovered the expansion of the universe, and Anonymous (1941) is the source of Payne-Gaposchkin's closing quote.

4: GIVING THE GALAXY A SPIN

Berendzen, Hart, and Seeley (1976) discuss many of the developments in this chapter. The quote concerning Lowell Observatory is from page 131 of Tombaugh and Moore (1980). Slipher (1913) discovered the large radial veloc- ity of Andromeda, and Slipher (1914) discovered the rotation of the Sombrero Galaxy. Hetherington (1975) and Hoyt (1980) describe Slipher's work.

Vogel (1891, 1892, 1900) describes early radial-velocity work, and F (1908), Frost (1908), and Knobel (1893) describe Vogel's work. Maw (1906) and Trumpler (1938) discuss Campbell and Anonymous (1959) Wright. Camp- bell (1898a) describes the Lick spectrograph. The Lick catalogue of radial velocities is Campbell and Moore (1928). Shapley (1918d) plotted the distribu- tions of Cepheids and RR Lyraes.

Kapteyn (1905) discovered the two star streams. The Newcomb quote is from Anonymous (1922), and Eddington's is from Eddington (1922). Edding- ton (1917) and Hertzsprung (1917) describe Karl Schwarzschild's life and work.

High-velocity stars were reported by the following: Campbell (1895) de- scribes the radial velocity of Zeta Herculis, Campbell (1898b,c) Eta Cephei, and Campbell (1901a,b,c) stars with velocities near 100 kilometers per second. Curtis (1909) discovered the high radial velocity of Kapteyn's Star. Adams and Kohlschütter (1913) and Adams and Joy (1917) discovered stars with velocities exceeding 300 kilometers per second.

Boss (1918) and Adams and Joy (1919) noticed the asymmetry of the high-velocity stars' motions, and Strömberg (1924, 1925) suggested that Galac- tic rotation could explain it and discovered that the velocity dispersion in- creased as the V velocity became more negative. For an English version of Lindblad's 1925 Galactic rotation proposal, see page 550 of Lang and Ginger- ich (1979); for an English version of his star-stream solution, see Lindblad (1927). For accounts of Oort's life, see Blaauw and Schmidt (1993), Oort (1981), and van Woerden, Brouw, and van de Hulst (1980). Whitney's quote is from Whitney (1981). Oort (1927a) confirmed Galactic rotation, primarily through his use of radial velocities; Oort (1927b) analyzed proper motions, and Oort (1927c,d, 1928) provided additional evidence for Galactic rotation. Ac- counts of Trumpler's life and work appear by Gingerich (1985b), Irwin (1956), and Seeley and Berendzen (1972). Trumpler (1930a) discovered interstellar absorption; previous work is Trumpler (1929), and subsequent work Trumpler (1930b,c).

5: CITIZENS OF THE GALAXY

Iben (1991) and Chiosi, Bertelli, and Bressan (1992) give technical reviews of stellar evolution. Kaler's (1992) beautiful book is a good overview on a more popular level. DeVorkin (1977, 1978) and Strand (1977) describe the origin of the H-R diagram. Andersen (1991) reviews our knowledge of stellar masses. Leggett (1992) gives the latest data on hundreds of nearby red dwarfs. The following reported nearby red dwarfs: Innes (1915) discovered Proxima Centauri, Barnard (1916) discovered Barnard's Star, and Wolf (1918) discovered Wolf 359, whose distance was first measured by van Maanen, Brown, and Humason (1927). Van Biesbroeck (1944) discovered Van Biesbroeck's Star, and in 1925 Ross (1926) discovered Ross 154, Ross 248, and Ross 128. Croswell (1992e) explains why the stars we see at night differ from those which are most common in the Galaxy. Nather (1990) gives a spirited popular discussion of white dwarfs. The Russell quote on white dwarfs is from Russell (1977). Adams (1914) reported on the spectral type of Omicron^2 Eridani B, and Adams (1915) discovered the spectral type of Sirius B. Trimble (1991a) discusses neutron stars and black holes, as does Greenstein (1983) on a more popular level; Croswell (1992a) describes black holes and Croswell (1992c) type Ia supernovae.

6: THE DEMOGRAPHICS OF THE MILKY WAY

For reviews of stellar populations, see King (1971), van den Bergh (1975), Mould (1982), and Sandage (1986). The opening quote is from Sandage (1986). For a thorough discussion of the Sun's motion perpendicular to the Galactic plane, see Bahcall and Bahcall (1985).

Unfortunately, Baade published little during his life; the best account of his thoughts is Baade (1963), which was published after his death and consists of tape recordings of lectures he delivered at Harvard in 1958. This is the source of most of Baade's quotes in this chapter. For accounts of Baade's life, see Arp (1961), Jackson (1954), Osterbrock (1993), Sandage (1961), and O. C. Wilson (1955); Osterbrock's quote is from Osterbrock (1993). Hubble's galaxy classification scheme is Hubble (1926b); his investigations of Andromeda and M33 are Hubble (1929a) and Hubble (1926a), and of elliptical galaxies Hubble (1930). Early work on galaxy colors appears in Seares (1916a,b). Sculptor and Fornax were reported by Shapley (1938a,b, 1939); see Baade and Hubble (1939) for their investigation. Baade's landmark papers on stellar populations are Baade (1944a,b). The discovery of the main sequence in the globular clusters was made by Arp, Baum, and Sandage (1952, 1953) and Sandage (1953). Hints that population I Cepheids differ from population II Cepheids appear in Baade (1948). For Baade's doubling of galactic distances, see Hoyle (1954b).

7: A SPIRAL GALAXY

Most of this chapter's quotes are from Morgan (1993) and Osterbrock (1993). Gingerich (1985a) describes the discovery of the spiral arms. Speculations that the Milky Way is a spiral appear in Alexander (1852), Proctor (1869), Easton (1900, 1913), and Campbell (1926); see Seeliger (1900) for an argument against Easton (1900). Bok's quote is spliced together from page 269 of Bok and Bok (1981) and Bok (1969). Levy's (1993) book is a biography of Bok. Baade and Mayall (1951) discuss Andromeda.

Additional background on Morgan's early life appears in Morgan (1988) and in an interesting footnote to Morgan (1984): "This paper is dedicated to my father, William Thomas Morgan (1877–19??). You will never know what I owe you." Note the uncertain year of death. A report on the work of Morgan and Nassau appears in Anonymous (1950). More details on the Greenstein-Henyey camera—which Sharpless and Osterbrock used to find H II regions—appear in Struve (1951); also see Greenstein (1984). Nassau and Morgan (1951) studied OB stars, and Sharpless and Osterbrock (1952) describe nearby H II regions. The abstract of the discovery of the spiral arms appeared in Morgan, Sharpless, and Osterbrock (1952), and a better account was Anonymous (1952). The Sagittarius arm was confirmed by Morgan, Whitford, and Code (1953). The most vivid description of Morgan's 1951 talk is by Struve (1953).

For a detailed discussion of Jansky's work, see Sullivan (1978). Reber's first papers in *The Astrophysical Journal* were Reber (1940, 1944). The discovery reports of 21-centimeter radiation were, in order, Ewen and Purcell (1951), Muller and Oort (1951), and Pawsey (1951). The first radio maps showing a spiral pattern for the Milky Way were Oort (1953) and van de Hulst (1953), and a more detailed report appeared in van de Hulst, Muller, and Oort (1954). The 1958 radio map was by Oort, Kerr, and Westerhout (1958). The warp, discovered in 1956, was independently reported by Burke (1957) and Kerr (1957).

8: THE RISE OF THE SUBDWARFS

Aller's quotes are from Aller (1993), with additional information provided by Chamberlain (1993). The 1951 paper was Chamberlain and Aller (1951); an early report appeared by Chamberlain, Aller, and Sanford (1950). Even before that work, Martin Schwarzschild suspected that abundance differences existed between populations I and II; see Schwarzschild and Schwarzschild (1950), which was followed by Schwarzschild, Spitzer, and Wildt (1951). Sandage's quote is from page 432 of Sandage (1986). Aller (1943) claimed that Wolf-Rayet stars had abnormal abundances. Adams and Lasby (1911) discovered the high radial velocities of HD 19445 (also known as Lalande 5761) and HD 140283 (also known as Lalande 28607). Adams (1913) and Adams and Kohl-schütter (1914) commented on the abnormal spectra of HD 19445 and HD

140283. Adams and Joy (1922) included these two stars in a table of three odd
A stars, while Adams, Joy, Humason, and Brayton (1935) listed a total of six.
Kuiper (1939) coined the term "subdwarf." Popper (1943) measured the high
radial-velocity of CD −29°2277 in a follow-up to Popper (1942), and additional
details appear in Popper (1993). Carney and Latham (1987) found the current
radial-velocity champion, Giclas 233-27. The first report on the color of HD
19445 was by MacRae, Harris, and Rogerson (1952). Better abundance analy-
ses of HD 19445 and HD 140283 appear in Aller and Greenstein (1960).

Roman's quotes are from Roman (1993). Roman (1954) discovered the
ultraviolet excess, and Roman (1950, 1952) discovered the kinematic differences between disk stars having different line strengths. Her high-velocity
catalogue is Roman (1955).

The discovery of the white stars near the Milky Way was related by
Anonymous (1891); also see D (1897) and Tucker (1898). Pickering (1904,
1905) reported on the concentration of the B stars. Campbell (1910, 1911,
1912) and Kapteyn (1910) discovered the correlation of velocity with spectral
type. Eddington's quote is from page 37 of Eddington (1914). Early explanations in terms of energy equipartition appear, for example, in Halm (1911) and
Paddock (1913).

All quotes of Spitzer and Schwarzschild are from Spitzer (1993) and
Schwarzschild (1993). Their papers are Spitzer and Schwarzschild (1951,
1953). For a personal account, see Spitzer (1989). Osterbrock (1952) explored
the same concept.

9: THE ALCHEMY OF THE HEAVENS

The opening poem is by the author. All quotes are from Hoyle (1993), Fowler
(1993), E. M. Burbidge (1993), G. Burbidge (1993), and Cameron (1993).
B²FH is Burbidge, Burbidge, Fowler, and Hoyle (1957). Hoyle (1982a,b, 1986,
1994) discusses his life. Hoyle (1946) explored the synthesis of elements up to
iron in stars. For one account of big bang nucleosynthesis, see Alpher, Bethe,
and Gamow (1948). Hoyle (1954a) predicted the carbon-12 resonance, exactly
where Dunbar, Pixley, Wenzel, and Whaling (1953) found it; see also Hoyle,
Dunbar, Wenzel, and Whaling (1953). Fowler (1992) talks about his life, and
Burbidge and Burbidge (1982) and Burbidge (1994) offer further insights.
Fowler, Burbidge, and Burbidge (1955) began to investigate the s-process.
Merrill (1952) detected technetium as well as barium and zirconium in red
giants. Cameron (1954, 1955, 1960) proposed the carbon-13 and neon-22 neutron sources, and Cameron (1957, 1958) explored stellar nucleosynthesis in
general. Burbidge and Burbidge (1957) observed the barium star HD 46407.
Suess and Urey (1956) determined the abundances of the elements; for a
modern abundance table, see Anders and Grevesse (1989). Baade et al. (1956)
connect the supernova in IC 4182 with californium-254.

Bondi and Gold (1948) and Hoyle (1948) proposed the steady state cosmology; Hoyle (1950) coined "big bang." Hoyle (1983) discusses his true moti-
vations for studying nucleosynthesis. Bondi, Gold, and Hoyle (1955) and
Burbidge (1958) pointed out problems with helium. Hoyle and Tayler (1964)

worked out the primordial abundance of helium, and Wagoner, Fowler, and Hoyle (1967) extended this to other light nuclei. Reed (1983) describes the pulsar fiasco. Hoyle, Burbidge, and Narlikar (1993) present the new steady state cosmology, while Narlikar (1993) describes it at a more popular level, and Schramm (1993) attacks it.

1 0: ELS

All quotes are from Eggen (1993a), Lynden-Bell (1993), and Sandage (1993). ELS is Eggen, Lynden-Bell, and Sandage (1962), based on Eggen's (1962, 1964) catalogues. Eggen (1993b) talks about his life. The 1959 papers are Eggen and Sandage (1959) and Sandage and Eggen (1959). The Vatican conference is O'Connell (1958), and the split of the Milky Way into five populations appears in Oort (1958).

1 1: GALAXY IN CHAOS

Quotes in the opening pages are from Searle (1993) and Zinn (1993), and their paper is Searle and Zinn (1978). Most other quotes are from Sandage (1993), Lynden-Bell (1993), Norris (1993), Carney (1993), and Eggen (1993a).

Schmidt (1963) first determined the redshift of a quasar. Sandage (1969) found a metallicity gradient in the halo, as did Sandage (1981) and Sandage and Fouts (1987). Sandage and Wildey (1967) discovered NGC 7006's red horizontal branch. Toomre and Toomre (1972) investigated galactic collisions. Isobe (1974) and Yoshii and Saio (1979) argued for a longer timescale of Galactic collapse. Bolte (1989) determined the relative ages of NGC 288 and NGC 362, and precise metallicities were determined by Dickens, Croke, Cannon, and Bell (1991). Also, Gratton and Ortolani (1988) demonstrated that the globular Palomar 12 was relatively young. Sandage's latest data are contained in Sandage and Kowal (1986), Fouts and Sandage (1986), and Sandage and Fouts (1987); Sandage (1990) defended ELS against its critics.

Bond (1981) raised the population III specter. Norris, Bessel, and Pickles (1985) found metal-poor stars on circular orbits in the Bidelman and MacConnell (1973) catalogue, and Norris (1986) found no dependence of stellar V velocity on metallicity at low metallicities. Norris (1994) suggested the halo might contain two components, one conforming to ELS and the other to Searle and Zinn. Carney and Latham (1987) began a survey of high-velocity stars. Carney, Aguilar, Latham, and Laird (1990) found no metallicity gradient in the halo, and Carney, Latham, and Laird (1990) questioned ELS. Carney, Latham, Laird, and Aguilar (1994) extended the original Carney-Latham survey.

12: THE THICK DISK

Most quotes are from Gilmore (1993) and Bahcall (1993); most others are from Sandage (1993), Carney (1993), and Norris (1993). Gilmore and Reid (1983) proposed that the Galaxy had a thick disk; in contrast, Bahcall and Soneira's (1980, 1981b) Galaxy model had no thick disk. Bahcall and Soneira (1981a) pointed out the error in Allen's (1973) book. Gilmore (1984) and Gilmore, Reid, and Hewett (1985) presented evidence in favor of the thick disk, while Bahcall and Ratnatunga (1985), Bahcall and Soneira (1984), and Bahcall, Soneira, Morton, and Tritton (1983) disputed it. Bahcall (1986) reviewed the Bahcall-Soneira model.

Friel (1987) used spectra to determine that the Galaxy had a thick disk. Sandage and Fouts (1987) and Carney, Latham, and Laird (1989) also favored the thick disk. Casertano, Ratnatunga, and Bahcall (1990) conceded that the Galaxy had a population between the thin disk and halo.

Oort (1958) listed intermediate population II. Keenan and Keller (1951, 1953) determined the H-R diagram of what astronomers now know are thick disk stars; for further discussions of this, see Sandage (1986) and Sandage and Fouts (1987). Burstein (1979) coined the phrase "thick disk"; Hartkopf and Yoss (1982) saw the thick disk at the Galactic poles. Kinman (1959) and Morgan (1959) showed that the globulars split into halo and disk systems, following work by Mayall (1946) and Morgan (1956); but Woltjer (1975) and Harris (1976) disputed this result. Zinn (1985) definitively showed that the globulars do form two systems, one halo and the other disk. Armandroff (1989) gave a better scale height for the disk globulars. Norris (1987) argued that the thick disk was the tail of the thin disk and suggested that scattering might be the cause of the thick disk; Norris and Ryan (1991) argued that an ELS-type collapse may have formed the thick disk.

13: SHIFTS IN THE GALACTIC WIND

All quotes are from Sneden (1993), Beers (1993), and Woosley (1993). Wheeler, Sneden, and Truran (1989) and Trimble (1991b) give excellent reviews of recent developments in stellar abundances. Abundances in Table 13-1 are from Anders and Grevesse (1989); the lower (meteoritic) iron abundance is adopted. Gasson and Pagel (1966) and Conti et al. (1967) first observed oxygen in stars. Lambert, Sneden, and Ries (1974) detected oxygen in HD 122563, and Sneden, Lambert, and Whitaker (1979) studied oxygen in main-sequence stars. Marschall (1994) discusses supernovae. Cunha and Lambert (1992, 1994) studied oxygen and other elements in Orion. Spite and Spite (1978), Truran (1981), Sneden and Parthasarathy (1983), Sneden and Pilachowski (1985), and Gilroy, Sneden, Pilachowski, and Cowan (1988) discuss r-process elements in metal-poor stars.

Beers, Preston, and Shectman (1985, 1992) searched for metal-poor stars, and Bond (1981) raised the issue of the supposed absence of extremely metal-poor stars. Norris, Peterson, and Beers (1993) analyzed abundances of several extremely metal-poor stars. Domogatsky and Nadyozhin (1977) and Woosley (1977) explored the neutrino-process. Woosley and Haxton (1988) proposed that the neutrino-process could produce fluorine, and Woosley, Hartmann, Hoffman, and Haxton (1990) performed a more thorough study. Jorissen, Smith, and Lambert (1992) detected fluorine in giant stars.

14: THE GALACTIC METROPOLIS

Dickey and Lockman (1990) and Liszt (1992) discuss interstellar atomic hydrogen in the Milky Way, and Blitz (1992), Bronfman (1992), and Combes (1991) describe the Galaxy's molecular hydrogen. Kerr and Lynden-Bell (1986) give the "official" distance to the Galactic center, while Reid (1993) reviews recent determinations.

Quotes on the bulge are from Rich (1993). Recent reviews of the bulge include Whitford (1985), Frogel (1988), and Rich (1992). Baade (1946) detected RR Lyraes in the bulge. Morgan (1956) found Andromeda's bulge to be metal-rich and later found the Milky Way's bulge likewise, as discussed by Morgan and Osterbrock (1969). Nassau and Blanco (1958) discovered M-type giants in the bulge. Stebbins and Whitford (1947) made infrared observations of the bulge, as described further by Whitford (1986). Whitford and Rich (1983) and Rich (1988) explored the metallicities in the bulge. De Vaucouleurs (1964) suggested the Milky Way is a barred spiral.

Quotes concerning the Galactic center are from Yusef-Zadeh (1993). Recent reviews on the subject include Genzel and Townes (1987), Yusef-Zadeh and Wardle (1992), and Yusef-Zadeh (1994). Sullivan (1978) describes Jansky's detection of the Galactic center's radio waves, and Becklin and Neugebauer (1968) picked up the center's infrared radiation. Lynden-Bell and Rees (1971) suggested that the Galactic center harbors a massive black hole. Croswell (1992a) described black holes and Croswell (1993a) the likely nature of quasars. Eckart et al. (1992) picked up infrared radiation from the direction of Sagittarius A*, and Backer and Sramek (1982) measured its proper motion. Wardle and Yusef-Zadeh (1992) proposed using gravitational lensing to measure the black hole's mass. Eckart et al. (1993) presented recent infrared pictures of the Galactic center. Rieke and Rieke (1989) studied IRS 7, and Serabyn, Lacy, and Achtermann (1991) and Yusef-Zadeh and Morris (1991) discovered the star's cometary tail. Yusef-Zadeh and Melia (1992) pinpointed IRS 16 as the culprit responsible for IRS 7's tail. Yusef-Zadeh, Morris, and Chance (1984) discovered the magnetic filaments.

15: FRONTIERS

All quotes are from Pryor (1993) and Kuhn (1993b). Fich and Tremaine (1991) review methods for determining the Milky Way's mass, and Trimble (1987) explores dark matter in the Galaxy and beyond. Carney (1984) describes the outer halo. The conference proceedings edited by Haynes and Milne (1991) describe the Magellanic Clouds.

Dwarf galaxies are reviewed by Hodge and Michie (1969), Hodge (1971, 1989), Da Costa (1988), and Pryor (1992). Ursa Minor and Draco were discovered by A. G. Wilson (1955); Sculptor was first reported by Shapley (1938a); Sextans by Irwin et al. (1990); Carina by Cannon, Hawarden, and Tritton (1977); Fornax by Shapley (1938b); and Leo I and Leo II by Harrington and Wilson (1950). Mateo et al. (1993) and Lee et al. (1993) give the latest data on the dwarfs; Suntzeff et al. (1993) describe Sextans, Smecker-Hane, Stetson, Hesser, and Lehnert (1994) describe Carina, and Demers, Irwin, and Kunkel (1994) describe Fornax. These are the sources of most of the data in Table 15-1. Absolute magnitudes for the Magellanic Clouds are from van den Bergh (1992a).

Zaritsky et al. (1989) used the satellites to determine the Milky Way's mass. Rubin and Ford (1970) discovered the fast orbital speeds of stars in the outer disk of Andromeda, but Freeman (1970) first recognized the implications of such a find. Carney, Latham, and Laird (1988) and Leonard and Tremaine (1990) discussed recent determinations of the Galactic escape velocity. The current radial-velocity champion was reported by Carney and Latham (1987).

Aaronson (1983) determined the velocity dispersion of stars in Draco, and Faber and Lin (1983) and Lin and Faber (1983) interpreted the result in terms of dark matter. Kormendy (1987) noted the dwarfs' large central densities of dark matter. Kuhn and Miller (1989) and Kuhn (1993a) put forth other ways to explain the high velocity dispersion. Simonson (1975, 1993) reported on Snickers, and Pogge (1990) discussed the fate of this purported galaxy. Mathewson, Cleary, and Murray (1974) discovered the Magellanic Stream, and Lin and Richer (1992) reported that Ruprecht 106 may have come from the Clouds. Chapter 7 of Binney and Tremaine (1987) described dynamical friction. Croswell et al. (1991) studied distant stars in the halo. Distant RR Lyrae stars were reported by Hawkins (1983, 1984), who found an extremely high radial velocity for one star, later corrected by Norris and Hawkins (1991); and Ciardullo, Jacoby, and Bond (1989) also found a distant RR Lyrae star. Alcock et al. (1993), Aubourg et al. (1993), and Udalski et al. (1994) reported the detection of MACHOs.

16: FOSSILS OF CREATION

Most quotes are from Schramm (1993b) and Gilmore (1991, 1993). The book edited by Thompson, Carney, and Karwowski (1990) gives a good overview of primordial nucleosynthesis, and Walker et al. (1991) give the latest constraints on the mass density of the universe. Hubble (1929b) discovered the expansion of the universe, and Penzias and Wilson (1965) the microwave background; Chown's (1993) book concentrated on the history of the latter. Trimble's quote concerning lithium is from Trimble (1991b). M. Spite and F. Spite (1982) and F. Spite and M. Spite (1982) discovered lithium in halo stars; Smith, Lambert, and Nissen (1993) discovered lithium-6 in a halo star, while Steigman et al. (1993) pointed out the importance of this discovery. Crawford and Schramm

(1982), Witten (1984), and Applegate and Hogan (1985) explored the inhomogeneous early universe, and Boyd and Kajino (1989) called attention to the possible importance of beryllium. Gilmore, Edvardsson, and Nissen (1991) discovered beryllium in halo stars, and Duncan, Lambert, and Lemke (1992) boron.

17: OLDER THAN THE UNIVERSE?

All quotes are from Sandage (1993), Bolte (1993), VandenBerg (1993), and G. Burbidge (1993). Hubble (1929b) discovered the expansion of the universe. Van den Bergh (1992b) gave distances for the nearby galaxies. Recent, and contradictory, determinations of the Hubble constant are by Sandage and Tammann (1993) and Sandage et al. (1994), and by de Vaucouleurs (1993). Attacks on IC 4182 were made by Sandage and Tammann (1982), Pierce, Ressler, and Shure (1992), Sandage et al. (1992), and Jacoby, Pierce, and Massey (1992). Carroll, Press, and Turner (1992) discuss the cosmological constant, as does Croswell (1993b). Bondi and Gold (1948) and Hoyle (1948) developed the original steady state cosmology; Hoyle, Burbidge, and Narlikar (1993) have revived it.

18: AN INTELLIGENT GALAXY

Swift (1990) and Drake and Sobel (1992) describe the investigation of Tau Ceti and Epsilon Eridani. Nearby star data come from various sources, including, but not limited to, Leggett (1992), Gliese (1969), and Gliese and Jahreiss (1979). Dahn et al. (1972) discovered that Giclas 51-15 is a nearby star. Luyten (1949) discovered that Luyten 726-8 B is a flare star. Croswell (1988) discusses the status of the planets around Barnard's Star. Wolsczan and Frail (1992) discovered at least two planets around PSR B1257+12; for a popular discussion of pulsar planets, see Croswell (1992b). Croswell (1993c) describes the evolution of Mars. Croswell (1991) discusses the prospects for life around Alpha Centauri. Wetherill's work on Jupiter and Saturn was first reported by Croswell (1992d).

BIBLIOGRAPHY

Aaronson, Marc, 1983. Accurate Radial Velocities for Carbon Stars in Draco and Ursa Minor: The First Hint of a Dwarf Spheroidal Mass-to-Light Ratio. *Astrophysical Journal Letters*, 266, L11.

Adams, W. S., 1913. Some Radial Velocity Results. *Publications of the Astronomical Society of the Pacific*, 25, 259.

———, 1914. An A-type Star of Very Low Luminosity. *Publications of the Astronomical Society of the Pacific*, 26, 198.

———, 1915. The Spectrum of the Companion of Sirius. *Publications of the Astronomical Society of the Pacific*, 27, 236.

Adams, W. S., and Joy, A. H., 1917. Two Stars with Remarkable Radial Velocities. *Publications of the Astronomical Society of the Pacific*, 29, 259.

———, 1919. The Motions in Space of Some Stars of High Radial Velocity. *Astrophysical Journal*, 49, 179.

———, 1922. A Spectroscopic Method of Determining the Absolute Magnitudes of A-Type Stars and the Parallaxes of 544 Stars. *Astrophysical Journal*, 56, 242.

Adams, Walter S., Joy, Alfred H., Humason, Milton L., and Brayton, Ada Margaret, 1935. The Spectroscopic Absolute Magnitudes and Parallaxes of 4179 Stars. *Astrophysical Journal*, 81, 187.

Adams, W. S., and Kohlschütter, A., 1913. A Star with a Remarkable Radial Velocity. *Publications of the Astronomical Society of the Pacific*, 25, 289.

————, 1914. The Radial Velocities of One Hundred Stars with Measured Parallaxes. *Astrophysical Journal*, 39, 341.

Adams, Walter S., and Lasby, Jennie B., 1911. Some Stars with Great Radial Velocities. *Publications of the Astronomical Society of the Pacific*, 23, 239.

Aitken, R. G., 1898. The Star with the Largest Known Proper Motion. *Publications of the Astronomical Society of the Pacific*, 10, 37.

Alcock, C., et al., 1993. Possible Gravitational Microlensing of a Star in the Large Magellanic Cloud. *Nature*, 365, 621.

Alexander, Stephen, 1852. On the Origin of the Forms and the Present Condition of Some of the Clusters of Stars and Several of the Nebulae. *Astronomical Journal*, 2, 97; 2, 105.

Allen, C. W., 1973. *Astrophysical Quantities*, Third Edition (London: Athlone Press).

Aller, Lawrence H., 1943. A Study of Emission-Line Intensities in Some Bright Northern Wolf-Rayet Stars. *Astrophysical Journal*, 97, 135.

————, 1993. Interviews with Ken Croswell: June 29, 1993, and July 26, 1993.

Aller, Lawrence H., and Greenstein, Jesse L., 1960. The Abundances of the Elements in G-Type Subdwarfs. *Astrophysical Journal Supplement Series*, 5, 139.

Alpher, R. A., Bethe, H., and Gamow, G., 1948. The Origin of Chemical Elements. *Physical Review*, 73, 803.

Anders, Edward, and Grevesse, Nicolas, 1989. Abundances of the Elements: Meteoritic and Solar. *Geochimica et Cosmochimica Acta*, 53, 197.

Andersen, J., 1991. Accurate Masses and Radii of Normal Stars. *Astronomy and Astrophysics Review*, 3, 91.

Anonymous, 1891. The Milky Way Is a Separate System of Stars. *Publications of the Astronomical Society of the Pacific*, 3, 372.

————, 1921. The Astronomical Laboratory at Groningen. *Observatory*, 44, 28.

————, 1922. From an Oxford Note-Book. *Observatory*, 45, 271.

————, 1941. Harlow Shapley Honored. *Sky and Telescope*, 1, No. 2 (December 1941), 12.

————, 1950. Symposium on the Galaxy. *Sky and Telescope*, 9, 243 (August 1950).

————, 1952. Spiral Arms of the Galaxy. *Sky and Telescope*, 11, 138 (April 1952).

————, 1959. Former Lick Director Dies. *Sky and Telescope*, 18, 500 (July 1959).

Applegate, James H., and Hogan, Craig J., 1985. Relics of Cosmic Quark Condensation. *Physical Review D*, 31, 3037.

Armandroff, Taft E., 1989. The Properties of the Disk System of Globular Clusters. *Astronomical Journal*, 97, 375.

Arp, Halton C., 1961. Wilhelm Heinrich Walter Baade, 1893–1960. *Journal of the Royal Astronomical Society of Canada*, 55, 113.

Arp, H. C., Baum, W. A., and Sandage, A. R., 1952. The H-R Diagrams for the Globular Clusters M92 and M3. *Astronomical Journal*, 57, 4.

———, 1953. The Color-Magnitude Diagram of the Globular Cluster M92. *Astronomical Journal*, 58, 4.

Ashbrook, Joseph, 1974. The Story of Groombridge 1830. *Sky and Telescope*, 47, 296 (May 1974).

Aubourg, E., et al., 1993. Evidence for Gravitational Microlensing by Dark Objects in the Galactic Halo. *Nature*, 365, 623.

Baade, W., 1944a. The Resolution of Messier 32, NGC 205, and the Central Region of the Andromeda Nebula. *Astrophysical Journal*, 100, 137.

———, 1944b. NGC 147 and NGC 185, Two New Members of the Local Group of Galaxies. *Astrophysical Journal*, 100, 147.

———, 1946. A Search for the Nucleus of our Galaxy. *Publications of the Astronomical Society of the Pacific*, 58, 249.

———, 1948. A Program of Extragalactic Research for the 200-inch Hale Telescope. *Publications of the Astronomical Society of the Pacific*, 60, 230.

———, 1963. *Evolution of Stars and Galaxies*, edited by Cecilia Payne-Gaposchkin (Cambridge: Harvard University Press).

Baade, W., Burbidge, G. R., Hoyle, F., Burbidge, E. M., Christy, R. F., and Fowler, W. A., 1956. Supernovae and Californium 254. *Publications of the Astronomical Society of the Pacific*, 68, 296.

Baade, W., and Hubble, Edwin, 1939. The New Stellar Systems in Sculptor and Fornax. *Publications of the Astronomical Society of the Pacific*, 51, 40.

Baade, W., and Mayall, N. U., 1951. Distribution and Motions of Gaseous Masses in Spirals. In *Problems of Cosmical Aerodynamics* (Dayton: Central Air Documents Office), p. 165.

Backer, D. C., and Sramek, R. A., 1982. Apparent Proper Motions of the Galactic Center Compact Radio Source and PSR 1929+10. *Astrophysical Journal*, 260, 512.

Bahcall, John N., 1986. Star Counts and Galactic Structure. *Annual Review of Astronomy and Astrophysics*, 24, 577.

————, 1993. Interview with Ken Croswell: September 27, 1993.

Bahcall, John N., and Bahcall, Safi, 1985. The Sun's Motion Perpendicular to the Galactic Plane. *Nature*, 316, 706.

Bahcall, John N., and Ratnatunga, Kavan U., 1985. Comparisons of the Standard Galaxy Model with Observations in Two Fields. *Monthly Notices of the Royal Astronomical Society*, 213, 39P.

Bahcall, John N., and Soneira, Raymond M., 1980. The Universe at Faint Magnitudes. I. Models for the Galaxy and the Predicted Star Counts. *Astrophysical Journal Supplement Series*, 44, 73.

————, 1981a. The Distribution of Stars to V = 16th Magnitude near the North Galactic Pole: Normalization, Clustering Properties, and Counts in Various Bands. *Astrophysical Journal*, 246, 122.

————, 1981b. Predicted Star Counts in Selected Fields and Photometric Bands: Applications to Galactic Structure, the Disk Luminosity Function, and the Detection of a Massive Halo. *Astrophysical Journal Supplement Series*, 47, 357.

————, 1984. Comparisons of a Standard Galaxy Model with Stellar Observations in Five Fields. *Astrophysical Journal Supplement Series*, 55, 67.

Bahcall, John N., Soneira, Raymond M., Morton, Donald C., and Tritton, K. P., 1983. Some Constraints on the Color-Magnitude Diagram of Giants in the Galactic Spheroid. *Astrophysical Journal*, 272, 627.

Barnard, E. E., 1916. A Small Star with Large Proper-Motion. *Astronomical Journal*, 29, 181.

Becklin, E. E., and Neugebauer, G., 1968. Infrared Observations of the Galactic Center. *Astrophysical Journal*, 151, 145.

Beers, Timothy, 1993. Interview with Ken Croswell: October 4, 1993.

Beers, Timothy C., Preston, George W., and Shectman, Stephen A., 1985. A Search for Stars of Very Low Metal Abundance. I. *Astronomical Journal*, 90, 2089.

————, 1992. A Search for Stars of Very Low Metal Abundance. II. *Astronomical Journal*, 103, 1987.

Bennett, J. A., 1976. "On the Power of Penetrating into Space": The Telescopes of William Herschel. *Journal for the History of Astronomy*, 7, 75.

Berendzen, Richard, Hart, Richard, and Seeley, Daniel, 1976. *Man Discovers the Galaxies* (New York: Science History Publications).

Bessel, F. W., 1838. A Letter from Professor Bessel to Sir J. Herschel. *Monthly Notices of the Royal Astronomical Society*, 4, 152.

Bidelman, William P., and MacConnell, Darrell J., 1973. The Brighter Stars of Astrophysical Interest in the Southern Sky. *Astronomical Journal,* 78, 687.

Binney, James, and Tremaine, Scott, 1987. *Galactic Dynamics* (Princeton: Princeton University Press).

Blaauw, Adriaan, and Schmidt, Maarten (editors), 1965. *Galactic Structure* (Chicago: University of Chicago Press).

Blaauw, Adriaan, and Schmidt, Maarten, 1993. Jan Hendrik Oort (1900–1992). *Publications of the Astronomical Society of the Pacific,* 105, 681.

Blitz, Leo, 1992. Large Scale Properties of Giant Molecular Clouds. In *Evolution of Interstellar Matter and Dynamics of Galaxies,* edited by Jan Palouš, W. Butler Burton, and Per Olof Lindblad (Cambridge: Cambridge University Press), p. 158.

Bok, Bart J., 1969. The Spiral Structure of Our Galaxy—I. *Sky and Telescope,* 38, 392 (December 1969).

Bok, Bart J., and Bok, Priscilla F., 1981. *The Milky Way* (Cambridge: Harvard University Press).

Bolte, Michael, 1989. The Age of the Globular Cluster NGC 288, the Formation of the Galactic Halo, and the Second Parameter. *Astronomical Journal,* 97, 1688.

————, 1993. Interview with Ken Croswell: November 15, 1993.

Bond, Howard E., 1981. Where Is Population III? *Astrophysical Journal,* 248, 606.

Bondi, H., and Gold, T., 1948. The Steady-State Theory of the Expanding Universe. *Monthly Notices of the Royal Astronomical Society,* 108, 252.

Bondi, H., Gold, T., and Hoyle, F., 1955. Black Giant Stars. *Observatory,* 75, 80.

Boss, Benjamin, 1918. Real Stellar Motions. *Publications of the American Astronomical Society,* 4, 11.

Boyd, R. N., and Kajino, T., 1989. Can ^{9}Be Provide a Test of Cosmological Theories? *Astrophysical Journal Letters,* 336, L55.

Bronfman, Leonardo, 1992. Molecular Clouds and Young Massive Stars in the Galactic Disk. In *The Center, Bulge, and Disk of the Milky Way,* edited by Leo Blitz (Dordrecht: Kluwer Academic Publishers), p. 131.

Burbidge, E. Margaret, 1993. Interview with Ken Croswell: October 21, 1993.

————, 1994. Watcher of the Skies. *Annual Review of Astronomy and Astrophysics,* 32, 1.

Burbidge, E. Margaret, and Burbidge, G. R., 1957. Chemical Composition of the Ba II Star HD 46407 and Its Bearing on Element Synthesis in Stars. *Astrophysical Journal*, 126, 357.

―――, 1982. Nucleosynthesis in Galaxies. In *Essays in Nuclear Astrophysics*, edited by C. A. Barnes, D. D. Clayton, and D. N. Schramm (Cambridge: Cambridge University Press), p. 11.

Burbidge, E. Margaret, Burbidge, G. R., Fowler, William A., and Hoyle, F., 1957. Synthesis of the Elements in Stars. *Reviews of Modern Physics*, 29, 547.

Burbidge, G. R., 1958. Nuclear Energy Generation and Dissipation in Galaxies. *Publications of the Astronomical Society of the Pacific*, 70, 83.

―――, 1993. Interview with Ken Croswell: August 3, 1993.

Burke, Bernard F., 1957. Systematic Distortion of the Outer Regions of the Galaxy. *Astronomical Journal*, 62, 90.

Burnham, Robert Jr., 1978. *Burnham's Celestial Handbook* (New York: Dover).

Burstein, David, 1979. Structure and Origin of S0 Galaxies. III. The Luminosity Distribution Perpendicular to the Plane of the Disks in S0's. *Astrophysical Journal*, 234, 829.

Cameron, A. G. W., 1954. Origin of Anomalous Abundances of the Elements in Giant Stars. *Physical Review*, 93, 932.

―――, 1955. Origin of Anomalous Abundances of the Elements in Giant Stars. *Astrophysical Journal*, 121, 144.

―――, 1957. Nuclear Reactions in Stars and Nucleogenesis. *Publications of the Astronomical Society of the Pacific*, 69, 201.

―――, 1958. *Stellar Evolution, Nuclear Astrophysics, and Nucleogenesis*, CRL-41, second edition (Chalk River, Ontario: Atomic Energy of Canada Limited)

―――, 1960. New Neutron Sources of Possible Astrophysical Importance. *Astronomical Journal*, 65, 485.

―――, 1993. Interview with Ken Croswell: August 9, 1993.

Campbell, W. W., 1895. The Velocity of Zeta Herculis in the Line of Sight. *Publications of the Astronomical Society of the Pacific*, 7, 67.

―――, 1898a. The Mills Spectrograph of the Lick Observatory. *Astrophysical Journal*, 8, 123.

―――, 1898b. Some Stars with Great Velocities in the Line of Sight. *Astrophysical Journal*, 8, 157.

―――, 1898c. The Motion of Eta Cephei in the Line of Sight. *Publications of the Astronomical Society of the Pacific*, 10, 184.

——, 1901a. Some Stars with Large Radial Velocities. *Astrophysical Journal*, 13, 98.

——, 1901b. The Velocity of Groombridge 1830 in the Line of Sight. *Publications of the Astronomical Society of the Pacific*, 13, 70.

——, 1901c. Some Stars with Large Radial Velocities. *Publications of the Astronomical Society of the Pacific*, 13, 72.

——, 1910. Reports of Observatories. *Publications of the Astronomical Society of the Pacific*, 22, 216.

——, 1911. Some Peculiarities in the Motions of the Stars. *Lick Observatory Bulletin*, 6, 125 (No. 196).

——, 1912. Biennial Report of the Director of Lick Observatory. *Publications of the Astronomical Society of the Pacific*, 24, 237.

——, 1926. Do We Live in a Spiral Nebula? *Publications of the Astronomical Society of the Pacific*, 38, 75.

Campbell, William Wallace, and Moore, Joseph Haines, 1928. Radial Velocities of Stars Brighter than Visual Magnitude 5.51 as Determined at Mount Hamilton and Santiago. *Publications of the Lick Observatory*, 16, 1.

Cannon, R. D., Hawarden, T. G., and Tritton, S. B., 1977. A New Sculptor-type Dwarf Elliptical Galaxy in Carina. *Monthly Notices of the Royal Astronomical Society*, 180, 81P.

Carney, Bruce W., 1984. Probing the Outer Galactic Halo. *Publications of the Astronomical Society of the Pacific*, 96, 841.

——, 1993. Interview with Ken Croswell: September 3, 1993.

Carney, Bruce W., Aguilar, Luis, Latham, David W., and Laird, John B., 1990. A Survey of Proper Motion Stars. IX. The Galactic Halo's Metallicity Gradient. *Astronomical Journal*, 99, 201.

Carney, Bruce W., and Latham, David W., 1987. A Survey of Proper-Motion Stars. I. UBV Photometry and Radial Velocities. *Astronomical Journal*, 93, 116.

Carney, Bruce W., Latham, David W., and Laird, John B., 1988. A Survey of Proper-Motion Stars. V. Extreme-Velocity Stars and the Local Galactic Escape Velocity. *Astronomical Journal*, 96, 560.

——, 1989. A Survey of Proper-Motion Stars. VIII. On the Galaxy's Third Population. *Astronomical Journal*, 97, 423.

——, 1990. A Survey of Proper-Motion Stars. X. The Early Evolution of the Galaxy's Halo. *Astronomical Journal*, 99, 572.

Carney, Bruce W., Latham, David W., Laird, John B., and Aguilar, Luis A., 1994. A Survey of Proper Motion Stars. XII. An Expanded Sample. *Astronomical Journal*, 107, 2240.

Carroll, Sean M., Press, William H., and Turner, Edwin L., 1992. The Cosmological Constant. *Annual Review of Astronomy and Astrophysics*, 30, 499.

Casertano, Stefano, Ratnatunga, Kavan U., and Bahcall, John N., 1990. Kinematic Modeling of the Galaxy. II. Two Samples of High Proper Motion Stars. *Astrophysical Journal*, 357, 435.

Chamberlain, Joseph W., 1993. Interview with Ken Croswell: June 30, 1993.

Chamberlain, Joseph W., and Aller, Lawrence H., 1951. The Atmospheres of A-Type Subdwarfs and 95 Leonis. *Astrophysical Journal*, 114, 52.

Chamberlain, Joseph W., Aller, Lawrence H., and Sanford, Roscoe F., 1950. Curves of Growth for A-Type Sub-dwarfs. *Astronomical Journal*, 55, 167.

Chiosi, Cesare, Bertelli, Gianpaolo, and Bressan, Alessandro, 1992. New Developments in Understanding the HR Diagram. *Annual Review of Astronomy and Astrophysics*, 30, 235.

Chown, Marcus, 1993. *Afterglow of Creation* (London: Arrow Books Limited).

Ciardullo, Robin, Jacoby, George H., and Bond, Howard E., 1989. Serendipitous Discovery of a Distant Galactic RR Lyrae Star in Front of the Virgo Elliptical NGC 4472. *Astronomical Journal*, 98, 1648.

Combes, Françoise, 1991. Distribution of CO in the Milky Way. *Annual Review of Astronomy and Astrophysics*, 29, 195.

Conti, Peter S., Greenstein, Jesse L., Spinrad, Hyron, Wallerstein, George, and Vardya, M. S., 1967. Neutral Oxygen in Late-type Stars. *Astrophysical Journal*, 148, 105.

Crawford, Matt, and Schramm, David N., 1982. Spontaneous Generation of Density Perturbations in the Early Universe. *Nature*, 298, 538.

Croswell, Ken, 1988. Does Barnard's Star Have Planets? *Astronomy*, 16, No. 3 (March 1988), p. 6.

———, 1991. Does Alpha Centauri Have Intelligent Life? *Astronomy*, 19, No. 4 (April 1991), p. 28.

———, 1992a. The Best Black Hole in the Galaxy. *Astronomy*, 20, No. 3 (March 1992), p. 30.

———, 1992b. Puzzle of the Pulsar Planets. *New Scientist*, 135, No. 1830 (July 18, 1992), p. 40.

———, 1992c. When Little Stars Explode. *Star Date*, 20, No. 5 (September/October 1992), p. 4.

———, 1992d. Why Intelligent Life Needs Giant Planets. *New Scientist*, 136, No. 1844 (October 24, 1992), p. 18.

————, 1992e. The Grand Illusion. *Astronomy*, 20, No. 11 (November 1992), p. 44.

————, 1993a. Have Astronomers Solved the Quasar Enigma? *Astronomy*, 21, No. 2 (February 1993), p. 28.

————, 1993b. The Quest for the Cosmological Constant. *New Scientist*, 137, No. 1861 (February 20, 1993), p. 23.

————, 1993c. Death of Mars. *Star Date*, 21, No. 2 (March/April 1993), p. 4.

Croswell, Ken, Latham, David W., Carney, Bruce W., Schuster, William, and Aguilar, Luis, 1991. A Search for Distant Stars in the Milky Way Galaxy's Halo and Thick Disk. *Astronomical Journal*, 101, 2078.

Cunha, Katia, and Lambert, David L., 1992. Chemical Evolution of the Orion Association. I. The Oxygen Abundance of Main-sequence B Stars. *Astrophysical Journal*, 399, 586.

————, 1994. Chemical Evolution of the Orion Association. II. The Carbon, Nitrogen, Oxygen, Silicon, and Iron Abundances of Main-Sequence B Stars. *Astrophysical Journal*, 426, 170.

Curtis, Heber D., 1909. Three Stars of Great Radial Velocity. *Lick Observatory Bulletin*, 5, 133 (No. 162).

————, 1913. Address of the Retiring President of the Society, in Awarding the Bruce Medal to Professor J. C. Kapteyn. *Publications of the Astronomical Society of the Pacific*, 25, 15.

D., F. W., 1897. The Cape Photographic Durchmusterung. *Monthly Notices of the Royal Astronomical Society*, 57, 297.

Da Costa, G. S., 1988. Dwarf Spheroidal Galaxies and Globular Clusters. In *The Harlow-Shapley Symposium on Globular Cluster Systems in Galaxies*, IAU Symposium 126, edited by Jonathan E. Grindlay and A. G. Davis Philip (Dordrecht: Kluwer Academic Publishers), p. 217.

Da Costa, G. S., Hatzidimitriou, D., Irwin, M. J., and McMahon, R. G., 1991. The Radial Velocity and Metal Abundance of the Sextans Dwarf Spheroidal Galaxy. *Monthly Notices of the Royal Astronomical Society*, 249, 473.

Dahn, C. C., Behall, A. L., Guetter, H. H., Priser, J. B., Harrington, R. S., Strand, K. Aa., and Riddle, R. K., 1972. The Low-Luminosity Star G51-15. *Astrophysical Journal Letters*, 174, L87.

Demers, Serge, Irwin, M. J., and Kunkel, W. E., 1994. Large-scale Photographic Survey of Fornax. *Astronomical Journal*, 108, 1648.

De Vaucouleurs, G., 1964. Interpretation of Velocity Distribution of the Inner Regions of the Galaxy. In *The Galaxy and the Magellanic Clouds*, IAU Symposium 20, edited by F. J. Kerr and A. W. Rodgers (Canberra: Australian Academy of Science), p. 195.

————, 1987. Discovering M31's Spiral Shape. *Sky and Telescope*, 74, 595 (December 1987).

————, 1993. The Extragalactic Distance Scale. VIII. A Comparison of Distance Scales. *Astrophysical Journal*, 415, 10.

DeVorkin, David H., 1977. The Origins of the Hertzsprung-Russell Diagram. In *In Memory of Henry Norris Russell*, edited by A. G. Davis Philip and David H. DeVorkin (Albany, New York: Dudley Observatory), p. 61.

————, 1978. Steps Toward the Hertzsprung-Russell Diagram. *Physics Today*, 31, No. 3 (March 1978), p. 32.

Dewhirst, David W., and Hoskin, Michael, 1991. The Rosse Spirals. *Journal for the History of Astronomy*, 22, 257.

Dickens, R. J., Croke, B. F. W., Cannon, R. D., and Bell, R. A., 1991. Evidence from Stellar Abundances for a Large Age Difference between Two Globular Clusters. *Nature*, 351, 212.

Dickey, John M., and Lockman, Felix J., 1990. H I in the Galaxy. *Annual Review of Astronomy and Astrophysics*, 28, 215.

Domogatsky, G. V., and Nadyozhin, D. K., 1977. Neutrino Induced Production of Bypassed Elements. *Monthly Notices of the Royal Astronomical Society*, 178, 33P.

Drake, Frank, and Sobel, Dava, 1992. *Is Anyone Out There?* (New York: Delacorte).

Dunbar, D. N. F., Pixley, R. E., Wenzel, W. A., and Whaling, W., 1953. The 7.68-Mev State in C^{12}. *Physical Review*, 92, 649.

Duncan, Douglas K., Lambert, David L., and Lemke, Michael, 1992. The Abundance of Boron in Three Halo Stars. *Astrophysical Journal*, 401, 584.

Duncan, J. C., 1909. The Orbits of the Cepheid Variables Y Sagittarii and RT Aurigae; with a Discussion of the Possible Causes of This Type of Stellar Variation. *Publications of the Astronomical Society of the Pacific*, 21, 119.

Easton, C., 1900. A New Theory of the Milky Way. *Astrophysical Journal*, 12, 136.

————, 1913. A Photographic Chart of the Milky Way and the Spiral Theory of the Galactic System. *Astrophysical Journal*, 37, 105.

Eckart, A., Genzel, R., Hofmann, R., Sams, B. J., and Tacconi-Garman, L. E., 1993. Near-Infrared 0."15 Resolution Imaging of the Galactic Center. *Astrophysical Journal Letters*, 407, L77.

Eckart, A., Genzel, R., Krabbe, A., Hofmann, R., van der Werf, P. P., and Drapatz, S., 1992. Spatially Resolved Near-Infrared Emission and a Bubble of Hot Gas in the Central Active Region of the Galaxy. *Nature*, 355, 526.

Eddington, A. S., 1914. *Stellar Movements and the Structure of the Universe* (London: MacMillan)

————, 1917. Karl Schwarzschild. *Monthly Notices of the Royal Astronomical Society*, 77, 314.

————, 1922. Jacobus Cornelius Kapteyn. *Observatory*, 45, 261.

Eggen, Olin J., 1962. Space-Velocity Vectors for 3483 Stars with Accurately Determined Proper Motion and Radial Velocity. *Royal Observatory Bulletins*, No. 51.

————, 1964. A Catalogue of High-Velocity Stars. *Royal Observatory Bulletins*, No. 84.

————, 1993a. Interview with Ken Croswell: September 7, 1993.

————, 1993b. Notes from a Life in the Dark. *Annual Review of Astronomy and Astrophysics*, 31, 1.

Eggen, O. J., Lynden-Bell, D., and Sandage, A. R., 1962. Evidence from the Motions of Old Stars that the Galaxy Collapsed. *Astrophysical Journal*, 136, 748.

Eggen, Olin J., and Sandage, Allan R., 1959. Stellar Groups, IV. The Groombridge 1830 Group of High Velocity Stars and its Relation to the Globular Clusters. *Monthly Notices of the Royal Astronomical Society*, 119, 255.

Ewen, H. I., and Purcell, E. M., 1951. Radiation from Galactic Hydrogen at 1,420 Mc./sec. *Nature*, 168, 356.

F., A., 1908. Hermann Carl Vogel. *Monthly Notices of the Royal Astronomical Society*, 68, 254.

Faber, S. M., and Lin, D. N. C., 1983. Is There Nonluminous Matter in Dwarf Spheroidal Galaxies? *Astrophysical Journal Letters*, 266, L17.

Fich, Michel, and Tremaine, Scott, 1991. The Mass of the Galaxy. *Annual Review of Astronomy and Astrophysics*, 29, 409.

Fouts, Gary, and Sandage, Allan, 1986. New Subdwarfs. V. Radial Velocities for 889 High-Proper-Motion Stars Measured with the Mount Wilson 100 Inch Reflector. *Astronomical Journal*, 91, 1189.

Fowler, William A., 1992. From Steam to Stars to the Early Universe. *Annual Review of Astronomy and Astrophysics*, 30, 1.

————, 1993. Interview with Ken Croswell: July 14, 1993.

Fowler, W. A., Burbidge, G. R., and Burbidge, E. Margaret, 1955. Stellar Evolution and the Synthesis of the Elements. *Astrophysical Journal*, 122, 271.

Freeman, K. C., 1970. On the Disks of Spiral and S0 Galaxies. *Astrophysical Journal*, 160, 811.

Friel, Eileen D., 1987. A Spectroscopic Survey of High-Latitude Fields and Comparison to Galaxy Models. *Astronomical Journal*, 93, 1388.

Frogel, Jay A., 1988. The Galactic Nuclear Bulge and the Stellar Content of Spheroidal Systems. *Annual Review of Astronomy and Astrophysics*, 26, 51.

Frost, Edwin B., 1908. Hermann Carl Vogel. *Astrophysical Journal*, 27, 1.

Galloway, J. D., 1902. Kapteyn's Contributions to Our Knowledge of the Stars. *Publications of the Astronomical Society of the Pacific*, 14, 97.

Gasson, R. E. M., and Pagel, B. E. J., 1966. Forbidden Lines in the Spectra of the Sun and Arcturus. *Observatory*, 86, 196.

Genzel, R., and Townes, C. H., 1987. Physical Conditions, Dynamics, and Mass Distribution in the Center of the Galaxy. *Annual Review of Astronomy and Astrophysics*, 25, 377.

Gilman, Carolyn, 1978. John Goodricke and His Variable Stars. *Sky and Telescope*, 56, 400 (November 1978).

Gilmore, Gerard, 1984. New Light on Faint Stars—VI. Structure and Evolution of the Galactic Spheroid. *Monthly Notices of the Royal Astronomical Society*, 207, 223.

———, 1991. Interview with Ken Croswell: November 13, 1991.

———, 1993. Interview with Ken Croswell: September 15, 1993.

Gilmore, Gerard, Edvardsson, Bengt, and Nissen, P. E., 1991. First Detection of Beryllium in a Very Metal Poor Star: A Test of the Standard Big Bang Model. *Astrophysical Journal*, 378, 17.

Gilmore, Gerard, King, Ivan R., and van der Kruit, Pieter C., 1990. *The Milky Way as a Galaxy* (Mill Valley, California: University Science Books).

Gilmore, Gerard, and Reid, Neill, 1983. New Light on Faint Stars—III. Galactic Structure towards the South Pole and the Galactic Thick Disc. *Monthly Notices of the Royal Astronomical Society*, 202, 1025.

Gilmore, Gerard, Reid, Neill, and Hewett, Paul, 1985. New Light on Faint Stars—VII. Luminosity and Mass Distributions in Two High Galactic Latitude Fields. *Monthly Notices of the Royal Astronomical Society*, 213, 257.

Gilroy, Kalpana Krishnaswamy, Sneden, Christopher, Pilachowski, Catherine A., and Cowan, John J., 1988. Abundances of Neutron Capture Elements in Population II Stars. *Astrophysical Journal*, 327, 298.

Gingerich, Owen, 1984. Herschel's 1784 Autobiography. *Sky and Telescope*, 68, 317 (October 1984).

———, 1985a. The Discovery of the Spiral Arms of the Milky Way. In *The Milky Way Galaxy*, IAU Symposium 106, edited by Hugo van Woerden, Ronald J. Allen, and W. Butler Burton (Dordrecht: D. Reidel), p. 59.

———, 1985b. Robert Trumpler and the Dustiness of Space. *Sky and Telescope*, 70, 213 (September 1985).

————, 1988. How Shapley Came to Harvard Or, Snatching the Prize from the Jaws of Debate. *Journal for the History of Astronomy*, 19, 201.

————, 1990. Through Rugged Ways to the Galaxies. *Journal for the History of Astronomy*, 21, 76.

Gingerich, Owen, and Welther, Barbara, 1985. Harlow Shapley and the Cepheids. *Sky and Telescope*, 70, 540 (December 1985).

Glaisher, J. W. L., 1902. Address on Presenting the Gold Medal of the Society to Professor J. C. Kapteyn. *Monthly Notices of the Royal Astronomical Society*, 62, 334.

Gliese, W., 1969. *Catalogue of Nearby Stars*, Veröffentlichungen des Astronomischen Rechen-Instituts, Heidelberg, No. 22.

Gliese, W., and Jahreiss, H., 1979. Nearby Star Data Published 1969–1978. *Astronomy and Astrophysics Supplement Series*, 38, 423.

Goodricke, C. A., 1912. Gift to the Society of a Portrait of John Goodricke. *Monthly Notices of the Royal Astronomical Society*, 73, 3.

Gratton, R. G., and Ortolani, S., 1988. Deep Photometry of Globular Clusters. XI. Palomar 12: the Youngest Galactic Globular Cluster? *Astronomy and Astrophysics Supplement Series*, 73, 137.

Greenstein, George, 1983. *Frozen Star* (New York: Freundlich Books).

Greenstein, Jesse L., 1984. An Astronomical Life. *Annual Review of Astronomy and Astrophysics*, 22, 1.

Halley, E., 1718. Considerations on the Change of the Latitudes of Some of the Principal Fixt Stars. *Philosophical Transactions*, 30, 736.

Halm, J., 1911. Further Considerations Relating to the Systematic Motions of the Stars. *Monthly Notices of the Royal Astronomical Society*, 71, 610.

Harrington, R. G., and Wilson, A. G., 1950. Two New Stellar Systems in Leo. *Publications of the Astronomical Society of the Pacific*, 62, 118.

Harris, William E., 1976. Spatial Structure of the Globular Cluster System and the Distance to the Galactic Center. *Astronomical Journal*, 81, 1095.

Hartkopf, William I., and Yoss, Kenneth M., 1982. A Kinematic and Abundance Survey at the Galactic Poles. *Astronomical Journal*, 87, 1679.

Hartmann, J., 1904. Investigations on the Spectrum and Orbit of Delta Orionis. *Astrophysical Journal*, 19, 268.

Hawkins, M. R. S., 1983. Direct Evidence for a Massive Galactic Halo. *Nature*, 303, 406.

————, 1984. A Study of the Galactic Halo from a Complete Sample of RR Lyrae Variables to B = 21. *Monthly Notices of the Royal Astronomical Society*, 206, 433.

Haynes, Raymond, and Milne, Douglas (editors), 1991. *The Magellanic Clouds*, IAU Symposium 148 (Dordrecht: Kluwer Academic Publishers).

Hertzsprung, E., 1913. Über die räumliche Verteilung der Veränderlichen vom Delta Cephei-Typus. *Astronomische Nachrichten*, 196, 201.

———, 1917. Karl Schwarzschild. *Astrophysical Journal*, 45, 285.

Hetherington, Norriss S., 1975. The Simultaneous 'Discovery' of Internal Motions in Spiral Nebulae. *Journal for the History of Astronomy*, 6, 115.

Hodge, Paul W., 1971. Dwarf Galaxies. *Annual Review of Astronomy and Astrophysics*, 9, 35.

———, 1989. Populations in Local Group Galaxies. *Annual Review of Astronomy and Astrophysics*, 27, 139.

Hodge, Paul W., and Michie, Richard W., 1969. The Structure of Dwarf Elliptical Galaxies of the Local Group. *Astronomical Journal*, 74, 587.

Hoskin, Michael, 1970. The Cosmology of Thomas Wright of Durham. *Journal for the History of Astronomy*, 1, 44.

———, 1976. The 'Great Debate': What Really Happened. *Journal for the History of Astronomy*, 7, 169.

———, 1979. Goodricke, Pigott and the Quest for Variable Stars. *Journal for the History of Astronomy*, 10, 23.

———, 1982a. *Stellar Astronomy: Historical Studies* (Bucks, England: Science History Publications).

———, 1982b. The First Drawing of a Spiral Nebula. *Journal for the History of Astronomy*, 13, 97.

———, 1987. John Herschel's Cosmology. *Journal for the History of Astronomy*, 18, 1.

———, 1988. Shapley's Debate. In *The Harlow-Shapley Symposium on Globular Cluster Systems in Galaxies*, IAU Symposium 126, edited by Jonathan E. Grindlay and A. G. Davis Philip (Dordrecht: Kluwer Academic Publishers), p. 3.

Hoskin, Michael, and Rochester, George D., 1992. Thomas Wright and the Royal Society. *Journal for the History of Astronomy*, 23, 167.

Hoyle, F., 1946. The Synthesis of the Elements from Hydrogen. *Monthly Notices of the Royal Astronomical Society*, 106, 343.

———, 1948. A New Model for the Expanding Universe. *Monthly Notices of the Royal Astronomical Society*, 108, 372.

———, 1950. *The Nature of the Universe* (New York: Harper and Brothers).

———, 1954a. On Nuclear Reactions Occurring in Very Hot Stars. I. The Synthesis of Elements from Carbon to Nickel. *Astrophysical Journal Supplement Series*, 1, 121.

————, 1954b. Commission des Nebuleuses Extragalactiques. *Transactions of the International Astronomical Union*, 8, 397.

————, 1982a. The Universe: Past and Present Reflections. *Annual Review of Astronomy and Astrophysics*, 20, 1.

————, 1982b. Two Decades of Collaboration with Willy Fowler. In *Essays in Nuclear Astrophysics*, edited by C. A. Barnes, D. D. Clayton, and D. N. Schramm (Cambridge: Cambridge University Press), p. 1.

————, 1983. Hoyle on Stars. *New Scientist*, 100, No. 1385, 607 (November 24, 1983).

————, 1986. Personal Comments on the History of Nuclear Astrophysics. *Quarterly Journal of the Royal Astronomical Society*, 27, 445.

————, 1993. Interview with Ken Croswell: August 5, 1993.

————, 1994. *Home Is Where the Wind Blows: Chapters from a Cosmologist's Life* (Mill Valley, California: University Science Books).

Hoyle, F., Burbidge, G., and Narlikar, J. V., 1993. A Quasi-Steady State Cosmological Model with Creation of Matter. *Astrophysical Journal*, 410, 437.

Hoyle, F., Dunbar, D. N. F., Wenzel, W. A., and Whaling, W., 1953. A State in C^{12} Predicted from Astrophysical Evidence. *Physical Review*, 92, 1095.

Hoyle, F., and Tayler, R. J., 1964. The Mystery of the Cosmic Helium Abundance. *Nature*, 203, 1108.

Hoyt, William Graves, 1980. *Planets X and Pluto* (Tucson: University of Arizona Press).

Hubble, Edwin, 1926a. A Spiral Nebula as a Stellar System, Messier 33. *Astrophysical Journal*, 63, 236.

————, 1926b. Extra-Galactic Nebulae. *Astrophysical Journal*, 64, 321.

————, 1929a. A Spiral Nebula as a Stellar System, Messier 31. *Astrophysical Journal*, 69, 103.

————, 1929b. A Relation between Distance and Radial Velocity among Extra-galactic Nebulae. *Proceedings of the National Academy of Sciences*, 15, 168.

————, 1930. Distribution of Luminosity in Elliptical Nebulae. *Astrophysical Journal*, 71, 231.

Iben, Icko, Jr., 1991. Single and Binary Star Evolution. *Astrophysical Journal Supplement Series*, 76, 55.

Innes, R., 1915. A Faint Star of Large Proper Motion. *Union Observatory Circular* No. 30.

Irwin, John, 1956. A California Gauger of Star Clusters. *Sky and Telescope*, 16, 13 (November 1956).

Irwin, Mike, 1993. Interview with Ken Croswell: December 16, 1993.

Irwin, M. J., Bunclark, P. S., Bridgeland, M. T., and McMahon, R. G., 1990. A New Satellite Galaxy of the Milky Way in the Constellation of Sextans. *Monthly Notices of the Royal Astronomical Society*, 244, 16P.

Isobe, Syuzo, 1974. Space Motion of Subdwarfs and Initial Contraction of the Galaxy. *Astronomy and Astrophysics*, 36, 333.

Jackson, John, 1954. The President's Address on the Award of the Gold Medal to Dr. Walter Baade. *Monthly Notices of the Royal Astronomical Society*, 114, 370.

Jacoby, G. H., Pierce, M. J., and Massey, P., 1992. Evidence for Significant Reddening of Young Stars in IC 4182. *Bulletin of the American Astronomical Society*, 24, 1255.

Jaki, Stanley L., 1971. The Milky Way before Galileo. *Journal for the History of Astronomy*, 2, 161.

————, 1972. *The Milky Way: An Elusive Road for Science* (New York: Science History Publications).

Jeans, James, 1923. Jacobus Cornelius Kapteyn. *Monthly Notices of the Royal Astronomical Society*, 83, 250.

Jones, Kenneth Glyn, 1971. The Observational Basis for Kant's *Cosmogony:* A Critical Analysis. *Journal for the History of Astronomy*, 2, 29.

Jorissen, A., Smith, V. V., and Lambert, D. L., 1992. Fluorine in Red Giant Stars: Evidence for Nucleosynthesis. *Astronomy and Astrophysics*, 261, 164.

Joy, Alfred H., 1937. Some Early Variable Star Observers. *Astronomical Society of the Pacific Leaflet* No. 99.

Kaler, James B., 1992. *Stars* (New York: W. H. Freeman).

Kapteyn, J. C., 1905. Star Streaming. *Report of the British Association for the Advancement of Science*, 237.

————, 1909a. On the Absorption of Light in Space. *Astrophysical Journal*, 29, 46.

————, 1909b. On the Absorption of Light in Space: Second Paper. *Astrophysical Journal*, 30, 284 [erratum: *Astrophysical Journal*, 30, 398].

————, 1910. On Certain Statistical Data Which May Be Valuable in the Classification of the Stars in the Order of Their Evolution. *Astrophysical Journal*, 31, 258.

————, 1914a. The "Plan of Selected Areas." *Monthly Notices of the Royal Astronomical Society*, 74, 348.

————, 1914b. On the Change of Spectrum and Color Index with Distance and Absolute Brightness. Present State of the Question. *Astrophysical Journal*, 40, 187.

————, 1922. First Attempt at a Theory of the Arrangement and Motion of the Sidereal System. *Astrophysical Journal*, 55, 302.

Kapteyn, J. C., and van Rhijn, P. J., 1920. On the Distribution of the Stars in Space Especially in the High Galactic Latitudes. *Astrophysical Journal*, 52, 23.

————, 1922a. On the Upper Limit of Distance to Which the Arrangement of Stars in Space Can at Present Be Determined with Some Confidence. *Astrophysical Journal*, 55, 242.

————, 1922b. The Proper Motions of Delta Cephei Stars and the Distances of the Globular Clusters. *Bulletin of the Astronomical Institutes of the Netherlands*, 1, 37.

Keenan, Philip C., and Keller, Geoffrey, 1951. The Spectrum-Luminosity Diagram for the High-Velocity Stars. *Astronomical Journal*, 56, 131.

————, 1953. Spectral Classification of the High-Velocity Stars. *Astrophysical Journal*, 117, 241.

Kerr, F. J., 1957. A Magellanic Effect on the Galaxy. *Astronomical Journal*, 62, 93.

Kerr, F. J., and Lynden-Bell, D., 1986. Review of Galactic Constants. *Monthly Notices of the Royal Astronomical Society*, 221, 1023.

King, Ivan R., 1971. Stellar Populations in Galaxies. *Publications of the Astronomical Society of the Pacific*, 83, 377.

Kinman, T. D., 1959. Globular Clusters, II. The Spectral Types of Individual Stars and of the Integrated Light. *Monthly Notices of the Royal Astronomical Society*, 119, 538.

Knobel, E. B., 1893. Address on Presenting the Gold Medal to Professor H. C. Vogel. *Monthly Notices of the Royal Astronomical Society*, 53, 310.

Kormendy, John, 1987. Dark Matter in Dwarf Galaxies. In *Dark Matter in the Universe*, IAU Symposium 117, edited by J. Kormendy and G. R. Knapp (Dordrecht: Kluwer Academic Publishers), p. 139.

Kuhn, Jeffrey R., 1993a. Unbound Dwarf Spheroidal Galaxies and the Mass of the Milky Way. *Astrophysical Journal Letters*, 409, L13.

————, 1993b. Interview with Ken Croswell: September 17, 1993.

Kuhn, J. R., and Miller, R. H., 1989. Dwarf Spheroidal Galaxies and Resonant Orbital Coupling. *Astrophysical Journal Letters*, 341, L41.

Kuiper, G. P., 1939. Two New White Dwarfs; Notes on Proper Motion Stars. *Astrophysical Journal*, 89, 548.

Lambert, D. L., Sneden, C., and Ries, L. M., 1974. The Oxygen Abundance in the Metal-Deficient Star HD 122563. *Astrophysical Journal*, 188, 97.

Lang, Kenneth R., and Gingerich, Owen, 1979. *A Source Book in Astronomy and Astrophysics, 1900–1975* (Cambridge: Harvard University Press).

Leavitt, Henrietta S., 1907. 1777 Variables in the Magellanic Clouds. *Annals of Harvard College Observatory*, 60, 87.

———, 1912. Periods of 25 Variable Stars in the Small Magellanic Cloud. *Harvard College Observatory Circular* No. 173.

Lee, Myung Gyoon, Freedman, Wendy, Mateo, Mario, Thompson, Ian, Roth, Miguel, and Ruiz, Maria-Teresa, 1993. Leo I: The Youngest Milky Way Dwarf Spheroidal Galaxy? *Astronomical Journal*, 106, 1420.

Leggett, S. K., 1992. Infrared Colors of Low-Mass Stars. *Astrophysical Journal Supplement Series*, 82, 351.

Leonard, Peter J. T., and Tremaine, Scott, 1990. The Local Galactic Escape Speed. *Astrophysical Journal*, 353, 486.

Levy, David H., 1993. *The Man Who Sold the Milky Way: A Biography of Bart Bok* (Tucson: University of Arizona Press).

Lin, D. N. C., and Faber, S. M., 1983. Some Implications of Nonluminous Matter in Dwarf Spheroidal Galaxies. *Astrophysical Journal Letters*, 266, L21.

Lin, D. N. C., and Richer, Harvey B., 1992. Young Globular Clusters in the Milky Way Galaxy. *Astrophysical Journal Letters*, 388, L57.

Lindblad, Bertil, 1927. On the State of Motion in the Galactic System. *Monthly Notices of the Royal Astronomical Society*, 87, 553.

Liszt, Harvey S., 1992. H I in the Inner Galaxy. In *The Center, Bulge, and Disk of the Milky Way*, edited by Leo Blitz (Dordrecht: Kluwer Academic Publishers), p. 111.

Luyten, Willem J., 1922. On the Distances of the Cepheids. *Publications of the Astronomical Society of the Pacific*, 34, 166.

———, 1949. A New Star of Large Proper Motion (L 726-8). *Astrophysical Journal*, 109, 532.

Lynden-Bell, Donald, 1993. Interview with Ken Croswell: September 22, 1993.

Lynden-Bell, D., and Rees, M. J., 1971. On Quasars, Dust and the Galactic Centre. *Monthly Notices of the Royal Astronomical Society*, 152, 461.

MacRae, Donald A., Harris, D. L., and Rogerson, J. B., 1952. Photoelectric Photometry at the Warner and Swasey Observatory. *Astronomical Journal*, 57, 19.

Marschall, Laurence A., 1994. *The Supernova Story* (Princeton: Princeton University Press).

Mateo, Mario, Olszewski, Edward W., Pryor, Carlton, Welch, Douglas L., and Fischer, Philippe, 1993. The Carina Dwarf Spheroidal Galaxy: How Dark is it? *Astronomical Journal*, 105, 510.

Mathewson, D. S., Cleary, M. N., and Murray, J. D., 1974. The Magellanic Stream. *Astrophysical Journal*, 190, 291.

Maw, W. H., 1906. Address on Presenting the Gold Medal of the Society to Professor W. W. Campbell. *Monthly Notices of the Royal Astronomical Society*, 66, 245.

Mayall, N. U., 1946. The Radial Velocities of Fifty Globular Star Clusters. *Astrophysical Journal*, 104, 290.

Merrill, Paul W., 1952. Spectroscopic Observations of Stars of Class S. *Astrophysical Journal*, 116, 21.

Mihalas, Dimitri, and Binney, James, 1981. *Galactic Astronomy: Structure and Kinematics* (San Francisco: W. H. Freeman).

Morgan, W. W., 1956. The Integrated Spectral Types of Globular Clusters. *Publications of the Astronomical Society of the Pacific*, 68, 509.

———, 1959. The Integrated Spectra of Globular Clusters. *Astronomical Journal*, 64, 432.

———, 1984. The MK System and the MK Process. In *The MK Process and Stellar Classification*, edited by R. F. Garrison (Toronto: David Dunlap Observatory), p. 18.

———, 1988. A Morphological Life. *Annual Review of Astronomy and Astrophysics*, 26, 1.

———, 1993. Interview with Ken Croswell: May 12, 1993.

Morgan, W. W., and Osterbrock, D. E., 1969. On the Classification of the Forms and the Stellar Content of Galaxies. *Astronomical Journal*, 74, 515.

Morgan, W. W., Sharpless, Stewart, and Osterbrock, Donald, 1952. Some Features of Galactic Structure in the Neighborhood of the Sun. *Astronomical Journal*, 57, 3.

Morgan, W. W., Whitford, A. E., and Code, A. D., 1953. Studies in Galactic Structure. I. A Preliminary Determination of the Space Distribution of the Blue Giants. *Astrophysical Journal*, 118, 318.

Mould, J. R., 1982. Stellar Populations in the Galaxy. *Annual Review of Astronomy and Astrophysics*, 20, 91.

Muller, C. A., and Oort, J. H., 1951. The Interstellar Hydrogen Line at 1,420 Mc.sec., and an Estimate of Galactic Rotation. *Nature*, 168, 357.

Narlikar, Jayant, 1993. Challenge for the Big Bang. *New Scientist*, 138, No. 1878 (June 19, 1993), p. 27.

Nassau, J. J., and Blanco, V. M., 1958. M-type Stars and Red Variables in the Galactic Center. *Astrophysical Journal*, 128, 46.

Nassau, J. J., and Morgan, W. W., 1951. A Finding List of O and B Stars of High Luminosity. *Astrophysical Journal*, 113, 141.

Nather, R. Edward, 1990. Learning from White Dwarf Stars. *Star Date*, 18, No. 3 (May/June 1990), p. 4.

Norris, John, 1986. Population Studies. II. Kinematics as a Function of Abundance and Galactocentric Position for [Fe/H] $\leq$ −0.6. *Astrophysical Journal Supplement Series*, 61, 667.

―――, 1987. Population Studies: the Nature of the Thick Disk. *Astrophysical Journal Letters*, 314, L39.

―――, 1993. Interview with Ken Croswell: August 30, 1993.

―――, 1994. Population Studies. XII. The Duality of the Galactic Halo. *Astrophysical Journal*, 431, 645.

Norris, John, Bessell, M. S., and Pickles, A. J., 1985. Population Studies. I. The Bidelman-MacConnell "Weak-Metal" Stars. *Astrophysical Journal Supplement Series*, 58, 463.

Norris, John E., and Hawkins, M. R. S., 1991. Population Studies. X. Constraints on the Mass and Extent of the Galaxy's Dark Corona. *Astrophysical Journal*, 380, 104.

Norris, John E., Peterson, Ruth C., and Beers, Timothy C., 1993. Abundances of Four Ultra-Metal-Deficient Stars. *Astrophysical Journal*, 415, 797.

Norris, John E., and Ryan, Sean G., 1991. Population Studies. XI. The Extended Disk, Halo Configuration. *Astrophysical Journal*, 380, 403.

O'Connell, D. J. K., 1958. *Stellar Populations* (New York: Interscience Publishers).

Oort, J. H., 1927a. Observational Evidence Confirming Lindblad's Hypothesis of a Rotation of the Galactic System. *Bulletin of the Astronomical Institutes of the Netherlands*, 3, 275.

―――, 1927b. Investigations Concerning the Rotational Motion of the Galactic System, Together with New Determinations of Secular Parallaxes, Precession and Motion of the Equinox. *Bulletin of the Astronomical Institutes of the Netherlands*, 4, 79.

―――, 1927c. Additional Notes Concerning the Rotation of the Galactic System. *Bulletin of the Astronomical Institutes of the Netherlands*, 4, 91.

―――, 1927d. Summary of the Principal Radial Velocity Data Used for the Results of B. A. N. 120 and 132. *Bulletin of the Astronomical Institutes of the Netherlands*, 4, 93.

————, 1928. Dynamics of the Galactic System in the Vicinity of the Sun. *Bulletin of the Astronomical Institutes of the Netherlands*, 4, 269.

————, 1953. L'Hydrogène Interstellaire. *Ciel et Terre*, 69, 117.

————, 1958. Summary—from the Astronomical Point of View. In *Stellar Populations*, edited by D. J. K. O'Connell (New York: Interscience Publishers).

————, 1981. Some Notes on My Life as an Astronomer. *Annual Review of Astronomy and Astrophysics*, 19, 1.

Oort, J. H., Kerr, F. T., and Westerhout, G., 1958. The Galactic System as a Spiral Nebula. *Monthly Notices of the Royal Astronomical Society*, 118, 379.

Osterbrock, Donald E., 1952. The Time of Relaxation for Stars in a Fluctuating Density Field. *Astrophysical Journal*, 116, 164.

————, 1993. Interview with Ken Croswell: May 17, 1993.

————, 1995. I. The Preparation: 1893–1935. Walter Baade, Observational Astrophysicist. *Journal for the History of Astronomy*, 26, 1.

Paczyński, Bohdan, 1986. Gravitational Microlensing by the Galactic Halo. *Astrophysical Journal*, 304, 1.

Paddock, G. F., 1913. The Relation of Stellar Velocities and Masses. *Publications of the Astronomical Society of the Pacific*, 25, 221.

Paul, Erich Robert, 1993. *The Milky Way Galaxy and Statistical Cosmology, 1890–1924* (Cambridge: Cambridge University Press).

Pawsey, J. L., 1951. *Nature*, 168, 358.

Penzias, A. A., and Wilson, R. W., 1965. A Measurement of Excess Antenna Temperature at 4080 Mc/s. *Astrophysical Journal*, 142, 419.

Pickering, Edward C., 1901. Sixty-four New Variable Stars. *Harvard College Observatory Circular* No. 54.

————, 1904. Distribution of Stellar Spectra. *Annals of Harvard College Observatory*, 56, 1.

————, 1905. Stars Having Spectra of Class B. *Annals of Harvard College Observatory*, 56, 27.

Pierce, Michael J., Ressler, Michael E., and Shure, Mark S., 1992. An Absolute Calibration of Type Ia Supernovae and the Value of H_0. *Astrophysical Journal Letters*, 390, L45.

Pogge, Richard W., 1990. Whatever Became of "Snickers"? Private communication.

Popper, Daniel M., 1942. Radial Velocities of Proper-Motion Stars. *Astrophysical Journal*, 95, 307.

———, 1943. Radial Velocities of Proper-Motion Stars. II. *Astrophysical Journal*, 98, 209.

———, 1993. Letter to Ken Croswell: July 27, 1993.

Proctor, R. A., 1869. A New Theory of the Milky Way. *Monthly Notices of the Royal Astronomical Society*, 30, 50.

Pryor, C., 1992. Dark Matter in Dwarf Galaxies. In *Morphological and Physical Classification of Galaxies*, edited by G. Longo, M. Capaccioli, and G. Busarello (Dordrecht: Kluwer Academic Publishers), p. 163.

———, 1993. Interview with Ken Croswell: September 21, 1993.

Reber, Grote, 1940. Cosmic Static. *Astrophysical Journal*, 91, 621.

———, 1944. Cosmic Static. *Astrophysical Journal*, 100, 279.

Reed, George, 1983. The Discovery of Pulsars: was Credit Given Where it was Due? *Astronomy*, 11, No. 12 (December 1983), p. 24.

Reid, Mark J., 1993. The Distance to the Center of the Galaxy. *Annual Review of Astronomy and Astrophysics*, 31, 345.

Rich, R. Michael, 1988. Spectroscopy and Abundances of 88 K Giants in Baade's Window. *Astronomical Journal*, 95, 828.

———, 1992. The Evolution of the Galactic Bulge. In *The Center, Bulge, and Disk of the Milky Way*, edited by Leo Blitz (Dordrecht: Kluwer Academic Publishers), p. 47.

———, 1993. Interviews with Ken Croswell: October 5, 1993; October 15, 1993.

Rieke, G. H., and Rieke, M. J., 1989. Ionization of the Mass-Loss Wind of the M Supergiant IRS 7 by the Ultraviolet Flux in the Galactic Center. *Astrophysical Journal Letters*, 344, L5.

Roman, Nancy G., 1950. A Correlation between the Spectroscopic and Dynamical Characteristics of the Late F- and Early G-Type Stars. *Astrophysical Journal*, 112, 554.

———, 1952. The Spectra of the Bright Stars of Types F5-K5. *Astrophysical Journal*, 116, 122.

———, 1954. A Group of High Velocity F-Type Stars. *Astronomical Journal*, 59, 307.

———, 1955. A Catalogue of High-Velocity Stars. *Astrophysical Journal Supplement Series*, 2, 195.

———, 1993. Interview with Ken Croswell: July 1, 1993.

Ross, Frank E., 1926. New Proper Motion Stars. (Second List). *Astronomical Journal*, 36, 124.

Rubin, Vera C., and Ford, W. Kent, Jr., 1970. Rotation of the Andromeda Nebula from a Spectroscopic Survey of Emission Regions. *Astrophysical Journal*, 159, 379.

Russell, Henry Norris, 1977. Transcript of Colloquium Given at Princeton University Observatory—April 27, 1954. In *In Memory of Henry Norris Russell*, edited by A. G. Davis Philip and David H. DeVorkin (Albany, New York: Dudley Observatory), p. 97.

Sandage, Allan R., 1953. The Color-Magnitude Diagram for the Globular Cluster M3. *Astronomical Journal*, 58, 61.

———, 1961. Wilhelm Heinrich Walter Baade. *Quarterly Journal of the Royal Astronomical Society*, 2, 118.

———, 1969. New Subdwarfs. II. Radial Velocities, Photometry, and Preliminary Space Motions for 112 Stars with Large Proper Motion. *Astrophysical Journal*, 158, 1115.

———, 1981. New Subdwarfs. III. On Obtaining the Vertical Galactic Metallicity Gradient from the Kinematics of Nearby Stars. *Astronomical Journal*, 86, 1643.

———, 1986. The Population Concept, Globular Clusters, Subdwarfs, Ages, and the Collapse of the Galaxy. *Annual Review of Astronomy and Astrophysics*, 24, 421.

———, 1990. On the Formation and Age of the Galaxy. *Journal of the Royal Astronomical Society of Canada*, 84, 70.

———, 1993. Interview with Ken Croswell: September 15, 1993.

Sandage, Allan R., and Eggen, Olin J., 1959. On the Existence of Subdwarfs in the $(M_{Bol}, \text{Log } T_e)$-Diagram. *Monthly Notices of the Royal Astronomical Society*, 119, 278.

Sandage, Allan, and Fouts, Gary, 1987. New Subdwarfs. VI. Kinematics of 1125 High-Proper-Motion Stars and the Collapse of the Galaxy. *Astronomical Journal*, 93, 74.

Sandage, Allan, and Kowal, Charles, 1986. New Subdwarfs. IV. UBV Photometry of 1690 High-Proper-Motion Stars. *Astronomical Journal*, 91, 1140.

Sandage, Allan, Saha, A., Tammann, G. A., Panagia, Nino, and Macchetto, D., 1992. The Cepheid Distance to IC 4182: Calibration of M_V (Max) for SN Ia 1937C and the Value of H_0. *Astrophysical Journal Letters*, 401, L7.

Sandage, Allan, Saha, A., Tammann, G. A., Labhardt, Lukas, Schwengeler, Hans, Panagia, N., and Macchetto, F. D., 1994. The Cepheid Distance to NGC 5253: Calibration of M(max) for the Type Ia Supernovae SN 1972E and SN 1895B. *Astrophysical Journal Letters*, 423, L13.

Sandage, Allan, and Tammann, G. A., 1982. Steps toward the Hubble Constant. VIII. The Global Value. *Astrophysical Journal*, 256, 339.

————, 1993. The Hubble Diagram in V for Supernovae of Type Ia and the Value of H_0 Therefrom. *Astrophysical Journal*, 415, 1.

Sandage, Allan, and Wildey, Robert, 1967. The Anomalous Color-Magnitude Diagram of the Remote Globular Cluster NGC 7006. *Astrophysical Journal*, 150, 469.

Schmidt, M., 1963. 3C 273: a Star-like Object with Large Red-shift. *Nature*, 197, 1040.

Schramm, David, 1993a. The Big Bang Strikes Back. *New Scientist*, 138, No. 1878 (June 19, 1993), p. 31.

————, 1993b. Interview with Ken Croswell: November 10, 1993.

Schwarzschild, Martin, 1993. Interview with Ken Croswell: July 2, 1993.

Schwarzschild, Martin, and Schwarzschild, Barbara, 1950. A Spectroscopic Comparison between High- and Low-Velocity F Dwarfs. *Astrophysical Journal*, 112, 248.

Schwarzschild, M., Spitzer, L. Jr., and Wildt, R., 1951. On the Difference in Chemical Composition between High- and Low-Velocity Stars. *Astrophysical Journal*, 114, 398.

Seares, Frederick H., 1916a. Color-Photographs of Nebulae. *Publications of the Astronomical Society of the Pacific*, 28, 123.

————, 1916b. Preliminary Results on the Color of Nebulae. *Proceedings of the National Academy of Sciences*, 2, 553.

————, 1922. J. C. Kapteyn. *Publications of the Astronomical Society of the Pacific*, 34, 233.

Searle, Leonard, 1993. Interview with Ken Croswell: August 11, 1993.

Searle, Leonard, and Zinn, Robert, 1978. Compositions of Halo Clusters and the Formation of the Galactic Halo. *Astrophysical Journal*, 225, 357.

Seeds, Michael A., 1980. The Wink in the Demon's Eye. *Astronomy*, 8, No. 12 (December 1980), p. 66.

Seeley, D., and Berendzen, R., 1972. The Development of Research in Interstellar Absorption, c. 1900–1930. *Journal for the History of Astronomy*, 3, 52; 3, 75.

Seeliger, H., 1900. Remarks on Mr. Easton's Article "On a New Theory of the Milky Way" in the *Astrophysical Journal* for September. *Astrophysical Journal*, 12, 376.

Serabyn, E., Lacy, J. H., and Achtermann, J. M., 1991. A Gaseous Tail Ablated from the Supergiant IRS 7 near the Galactic Center. *Astrophysical Journal*, 378, 557.

Serio, Giorgia Foderà, 1990. Giuseppe Piazzi and the Discovery of the Proper Motion of 61 Cygni. *Journal for the History of Astronomy*, 21, 275.

Shapley, Harlow, 1916. Outline and Summary of a Study of Magnitudes in the Globular Cluster Messier 13. *Publications of the Astronomical Society of the Pacific*, 28, 171.

―――, 1918a. Studies Based on the Colors and Magnitudes in Stellar Clusters. Sixth Paper: On the Determination of the Distances of Globular Clusters. *Astrophysical Journal*, 48, 89.

―――, 1918b. Studies Based on the Colors and Magnitudes in Stellar Clusters. Seventh Paper: The Distances, Distribution in Space, and Dimensions of 69 Globular Clusters. *Astrophysical Journal*, 48, 154.

―――, 1918c. Globular Clusters and the Structure of the Galactic System. *Publications of the Astronomical Society of the Pacific*, 30, 42.

―――, 1918d. Studies Based on the Colors and Magnitudes in Stellar Clusters. Eighth Paper: The Luminosities and Distances of 139 Cepheid Variables. *Astrophysical Journal*, 48, 279.

―――, 1919. Studies Based on the Colors and Magnitudes in Stellar Clusters. Twelfth Paper: Remarks on the Arrangement of the Sidereal Universe. *Astrophysical Journal*, 49, 311.

―――, 1938a. A Stellar System of a New Type. *Harvard College Observatory Bulletin*, 908, 1.

―――, 1938b. Two Stellar Systems of a New Kind. *Nature*, 142, 715.

―――, 1939. Galactic and Extragalactic Studies, II. Notes on the Peculiar Stellar Systems in Sculptor and Fornax. *Proceedings of the National Academy of Sciences*, 25, 565.

Sharpless, Stewart, and Osterbrock, Donald, 1952. The Nearest H II Regions. *Astrophysical Journal*, 115, 89.

Simonson, S. Christian III, 1975. A New Milky Way Satellite Found in 21-centimeter Line Observations. *Astrophysical Journal Letters*, 201, L103.

―――, 1993. The Mass of the Milky Way and the Distance to the Small Galaxy at l = 197, b = +2. *Bulletin of the American Astronomical Society*, 25, 892.

Slipher, V. M., 1909. Peculiar Star Spectra Suggestive of Selective Absorption of Light in Space. *Lowell Observatory Bulletin* No. 51.

―――, 1913. The Radial Velocity of the Andromeda Nebula. *Lowell Observatory Bulletin* No. 58.

―――, 1914. The Detection of Nebular Rotation. *Lowell Observatory Bulletin* No. 62.

Smecker-Hane, Tammy A., Stetson, Peter B., Hesser, James E., and Lehnert, Matthew D., 1994. The Stellar Populations of the Carina Dwarf Spheroidal Galaxy. I. A New Color-Magnitude Diagram for the Giant and Horizontal Branches. *Astronomical Journal*, 108, 507.

Smith, Robert W., 1983. The Great Debate Revisited. *Sky and Telescope*, 65, 28 (January 1983).

Smith, Verne V., Lambert, David L., and Nissen, Poul E., 1993. The ^{6}Li/^{7}Li Ratio in the Metal-Poor Halo Dwarfs HD 19445 and HD 84937. *Astrophysical Journal*, 408, 262.

Sneden, Christopher, 1993. Interview with Ken Croswell: September 16, 1993.

Sneden, Christopher, Lambert, David L., and Whitaker, Rodney W., 1979. The Oxygen Abundance in Metal-Poor Stars. *Astrophysical Journal*, 234, 964.

Sneden, Christopher, and Parthasarathy, M., 1983. The r- and s-Process Nuclei in the Early History of the Galaxy: HD 122563. *Astrophysical Journal*, 267, 757.

Sneden, Christopher, and Pilachowski, Catherine A., 1985. An Extremely Metal-Poor Star with r-Process Overabundances. *Astrophysical Journal Letters*, 288, L55.

Spite, F., and Spite, M., 1982. Abundance of Lithium in Unevolved Halo Stars and Old Disk Stars: Interpretation and Consequences. *Astronomy and Astrophysics*, 115, 357.

Spite, M., and Spite, F., 1978. Nucleosynthesis in the Galaxy and the Chemical Composition of Old Halo Stars. *Astronomy and Astrophysics*, 67, 23.

———, 1982. Lithium Abundance at the Formation of the Galaxy. *Nature*, 297, 483.

Spitzer, Lyman Jr., 1989. Dreams, Stars, and Electrons. *Annual Review of Astronomy and Astrophysics*, 27, 1.

———, 1993. Interview with Ken Croswell: July 16, 1993.

Spitzer, Lyman Jr., and Schwarzschild, Martin, 1951. The Possible Influence of Interstellar Clouds on Stellar Velocities. *Astrophysical Journal*, 114, 385.

———, 1953. The Possible Influence of Interstellar Clouds on Stellar Velocities. II. *Astrophysical Journal*, 118, 106.

Stebbins, Joel, and Whitford, A. E., 1947. Six-Color Photometry of Stars. V. Infrared Radiation from the Region of the Galactic Center. *Astrophysical Journal*, 106, 235.

Steigman, Gary, Fields, Brian D., Olive, Keith A., Schramm, David N., and Walker, Terry P., 1993. Population II ^{6}Li as a Probe of Nucleosynthesis and Stellar Structure and Evolution. *Astrophysical Journal Letters*, 415, L35.

Strand, K. Aa., 1977. Hertzsprung's Contributions to the HR Diagram. In *In Memory of Henry Norris Russell*, edited by A. G. Davis Philip and David H. DeVorkin (Albany, New York: Dudley Observatory), p. 55.

Strömberg, Gustaf, 1924. The Asymmetry in Stellar Motions and the Existence of a Velocity-Restriction in Space. *Astrophysical Journal*, 59, 228.

————, 1925. The Asymmetry in Stellar Motions as Determined from Radial Velocities. *Astrophysical Journal*, 61, 363.

Struve, Otto, 1951. Photography of the Counterglow. *Sky and Telescope*, 10, 215 (July 1951).

————, 1953. New Light on the Structure of the Galaxy Gained in 1952. *Astronomical Society of the Pacific Leaflet* No. 285.

Suess, Hans E., and Urey, Harold C., 1956. Abundances of the Elements. *Reviews of Modern Physics*, 28, 53.

Sullivan, Woodruff T. III, 1978. A New Look at Karl Jansky's Original Data. *Sky and Telescope*, 56, 101 (August 1978).

Suntzeff, Nicholas B., Mateo, Mario, Terndrup, Donald M., Olszewski, Edward W., Geisler, Doug, and Weller, W., 1993. Spectroscopy of Giants in the Sextans Dwarf Spheroidal Galaxy. *Astrophysical Journal*, 418, 208.

Swift, David W., 1990. *SETI Pioneers: Scientists Talk About Their Search for Extraterrestrial Intelligence* (Tucson: University of Arizona Press).

Thompson, William J., Carney, Bruce W., and Karwowski, Hugon J., 1990. *Workshop on Primordial Nucleosynthesis* (Singapore: World Scientific).

Tombaugh, Clyde W., and Moore, Patrick, 1980. *Out of the Darkness: The Planet Pluto* (Harrisburg: Stackpole).

Toomre, Alar, and Toomre, Juri, 1972. Galactic Bridges and Tails. *Astrophysical Journal*, 178, 623.

Trimble, Virginia, 1987. Existence and Nature of Dark Matter in the Universe. *Annual Review of Astronomy and Astrophysics*, 25, 425.

————, 1991a. Neutron Stars and Black Holes in Binary Systems. *Contemporary Physics*, 32, 103.

————, 1991b. The Origin and Abundances of the Chemical Elements Revisited. *Astronomy and Astrophysics Review*, 3, 1.

Trumpler, Robert J., 1929. Diameters and Distances of Open Star Clusters. *Publications of the Astronomical Society of the Pacific*, 41, 249.

————, 1930a. Preliminary Results on the Distances, Dimensions and Space Distribution of Open Star Clusters. *Lick Observatory Bulletin*, 14, 154.

————, 1930b. Absorption of Light in the Galactic System. *Publications of the Astronomical Society of the Pacific*, 42, 214.

————, 1930c. Spectrophotometric Measures of Interstellar Light Absorption. *Publications of the Astronomical Society of the Pacific*, 42, 267.

————, 1938. William Wallace Campbell. *The Sky*, 3, No. 2 (December 1938), 18.

Truran, J. W., 1981. A New Interpretation of the Heavy Element Abundances in Metal-Deficient Stars. *Astronomy and Astrophysics*, 97, 391.

Tucker, R. H., 1898. Correspondence of the Photographic Durchmusterung with the Visual. *Astrophysical Journal*, 7, 330.

Udalski, A., Szymański, M., Kałużny, J., Kubiak, M., Mateo, M., and Krzemiński, 1994. The Optical Gravitational Lensing Experiment: The Discovery of Three Further Microlensing Events in the Direction of the Galactic Bulge. *Astrophysical Journal Letters*, 426, L69.

Van Biesbroeck, G., 1944. The Star of Lowest Known Luminosity. *Astronomical Journal*, 51, 61.

Van de Hulst, H. C., 1953. The Galaxy Explored by Radio Waves. *Observatory*, 73, 129.

Van de Hulst, H. C., Muller, C. A., and Oort, J. H., 1954. The Spiral Structure of the Outer Part of the Galactic System Derived from the Hydrogen Emission at 21cm Wave Length. *Bulletin of the Astronomical Institutes of the Netherlands*, 12, 117.

VandenBerg, Don, 1993. Interview with Ken Croswell: November 24, 1993.

Van den Bergh, Sidney, 1975. Stellar Populations in Galaxies. *Annual Review of Astronomy and Astrophysics*, 13, 217.

————, 1992a. The Luminosity Function of the Local Group. *Astronomy and Astrophysics*, 264, 75.

————, 1992b. The Hubble Parameter. *Publications of the Astronomical Society of the Pacific*, 104, 861.

Van Maanen, A., 1922. J. C. Kapteyn, 1851–1922. *Astrophysical Journal*, 56, 145.

Van Maanen, A., Brown, J. A., and Humason, M. L., 1927. A Star of Extremely Low Luminosity. *Publications of the Astronomical Society of the Pacific*, 39, 173.

Van Rhijn, P. J., 1916. The Change of Color with Distance and Apparent Magnitude Together with a New Determination of the Mean Parallaxes of the Stars of Given Magnitude and Proper Motion. *Astrophysical Journal*, 43, 36.

Van Woerden, Hugo, Brouw, Willem N., and van de Hulst, Henk C. (editors), 1980. *Oort and the Universe* (Dordrecht: D. Reidel).

Vogel, H. C., 1891. On the Spectroscopic Method of Determining the Velocity of Stars in the Line of Sight. *Monthly Notices of the Royal Astronomical Society*, 52, 87.

————, 1892. List of the Proper Motions in the Line of Sight of Fifty-one Stars. *Monthly Notices of the Royal Astronomical Society*, 52, 541.

————, 1900. On the Progress Made in the Last Decade in the Determination of Stellar Motions in the Line of Sight. *Astrophysical Journal*, 11, 373.

Wagoner, Robert V., Fowler, William A., and Hoyle, F., 1967. On the Synthesis of Elements at Very High Temperatures. *Astrophysical Journal*, 148, 3.

Walker, Terry P., Steigman, Gary, Schramm, David N., Olive, Keith A., and Kang, Ho-Shik, 1991. Primordial Nucleosynthesis Redux. *Astrophysical Journal*, 376, 51.

Wardle, Mark, and Yusef-Zadeh, Farhad, 1992. Gravitational Lensing by a Massive Black Hole at the Galactic Center. *Astrophysical Journal Letters*, 387, L65.

Wheeler, J. Craig, Sneden, Christopher, and Truran, James W., Jr., 1989. Abundance Ratios as a Function of Metallicity. *Annual Review of Astronomy and Astrophysics*, 27, 279.

Whitford, Albert E., 1985. The Stellar Population of the Galactic Nuclear Bulge. *Publications of the Astronomical Society of the Pacific*, 97, 205.

————, 1986. A Half-Century of Astronomy. *Annual Review of Astronomy and Astrophysics*, 24, 1.

Whitford, A. E., and Rich, R. M., 1983. Metal Content of K Giants in the Nuclear Bulge of the Galaxy. *Astrophysical Journal*, 274, 723.

Whitney, Charles A., 1971. *The Discovery of Our Galaxy* (New York: Knopf).

————, 1981. Oort and the Universe. *Sky and Telescope*, 62, 149 (August 1981).

Williams, M. E. W., 1979. Flamsteed's Alleged Measurement of Annual Parallax for the Pole Star. *Journal for the History of Astronomy*, 10, 102.

Wilson, A. G., 1955. Sculptor-Type Systems in the Local Group of Galaxies. *Publications of the Astronomical Society of the Pacific*, 67, 27.

Wilson, O. C., 1955. The Award of the Bruce Gold Medal to Dr. Walter Baade. *Publications of the Astronomical Society of the Pacific*, 67, 57.

Witten, Edward, 1984. Cosmic Separation of Phases. *Physical Review D*, 30, 272.

Wolf, M., 1918. Zwei Sterne mit Großer Eigenbewegung in Leo. *Astronomische Nachrichten*, 206, 237.

Wolszczan, A., and Frail, D. A., 1992. A Planetary System around the Millisecond Pulsar PSR 1257+12. *Nature*, 355, 145.

Woltjer, L., 1975. The Galactic Halo: Globular Clusters. *Astronomy and Astrophysics*, 42, 109.

Woosley, S. E., 1977. Neutrino-induced Nucleosynthesis and Deuterium. *Nature*, 269, 42.

———, 1993. Interview with Ken Croswell: September 16, 1993.

Woosley, S. E., Hartmann, D. H., Hoffman, R. D., and Haxton, W. C., 1990. The ν-process. *Astrophysical Journal*, 356, 272.

Woosley, S. E., and Haxton, W. C., 1988. Supernova Neutrinos, Neutral Currents and the Origin of Fluorine. *Nature*, 334, 45.

Yoshii, Yuzuru, and Saio, Hideyuki, 1979. Kinematics of the Old Stars and Initial Contraction of the Galaxy. *Publications of the Astronomical Society of Japan*, 31, 339.

Yusef-Zadeh, Farhad, 1993. Interview with Ken Croswell: October 15, 1993.

———, 1994. The IRS 16 Complex: The Importance of Stellar Winds at the Galactic Center. In *Nuclei of Normal Galaxies: Lessons from the Galactic Center* (Dordrecht: Kluwer Academic Publishers).

Yusef-Zadeh, Farhad, and Melia, Fulvio, 1992. The Bow Shock Structure of IRS 7: Wind-Wind Collision near the Galactic Center. *Astrophysical Journal Letters*, 385, L41.

Yusef-Zadeh, F., and Morris, Mark, 1991. A Windswept Cometary Tail on the Galactic Center Supergiant IRS 7. *Astrophysical Journal Letters*, 371, L59.

Yusef-Zadeh, F., Morris, Mark, and Chance, D., 1984. Large, Highly Organized Radio Structures near the Galactic Centre. *Nature*, 310, 557.

Yusef-Zadeh, Farhad, and Wardle, Mark, 1992. A Coherent Picture of the Innermost Parsec of the Galaxy. In *The Center, Bulge, and Disk of the Milky Way*, edited by Leo Blitz (Dordrecht: Kluwer Academic Publishers), p. 1.

Zaritsky, Dennis, Olszewski, Edward W., Schommer, Robert A., Peterson, Ruth C., and Aaronson, Marc, 1989. Velocities of Stars in Remote Galactic Satellites and the Mass of the Galaxy. *Astrophysical Journal*, 345, 759.

Zinn, Robert, 1985. The Globular Cluster System of the Galaxy. IV. The Halo and Disk Subsystems. *Astrophysical Journal*, 293, 424.

———, 1993. Interview with Ken Croswell: August 17, 1993.

INDEX

References to figures are *italicized*.